Glyn A. S. Lloyd

Diagnostic Imaging of the Nose and Paranasal Sinuses

With 297 Figures

Springer-Verlag
London Berlin Heidelberg New York
Paris Tokyo

Glyn A. S. Lloyd, MA, DM, FRCR
Consultant Radiologist, Royal National Throat, Nose and Ear Hospital, Gray's Inn Road, London WC1X 8DA, UK.

ISBN 3-540-19518-1 Springer-Verlag Berlin Heidelberg New York
ISBN 0-387-19518-1 Springer-Verlag New York Berlin Heidelberg

Lloyd, Glyn A. S.
Diagnostic imaging of the nose and paranasal sinuses.
1. Man. Nose & paranasal sinuses. Changes in radiography
I. Title
616.2'12075
ISBN 3-540-19518-1

Library of Congress Cataloging-in-Publication Data
Lloyd, Glyn A. S.
Diagnostic imaging of the nose and paranasal sinuses.
Includes bibliographies and index.
1. Nose – Imaging. 2. Paranasal sinuses – Imaging. 3. Diagnostic imaging. I. Title. [DNLM: 1. Diagnostic Imaging. 2. Nose Diseases – Diagnosis. 3. Paranasal Sinus Diseases – diagnosis. WV 300 L793s]
RF345.L56 1988 617'.5230757 88-4620
ISBN 0-387-19518-1

Printed in Great Britain

Filmset, printed and bound by Butler & Tanner Ltd, Frome and London

2128/3830-543210

Preface

In the past two decades the radiological investigation and imaging of paranasal sinus disease has been revolutionised by the introduction of two new techniques: magnetic resonance and computerised tomography. These have made the diagnosis and localisation of neoplastic and non-neoplastic disease a far more exact process and they have completely replaced conventional pluridirectional tomography as a means of augmenting the initial plain radiographic examination of the paranasal sinuses. The recent introduction into clinical practice of the paramagnetic contrast agent gadolinium DTPA has extended further the range of magnetic resonance in the demonstration of naso-sinus disease. The account which follows is an attempt to summarise present knowledge in this division of otorhinolaryngology imaging.

The format of the chapters has been arranged in two halves: the initial nine chapters are concerned with investigative techniques and non-neoplastic disease, while the second half of the book is entirely devoted to benign and malignant tumours. The latter chapters are arranged according to the nomenclature proposed by my two ex-colleagues, Professor I. Friedmann and the late Dr D. A. Osborn, in their book *The Pathology of Granulomas and Neoplasms of the Nose and Paranasal Sinuses* (Churchill Livingstone, 1982).

Acknowledgements

The author is deeply grateful to his surgical and radiological colleagues for their cooperation and help in the preparation of the text. In this respect I am particularly indebted to Dr Peter Phelps, Professor D. F. N. Harrison, Miss V. J. Lund and Mr John Wright; many of the patients described presented initially at his Orbital Clinic at Moorfields Eye Hospital. I would also like to thank Mrs Joanne Evans and her staff of radiographers at the Royal National Throat, Nose and Ear Hospital and Mr Andrew Gardner of the photographic department for their assistance with the illustrations.

I am indebted to the Editors of the following journals for permission to reproduce previously published illustrations: *British Journal of Radiology* (Figs. 2.1, 2.2, 6.18, 7.5, 7.6, 7.10, 7.11, 7.12, 7.13, 7.14, 7.15, 8.5, 8.6, 8.7, 10.1, 10.2, 10.3 10.4, 10.5, 10.6, 10.7, 10.8, 10.9, 10.10, 10.22, 10.26, 13.10, 13.16, 14.4, 14.6, 15.2, 15.3, 15.14, 15.15, 18.11, 18.12); *Clinical Radiology* (Figs. 6.38, 6.40, 6.41, 6.42, 6.43, 6.44, 6.45, 15.12); *Clinical Otolaryngology* (Figs. 11.1, 11.2, 11.3, 11.4, 11.5, 11.6, 11.7, 11.8, 11.9, 11.10, 11.11, 11.12, 11.14, 11.15); *Journal of Laryngology and Otology* (Figs. 5.19, 6.26).

I am also grateful to W. B. Saunders and Co. for permission to reproduce Figs. 1.15, 4.12, 4.15, 5.6, 5.12, 5.17, 6.17, 6.37, 7.1, 10.11, 11.19 and 18.3; to H. K. Lewis and Co. for Figs. 3.5 and 5.23; and to Messrs Churchill Livingstone for Figs. 4.5, 4.6, 4.7, 4.8, 4.10, 4.11, 4.18, 10.13, 10.27 and 10.28.

Schering Health Care Ltd. kindly supplied the gadolinium DTPA for use with the magnetic resonance studies.

London, March 1988 G.A.S.L.

Contents

1 Basic Radiographic Technique and Normal Anatomy

Basic Radiographic Technique

Several technical factors are essential for good radiography of the sinuses. These include accurate coning of the incident beam, a fine focal spot X-ray tube, and a Potter Bucky or fine grid to obtain maximum contrast. The standard projections which may be employed are:

1. *Occipito-mental projection.* The subject sits facing the film and the radiographic base line is tilted to 45°. The incident beam is horizontal and is centred on the occipital bone 3 cm above the external occipital protuberance (Fig. 1.1). With most subjects a tube tilt is unnecessary since they can readily extend the head to the required position, but in older persons it may be necessary to tilt the tube slightly to compensate for any restricted extension of the head. This view shows the maxillary antra free of any overlap of the petrous bones, and if the mouth is kept open during the examination the sphenoid sinuses and nasopharynx can be seen through the open mouth (Fig. 1.2).

2. *Occipito-frontal projection.* The subject sits facing the film with the orbito-meatal line raised 20°, the incident beam horizontal and the tube centred to the nasion (Fig. 1.3). This projection demonstrates the fine detail of the frontal sinuses; the lateral walls of the antra are also seen, although the overlapping petrous temporal bones largely obscure the antra (Fig. 1.4).

3. *Lateral projection.* The subject sits with the radiographic base line horizontal and the sagittal plane parallel to the film. The incident beam is centred through the antrum (Fig. 1.5). The superimposition of the frontal sinuses and also of both maxillary antra detracts somewhat from the value of this projection. An alternative version of the lateral sinus projection, which better demonstrates the sinus and nasopharyngeal air spaces, is a high kilovolt lateral film using 150 kV or above and 3 mm of brass filtration (Fig. 1.6).

4. *Submentovertical projection.* The head is extended so that the vertex rests against the table top and the incident beam is centred between the angles of the jaw so that it is at right angles to the base line (Fig. 1.7). In middle-aged and elderly patients the tube has to be tilted upwards considerably as few patients in this age group can attain the degree of extension necessary to bring the skull base parallel to the film. This projection demonstrates the sphenoid sinuses and also the maxillary antra and orbital walls (Fig. 1.8).

5. *Oblique projection.* Rotation of the sagittal plane of the skull through an angle of 39° will enable the posterior ethmoid cells to be projected through the orbit and will show these cells largely clear of overlap shadows. The optic foramen is seen end-on in this projection.

Examination of the sinuses should always be made in the erect position with a horizontal X-ray beam to allow the demonstration of fluid levels within the sinus. Fluid in the sinus may be pus or mucopus occurring as a sequel of infection or allergy, but on occasion may be frank blood as a result of trauma. These different types of fluid cast shadows of identical density on the radiograph and the nature of the fluid cannot be recognised.

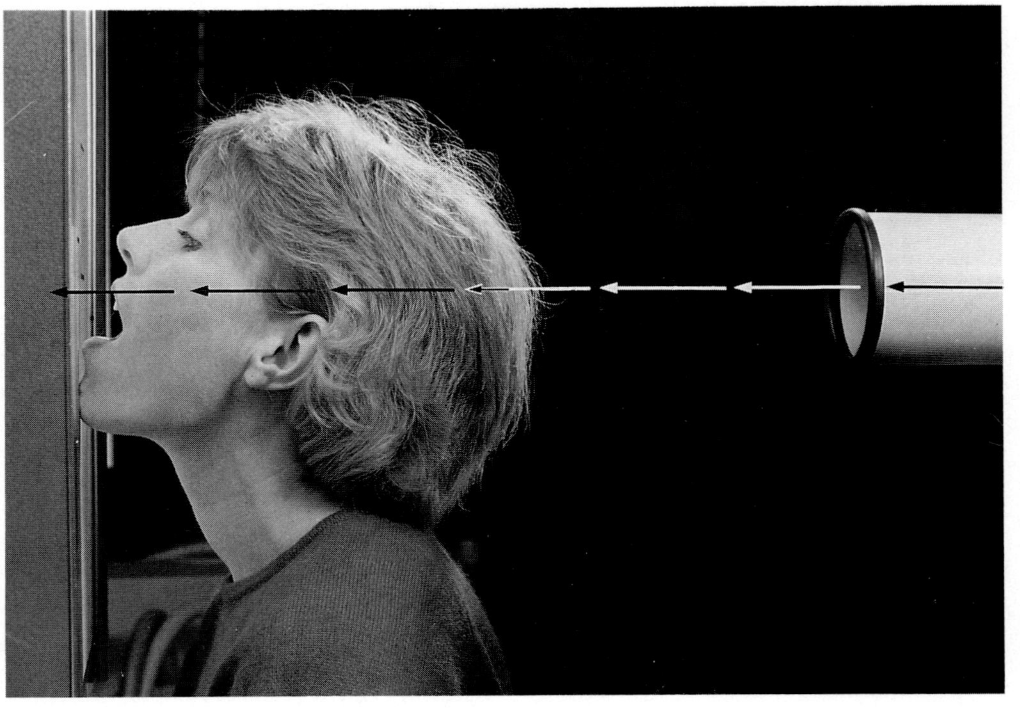

Fig. 1.1. Position for occipito-mental projection.

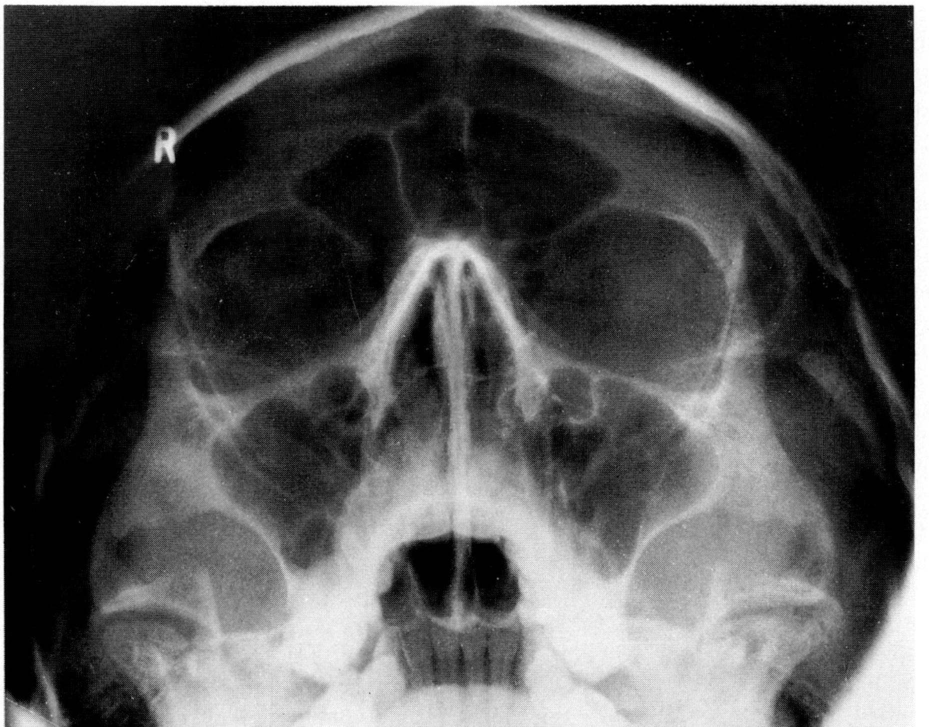

Fig. 1.2. Normal occipito-mental projection. The maxillary antra are shown free of any overlap of the petrous bones. The sphenoid sinuses are visible through the open mouth.

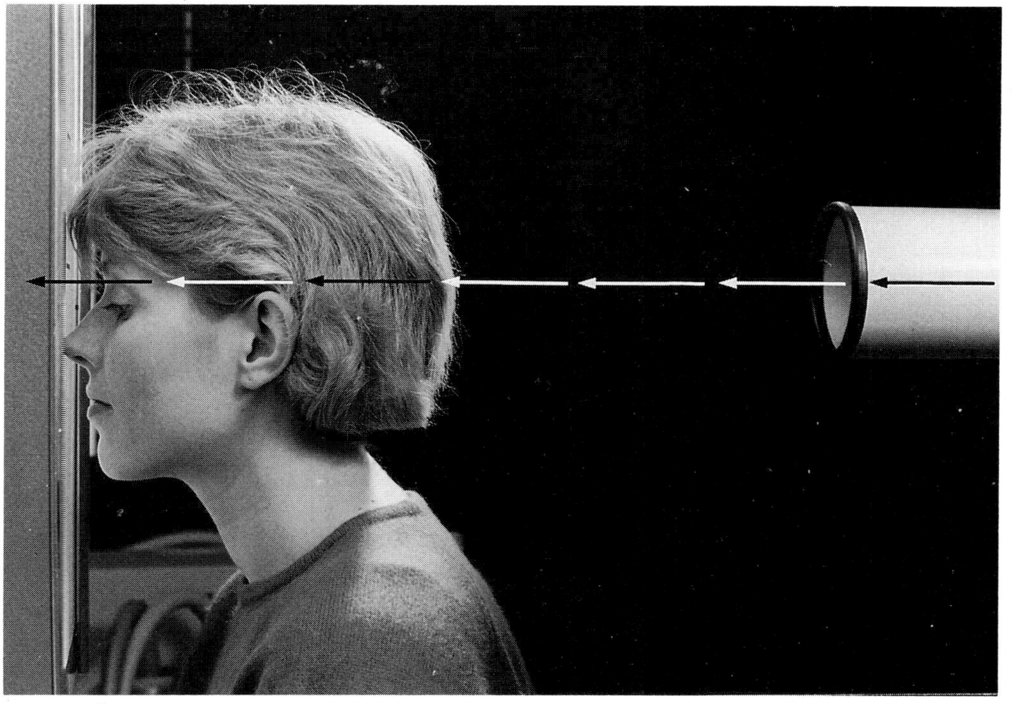

Fig. 1.3. Position for occipito-frontal projection.

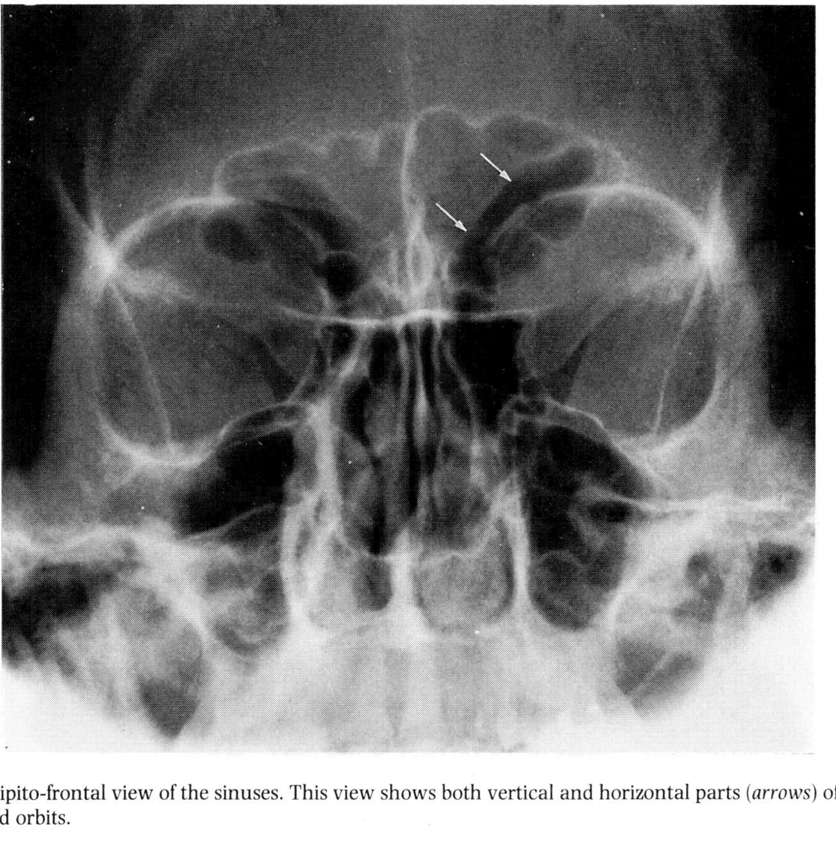

Fig. 1.4. Normal occipito-frontal view of the sinuses. This view shows both vertical and horizontal parts (*arrows*) of the frontal sinus, the ethmoid cells and orbits.

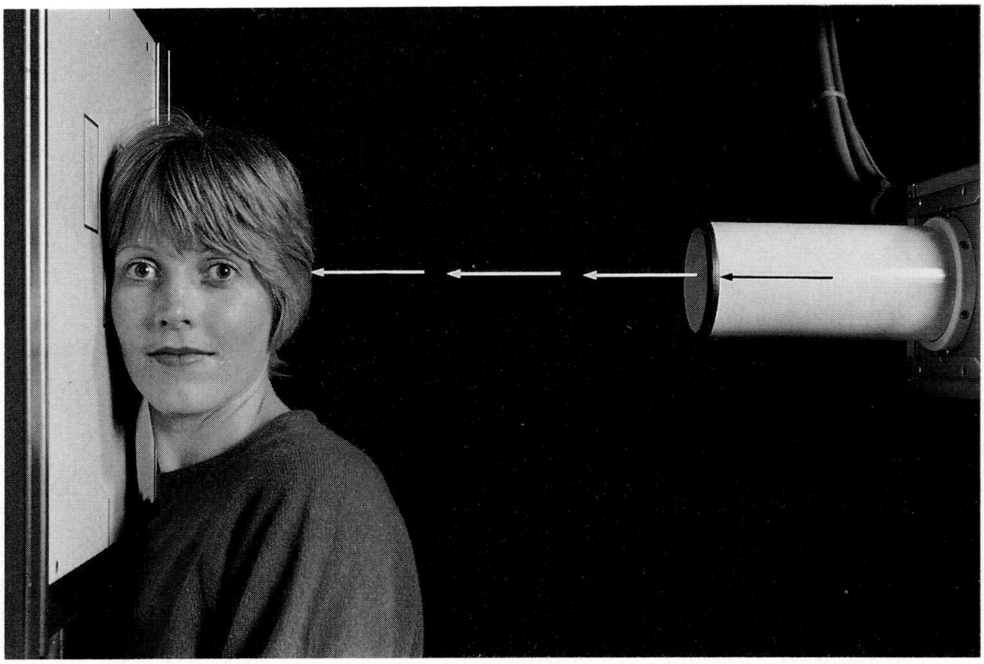

Fig. 1.5. Position for the lateral projection of the sinuses.

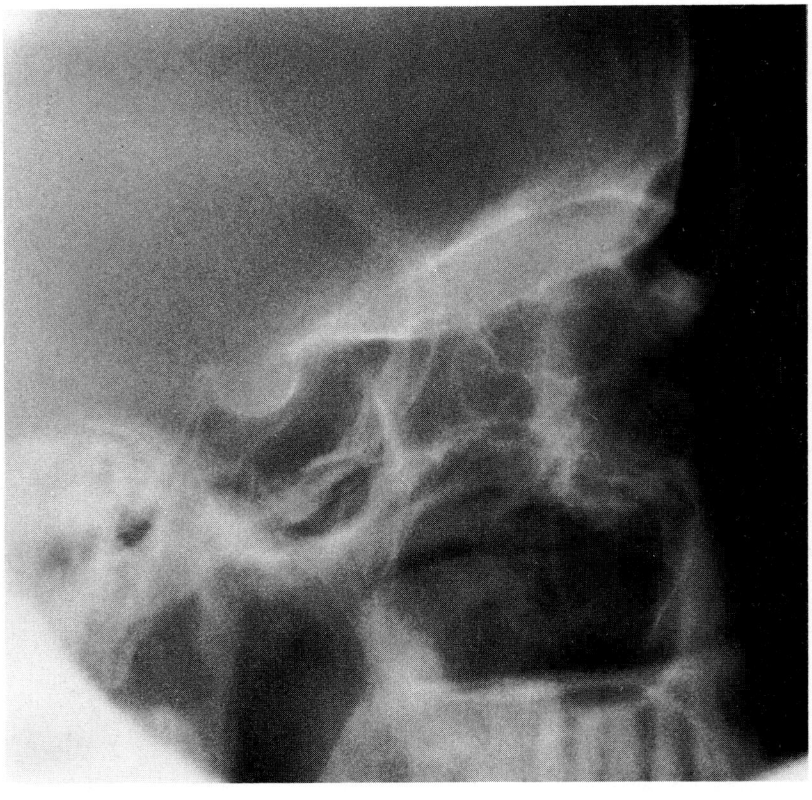

Fig. 1.6. Lateral high kilovolt view of the paranasal sinuses. The air spaces in the nose, sinuses and nasopharynx are better demonstrated by partial elimination of the bone structures.

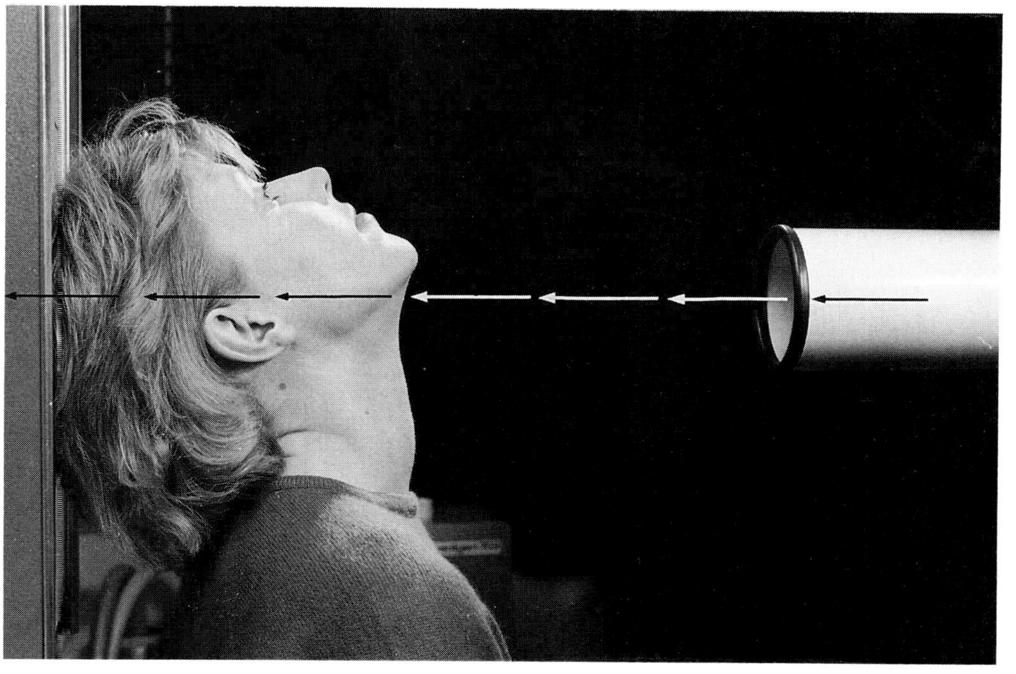

Fig. 1.7. Position for submentovertical projection.

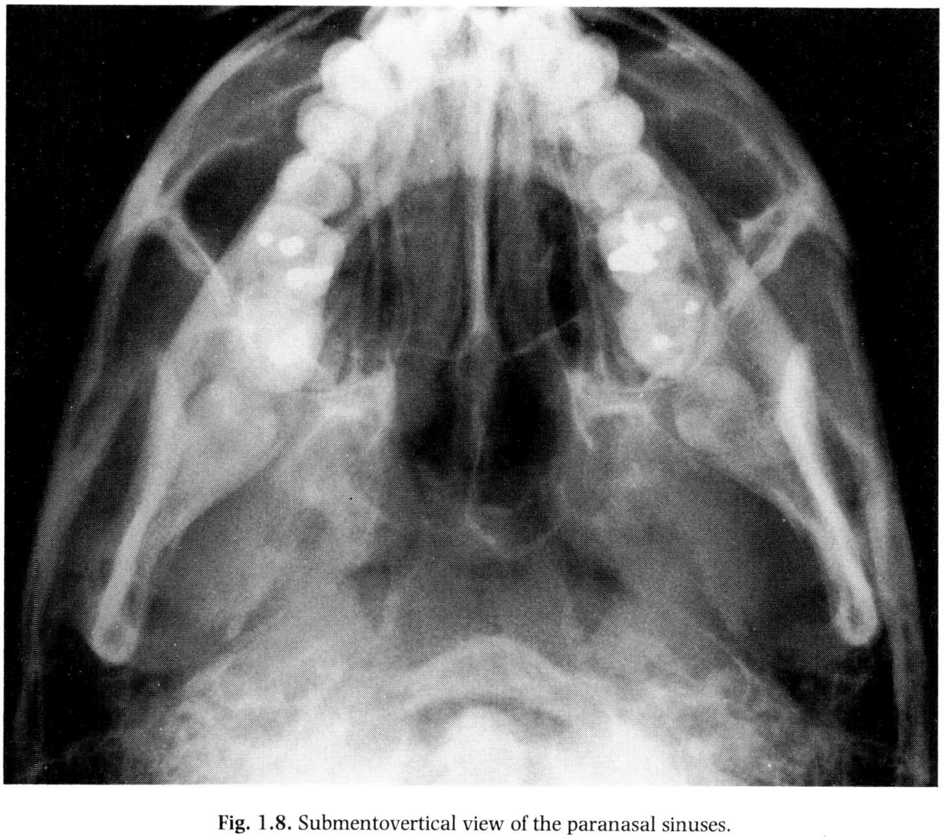

Fig. 1.8. Submentovertical view of the paranasal sinuses.

Nasal Bones

The nasal bones are shown in the occipito-mental view but bone detail is poor, and although fractures and displacements may be visible two further views are needed for their proper demonstration:

Lateral Projection

The examination is performed using non-screen film and with the patient seated. With the tube horizontal, the central ray is directed through the base of the nose to the centre of the film (Fig. 1.9).

Supero-inferior or Occlusal View

With the patient sitting, an occlusal film is placed lengthways between the teeth so that two thirds of the film projects forward. The central ray is directed along the vertical plane of the forehead and the tube centred in the mid-line through the nose (Fig. 1.10).

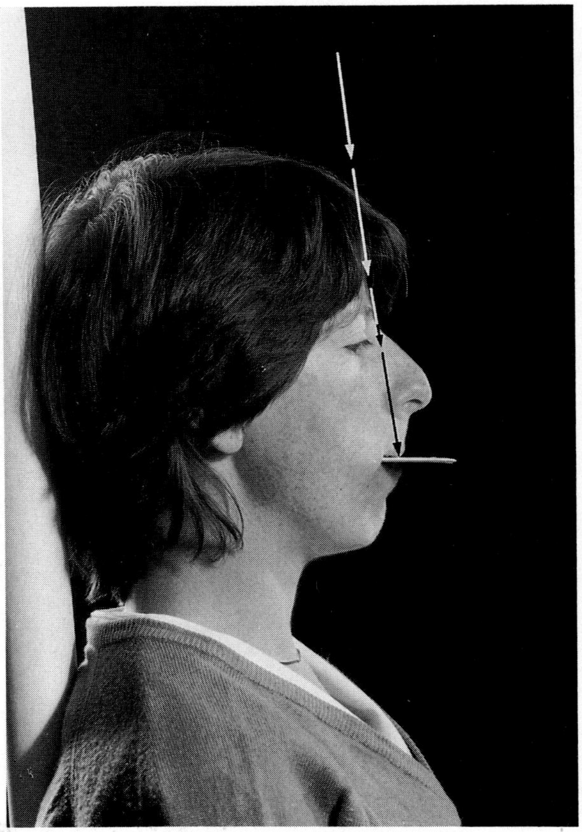

Fig. 1.10. Position for axial or supero-inferior occlusal view of the nasal bones.

Fig. 1.9. Lateral projection for the nasal bones.

Normal Anatomy

External Nose

The external nose is a pyramidal-shaped structure with its base directed inferiorly and supported by the facial bones. It consists of cartilage covered by skin and subcutaneous tissue and is lined by mucoperiosteum. The nasal cartilages which contribute to the structure of the nose consist of: (1) paired lateral cartilages superiorly; (2) two greater alar cartilages forming the tip of the nose and its lateral aspect; and (3) numerous lesser alar cartilages forming the lateral wall. These are supported by a mid-line triangular plate of thin cartilage – the mobile nasal septum – which is attached anteriorly along its inferior surface to the nasal anterior maxillary spine and to the perpendicular process of the ethmoid and vomer posteriorly. This cartilaginous plate is continuous posteriorly with the bony nasal septum. The nasal cartilages are attached superiorly to the paired nasal bones and supported on either side by the facial process of the maxilla.

The cartilaginous structure of the external nose is readily examined clinically, and radiology or other imaging methods are generally not required except to demonstrate changes in the nasal bones and anterior nasal spine.

Nasal Cavity

Superiorly the nasal cavity has a narrow roof formed by the cribriform plate of the ethmoid bone. Its lateral walls, which slope outwards and downwards, consist of the orbital plates of the ethmoid and palatine bones and the lacrimal bone, whilst its anterior margin is formed by the nasal process of the maxilla. The floor of the nasal cavity is formed by the palatal process of the maxilla. Posteriorly the nasal cavity opens into the nasopharynx by two oval openings – the posterior nares. Subdividing the nasal cavity is the bony nasal septum; this is formed by the vertical plate of the ethmoid in its upper two thirds, whilst the lower third is formed by the vomer.

Projecting from the lateral walls of the nose are the three paired turbinate bones running in a roughly horizontal direction, and dividing the nasal cavity into three sulci on each side: the superior, middle and inferior meatus. The superior meatus lies above the middle turbinate bone, and the posterior ethmoid and sphenoid cells drain into it. The middle meatus lies between the middle and inferior turbinates. On its lateral wall there is a curved depression – the hiatus semilunaris – into which drains the fronto-nasal duct anteriorly and behind it the ostium of the maxillary antrum. The inferior meatus is bounded above and medially by the inferior turbinate and laterally by the lateral wall of the nasal cavity. The naso-lacrimal duct drains into it.

Paranasal Sinuses

The paranasal sinuses consist of frontal sinuses, sphenoid sinuses, maxillary antra and the ethmoid complex of air cells. The sinuses lie within the facial bones and are grouped around the nasal cavities, which contain the superior, middle and inferior turbinate bones on their lateral wall.

Development

The sinuses are rudimentary at birth and the frontals not formed. They develop from buds of respiratory epithelium and grow rapidly from about the age of 7 until after puberty. The maxillary antrum is the first to develop and by the age of 2 years the outer wall of the antrum has extended laterally as far as the infraorbital foramen; by 8–10 years aeration is complete.

The frontal sinuses can seldom be demonstrated radiographically until they have extended into the vertical portion of the frontal bone, a process which varies considerably in its timing but which seldom occurs to any degree before the age of 2 years. By this time they generally measure less than 1 cm in height, reaching adult size by puberty. Pneumatisation of the sinus within the frontal bone proceeds in a postero-anterior direction; unilateral failure, or partial failure, of this process may lead to a thickening of the anterior wall of the adult sinus producing a relative loss of translucence on the affected side on plain sinus radiographs.

The sphenoid sinuses are absent at birth and do not extend below the planum sphenoidale until the age of 2–3 years. Pneumatisation inferior to the pituitary fossa can be seen by 4 years and is present in the majority at 8–10 years.

Frontal Sinuses

The fully developed frontal sinuses are two air-filled irregular cavities lying between the inner and outer tables of the frontal bone. They consist of a central cavity forming the main body of the sinus and two extensions: an extension upwards into the vertical part of the frontal bone; and a horizontal component, which extends backwards into the orbital plate of the frontal bone and pneumatises the orbital roof. The two frontal sinuses are separated by a bony or fibrous septum, which is seldom exactly mid-line in position; they communicate with the middle meatus of the nose by the fronto-nasal duct. This varies greatly in diameter, length and direction and is often very tortuous, being encroached upon by the fronto-ethmoidal air cells. Sometimes the ducts open directly into the ethmoid cells instead of into the nose.

The frontal sinuses are almost never symmetrical: one or both components of the sinus cavity may be congenitally absent or underdeveloped. Incomplete pneumatisation of the orbital roof is the commonest variant. Sometimes the frontal sinuses are totally absent – a feature usually associated with a persistent metopic suture.

Sphenoid Sinuses

The sphenoid sinuses are paired cavities occupying the body of the sphenoid bone. They are separated by a thin bony septum and each drains into the superior meatus of the same side with the posterior ethmoid cells. In the average subject pneumatisation extends posteriorly for approximately two thirds of the sphenoid bone, but the degree of pneumatisation varies widely. On occasion complete failure of pneumatisation may occur, or conversely the whole bone is completely pneumatised, the posterior clinoids being involved in the process. Sometimes the pneumatisation may involve the pterygoid processes, the greater wings of the sphenoid and exceptionally the basilar process of the sphenoid bone. The sphenoid sinuses may also encroach upon the antral cavity and extend into the ethmoidal field.

Maxillary Antra

The adult maxillary antra are a pair of air-filled cavities lying on either side in the body of the maxilla. Viewed in the postero-anterior plane they are pyramidal in shape with their apices directed downwards. The medial boundary of the antrum is formed from the lateral wall of the nasal cavity, and through it the antral ostium opens into the middle meatus. Anteriorly in the medial wall the naso-lacrimal duct runs in a bony canal into the inferior meatus. The roof of the antrum is the bony floor of the orbit, across which is a bony ridge enclosing the infraorbital nerve and vessels. Inferiorly the floor of the antrum is formed from that portion of the alveolar process of the maxilla containing the molar and pre-molar teeth. Beneath the cheek the antero-lateral wall is formed from the facial aspect of the maxilla, and behind, the postero-lateral wall separates the antrum from the pterygo-palatine fossa and the infratemporal fossa. The lateral wall is grooved by branches of the superior dental vessels and this may show as a small dehiscence on the radiograph.

Although they are the most symmetrical of the paranasal sinuses some variation in the size of the antra may occur in the normal subject. The commonest defect is a failure of pneumatisation of the alveolar recess, but not infrequently the whole air cavity is under-pneumatised (Fig. 1.11). The resultant increase in the thickness of the antral walls will produce a relative loss of translucence on the radiograph on the side affected, and may lead to misinterpretation of the change as sinus infection or mucosal thickening.

Sub-division of the antral cavities can occur by the formation of septa (Fig. 1.12), which may be bony or membranous, partial or complete. Not infrequently a posterior ethmoid cell grows downwards into the body of the maxilla and may encroach on the sinus, usually occupying the supero-medial part of the antral cavity. These aberrant cells may become infected without involvement of the antrum, or they may expand into the antral cavity as part of a spheno-ethmoidal mucocoele.

Ethmoid Labyrinth

The ethmoids are a complex of small air cells lying in the lateral wall of the nasal cavity between its upper part and the medial wall of the orbit and are separated from the anterior cranial fossa by the orbital plate or roof of the ethmoid bone. The medial wall is formed by a thin plate of bone to which are attached the superior and middle turbinates, and into which the air cells may sometimes extend (Fig. 1.13). The lateral wall of the ethmoid labyrinth consists of a thin lamina papyracea which separates the air cells from the orbit. Inferiorly the ethmoid cells articulate with the maxilla. The cells forming the ethmoid labyrinth are divided into anterior, middle and posterior groups. The posterior group of cells drain into the superior meatus in company with the sphenoid sinuses. The middle and anterior groups drain into the middle meatus; the anterior group into the infundibulum or the hiatus semilunaris; and the middle group posteriorly in the region of the bulla ethmoidalis.

The ethmoid cells may migrate during their development beyond the confines of the normal

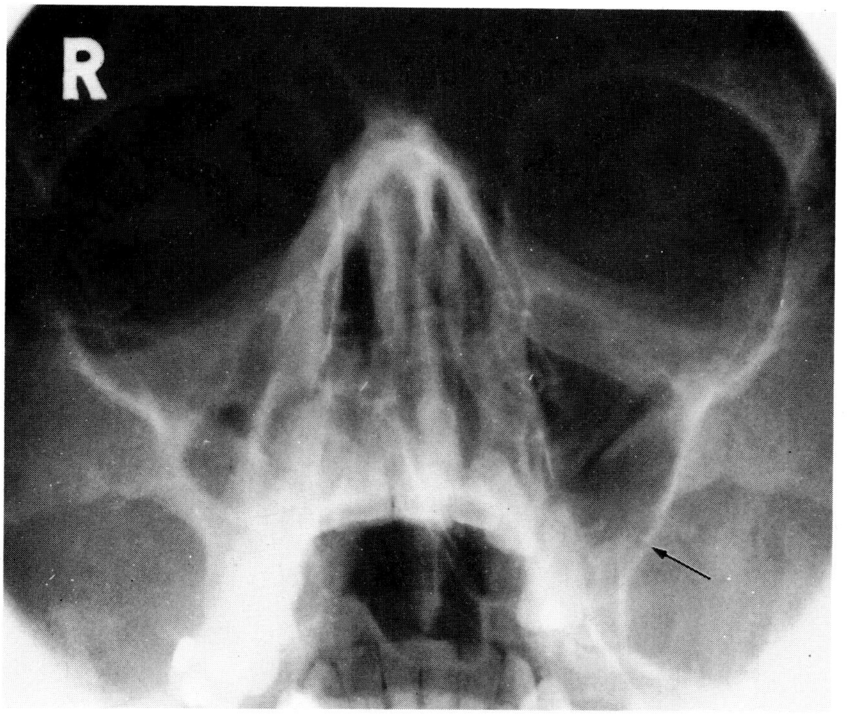

Fig. 1.11. Occipito-mental view of the maxillary antra showing underdevelopment of the sinus on the right side. On the left there is a bony groove on the lateral wall of the antrum caused by the superior dental vessels and nerves (*arrow*).

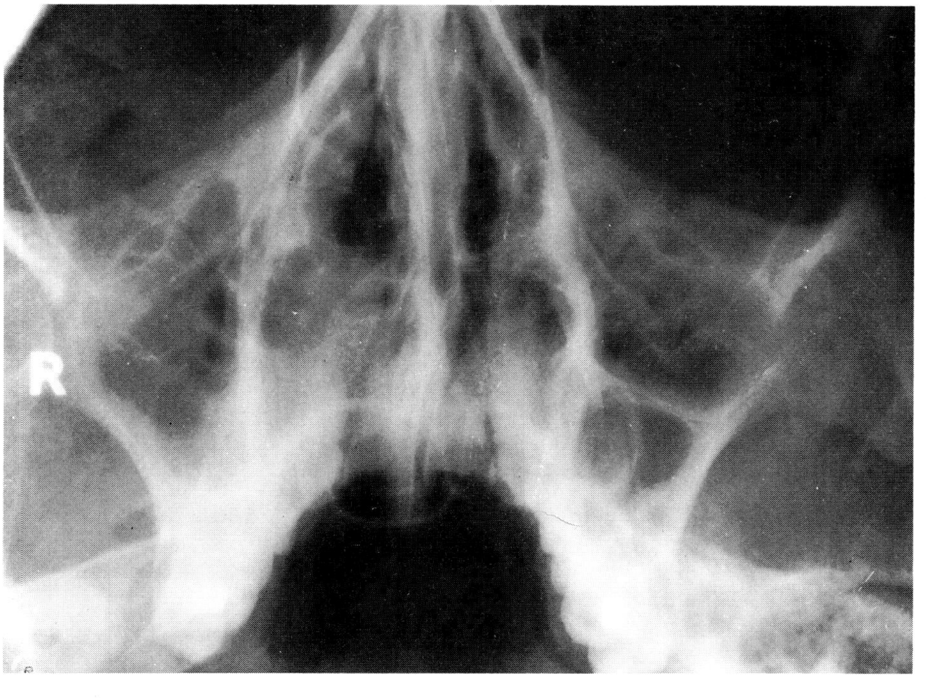

Fig. 1.12. Septate antrum. Occipito-mental view showing a bony septum traversing the antral cavity.

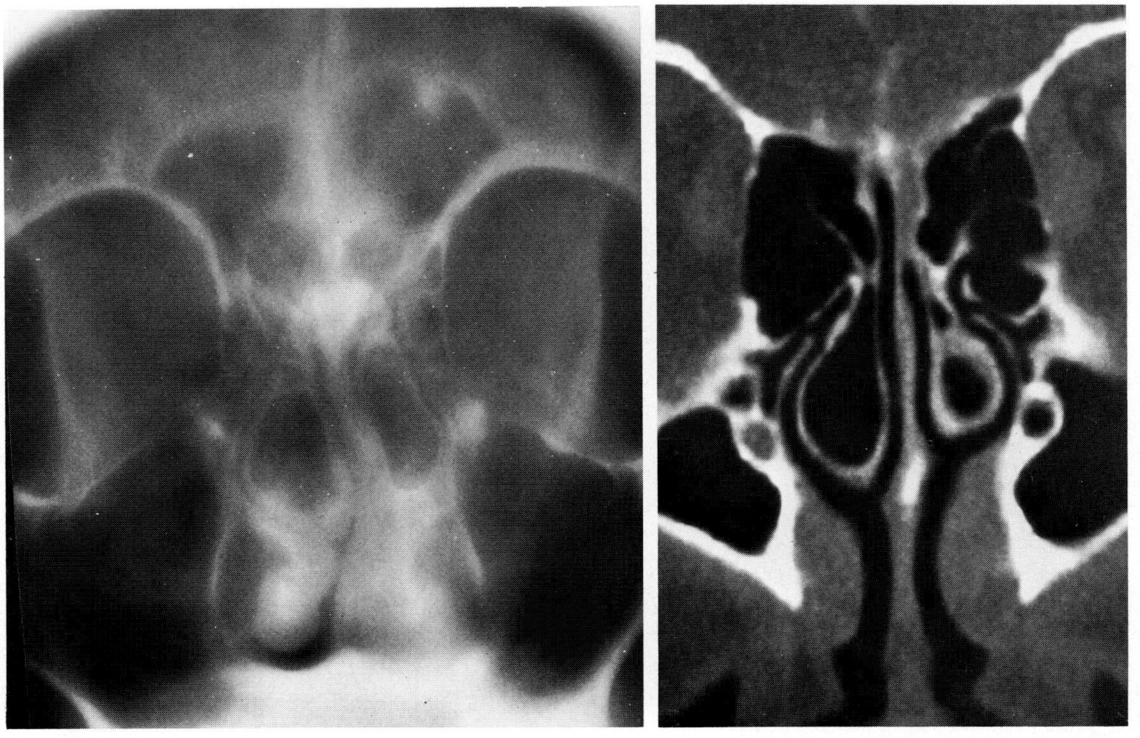

Fig. 1.13a, b. Pneumatisation of the middle turbinate shown on **a** hypocycloidal tomography and **b** computed tomographic scan in coronal sections.

ethmoid labyrinth. These are known as agger cells and may occur in the sphenoid bone, the frontal bone and the nasal bones, and may pneumatise the ethmoid conchae.

The normal anatomy of the sinuses as demonstrated by computerised axial tomography is shown by the series of scans in Fig. 1.14.

Developmental Variations

An increased pneumatisation of the frontal sinuses associated with overdevelopment of the supraorbital ridge is usually a prominent feature of acromegaly, and cerebral atrophy or agenesis of a cerebral hemisphere may also result in overpneumatisation of the frontal sinus. Congenital absence of the frontal sinus is said to occur in 5% of the normal population and is often associated with persistence of the metopic suture. Underdevelopment may also be seen in premature fusion of the cranial sutures (craniostenosis) and absence of the frontal sinuses occurs in the majority of patients with Down's syndrome. Agenesis or underdevelopment of the frontal sinuses is also a feature of Kartagener's syndrome. Kartagener (1933) described a triad of abnormalities consisting of chronic rhinosinusitis, bronchiectasis and situs inversus viscerum. This has recently been demonstrated to be the result of a genetic defect manifest in structural and functional abnormalities of the cilia (Imbrie 1981).

The ethmoid cells show increased width and overpneumatisation in hypertelorism from whatever cause, congenital or acquired: for example in the naso-encephalocoeles, or in untreated nasal polyposis (Lund and Lloyd 1983). Hypoplasia of the ethmoids is seen in hypotelorism, which is a narrowed space between the orbits and occurs in several congenital syndromes affecting the facial bones, notably the holoprosencephalies (Becker and McCarthy 1986).

Fig. 1.14. 1, Frontal sinuses: anterior and posterior walls; 2, crista galli; 3, anterior ethmoid cells; 4, lamina papyracea of the ethmoids; 5, sphenoid sinuses; 6, posterior ethmoid cells; 7, sphenoid intersinus septum; 8, nasal septum; 9, summit of the antrum; 10, pterygo-palatine fossa; 11, pneumatised pterygoid plates; 12, naso-lacrimal duct; 13, maxillary antrum; 14, infratemporal fossa and fat pad; 15, pterygoid laminae; 16, lateral pterygoid muscle. ▶

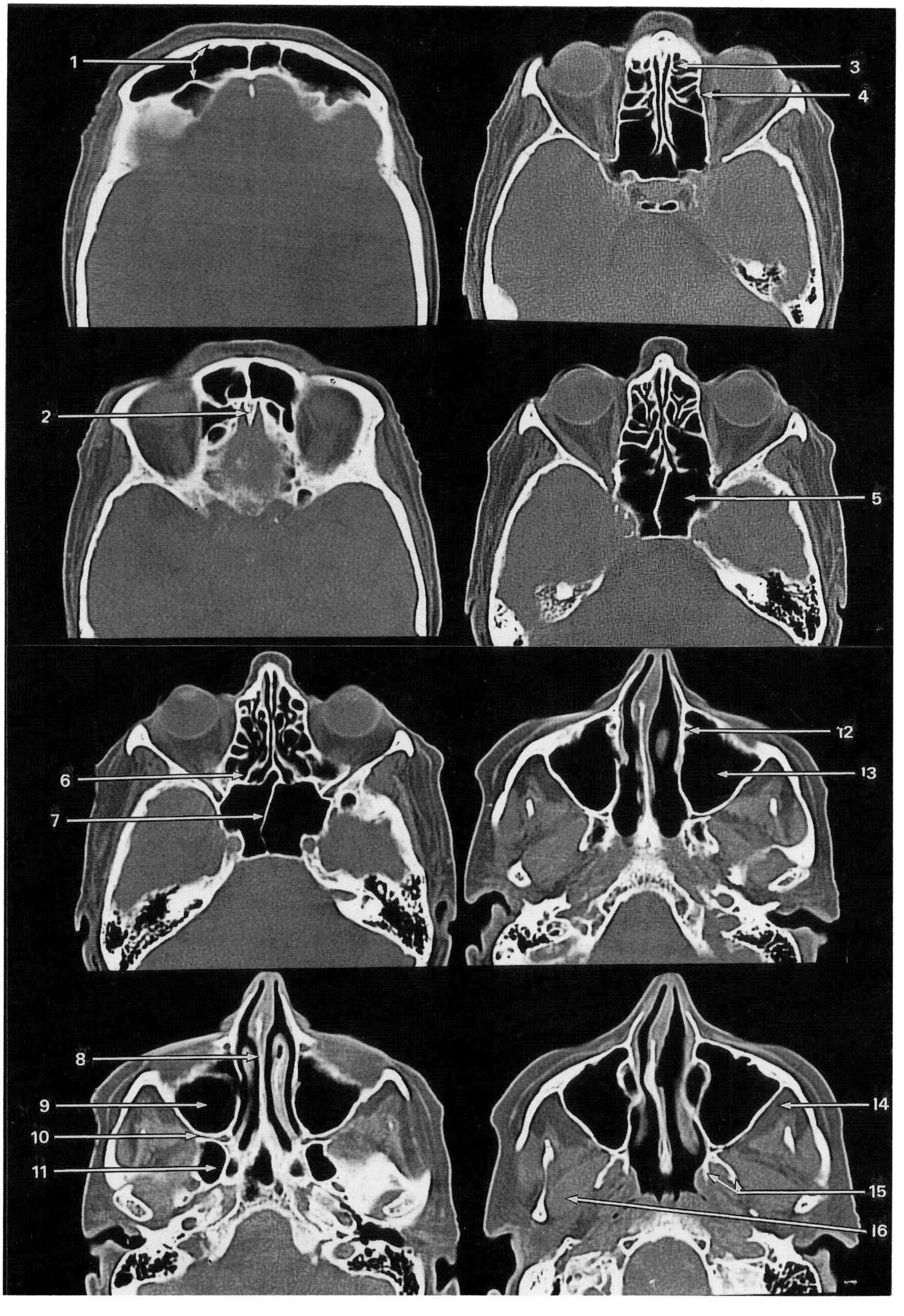
1
3
4
2
5
6
7
12
13
8
9
10
11
14
15
16

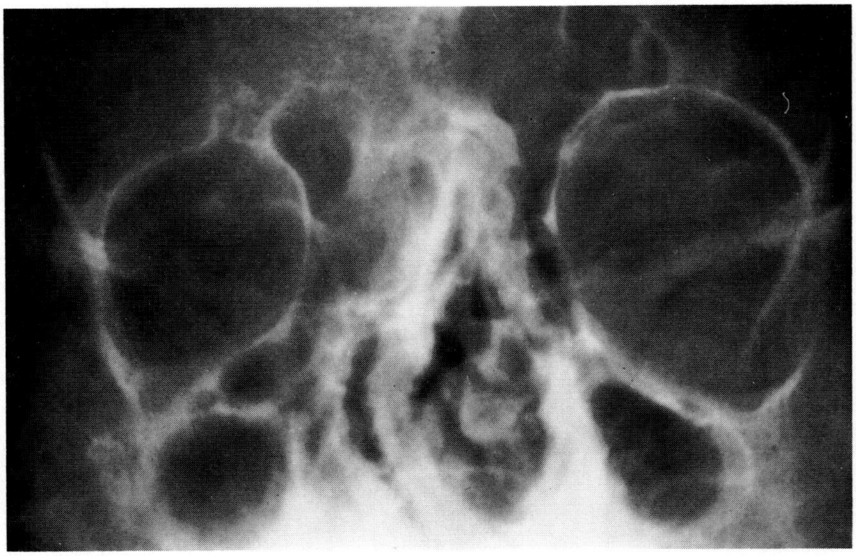

Fig. 1.15. Bilateral anophthalmos. The sinuses have overgrown to occupy the space left by the absent eyeballs.

The maxillary antra seldom show any appreciable over-pneumatisation, but relative over-development may occur in microcephaly (Samuel and Lloyd 1978) and may also be seen in congenital anophthalmos in which the ethmoid cells and to a lesser extent the maxillary antra overgrow, to occupy the space left by the absent eyeball (Fig. 1.15). Total absence of pneumatisation of the antra may be seen in sickle cell anaemia and in thalassaemia due to bone marrow hyperplasia.

Local over-pneumatisation of the paranasal sinuses (pneumosinus dilatans) may occur in the presence of fibro-osseous disease or as the response to a meningioma (Lloyd 1985) (see Chap. 6).

References

Becker MH, McCarthy JG (1986) Congenital abnormalities. Diagnostic imaging in ophthalmology. Springer, Berlin Heidelberg New York

Imbrie JD (1981) Kartagener's syndrome: a genetic defect affecting the function of cilia. Am J Otolaryngol 2:215–222

Kartagener M (1933) Zur Pathogenese der Bronkiektasien, Bronkiektasien bei situs viscerum inversus. Beitr Klin Tuberk 83:489–495

Lloyd GAS (1985) Orbital pneumosinus dilatans. Clin Radiol 36:381–386

Lund VJ, Lloyd GAS (1983) Radiological changes associated with benign nasal polyps. J Laryngol Otol 97:503–510

Samuel E, Lloyd GAS (1978) Clinical radiology of the ear, nose and throat. HK Lewis, London

2 Special Procedures

The initial radiological investigation of the paranasal sinuses is by the techniques described in Chap. 1. If the appearances on plain radiography are consistent with allergic or inflammatory disease and there is good correlation of the clinical findings, no further investigation is usually necessary. If, however, the evidence from plain radiography indicates an expanding or destructive lesion of the sinus walls or a tumour mass, and the clinical findings suggest a less benign process (for example pain and paraesthesia, epistaxis, facial swelling, nasal mass or orbital involvement), tomographic investigation is indicated. There are now three tomographic techniques available.

Conventional Tomography

Conventional tomography has been almost completely superseded by high-resolution computerised tomography (CT) and now plays only a minor role in the investigation of paranasal sinus disease. It has been reduced to functioning as a simple adjunct to plain radiography at initial examination. For example it is useful in showing whether there is a nasal mass present in association with an opaque sinus, particularly if the maxillary antrum is involved. Small nasal masses often go undetected on plain radiographs, and patients who are found to have unilateral total opacity of the antrum should have immediate tomography as a screening procedure.

Ideally conventional tomography should be performed using a machine capable of complex motion tomography rather than a simple linear movement. A full series of paranasal sinus tomograms consists of:

1. Coronal projections using an undertilted occipito-mental position of the skull angled 30°–35° cranially, or a projection corresponding to the occipito-frontal view described in Chap. 1, with the forehead placed in contact with the table top and the orbito-meatal line at right angles to it.
2. Lateral tomograms.
3. Axial tomography. For these films a specially designed wooden platform is placed on the table top so that the head may be hyperextended into the submentovertical position (Lloyd 1975).

Computerised Tomography

CT scanning has provided an important addition to the radiographic investigation of the paranasal sinuses, and has virtually replaced conventional tomography as a means of assessing tumours, mucocoeles and other expanding lesions in the sinuses. CT has the advantage of showing both bone destruction and the soft tissue extent of disease. In malignant disease it provides an accurate method of staging a tumour prior to radiotherapy or surgery and is important post-operatively to show recurrence of tumour. In addition it has extended the possibilities of differential diagnosis in the sinuses, not only by showing the soft tissue pattern of sinus disease, but by a more sensitive demonstration of calcification within a tumour, which in some instances may be characteristic. Occasionally CT may allow the radiologist to make the primary diagnosis of sinus malignancy prior to the clinician, and to indicate the best area for confirmatory biopsy.

Technique

Routine axial and coronal sections are obtained on all patients. Direct coronal scanning is necessary for adequate demonstration of sinus disease. Reformatted views should be reserved for sagittal sections, which are not directly obtainable with most scanner designs. They should only be used to provide coronal scans when direct scanning is for any reason impossible – for example if a patient cannot extend the head or cervical spine.

Axial Scans

Axial scans should be orientated in the same plane as those used for CT of the orbit (Lloyd 1979). The position of the patient's head is adjusted so that the scanning plane forms an angle of 16° caudally from the orbito-meatal line: in this way the plane of section will conform to the length of the optic nerve, and will also provide axial views of the optic canals and the adjacent posterior ethmoid cells and sphenoid sinuses (Fig. 2.1).

Coronal Scans

Coronal scans are performed by hyperextension of the patient's head and angulation of the gantry with the patient either prone or in the supine position. In some patients it is impossible to obtain true coronal scans, either because the patient cannot achieve sufficient extension of the neck, or because the angulation may need to be adjusted out of the coronal plane to avoid the effect of metallic dental fillings. These will degrade the image unless suitable computer software modification is available to overcome the problem.

For imaging of the sinuses 5-mm sections in both planes are generally adequate, with contiguous slices through the lesion. Imaging should include both wide window settings for bone detail and narrower window widths for good soft tissue contrast: generally the window settings should be within the range of 200–3000 Hounsfield units.

Contrast Medium

The CT attenuation values of both normal and abnormal tissues generally show an increase after the administration of intravenous contrast medium. In sinus neoplasia the degree of enhancement varies with tumours of different histology and there is also a considerable variation within the same histological type. Enhancement usually correlates closely with the vascularity of the tissue concerned, so that strong enhancement is to be expected for inflammatory tissue while retained secretion and uninfected mucocoeles should not enhance. By utilising any differential contrast enhancement a distinction can sometimes be made between tumour and adjacent normal or inflammatory tissue.

In practice these differences are often unclear, largely because of the wide range of tumour enhancement encountered, and this method of

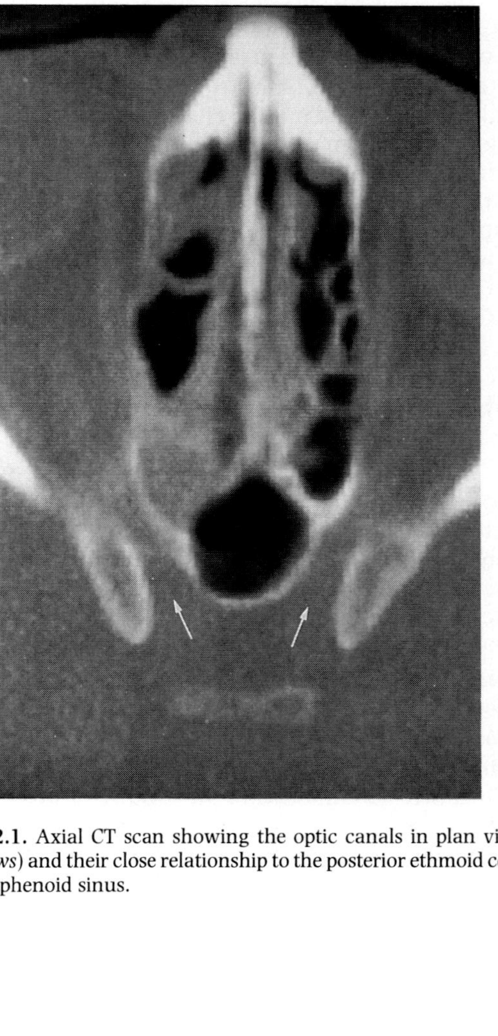

Fig. 2.1. Axial CT scan showing the optic canals in plan view (*arrows*) and their close relationship to the posterior ethmoid cells and sphenoid sinus.

assessing tumour extent and recurrence has now largely been replaced by the use of magnetic resonance tomography using paramagnetic contrast medium (see below). In these circumstances little or no added information is provided by giving contrast medium prior to CT of the nose and sinuses when magnetic resonance tomography is available. Intravenous contrast should be reserved for the following categories of patients:

1. Patients with vascular tumours such as angiofibroma. In these a bolus injection or drip infusion should be employed, scanning taking place during the actual administration of the contrast to catch the vascular phase of tumour enhancement.
2. Patients with suspected tumour spread into the anterior or middle cranial fossae, i.e. when the blood-brain barrier is involved. The tumour is then outlined against the non-enhancing brain tissue.
3. Patients with sinus infection in whom abscess formation is suspected either in the anterior fossa or in the orbit, when there is an associated orbital cellulitis. This also applies to pyocoeles, which may show a typical ring enhancement after contrast.

Magnetic Resonance Tomography

Principle and Method of Examination

Atomic nuclei with an odd number of protons or neutrons possess a magnetic moment and behave like spinning magnets. Hydrogen has such a nucleus and because of its wide distribution in the body is the element most commonly used for magnetic resonance tomography. When placed in a static magnetic field the hydrogen nuclei align their magnetic axes parallel to the field, the sum of these nuclear magnets giving rise to a weak magnetisation through the patient in the direction of the field, because more of these nuclear magnets align with the field than against it.

Application of an oscillating radio-frequency pulse of a specific frequency introduces energy into the patient and can cause the magnetisation to rotate into the transverse plane when a 90° pulse is used. Alternatively the magnetisation vector can be completely reversed in direction by a 180° pulse. When the pulse is discontinued the nuclei tend to return to their original orientation in the static magnetic field, while emitting the absorbed energy. When at 90° to the field, the transverse nuclear magnetisation will cut the receiver coils and induce a resonant signal voltage in the coil. The return to equilibrium for the component in the direction of the static magnetic field is exponential and is described by the time constant T_1. The return to equilibrium for the transverse component is also exponential and is described by the time constant T_2.

The intensity of the signal is related to the proton density, which refers to the distribution of resonating hydrogen nuclei within the patient. It must be emphasised, however, that not all protons give a magnetic resonance signal. The protons in large molecules such as proteins do not as a rule contribute to the signal, nor is there signal from solid structures such as bone. The distribution of the resonating protons is fairly uniform in the soft tissues and differences in density therefore slight. Contrast between areas of differing proton density can be enhanced if the scan is biased towards T_1 or T_2 relaxation characteristics. In practice, therefore, the resultant image is affected by the proton density and by one or other, or both, of these components.

Pulse Sequences

Three pulse sequences are commonly used in magnetic resonance tomography:

Saturation Recovery. This is the simplest pulse sequence. Ninety-degree radio-frequency pulses are repeatedly applied to the patient and the nuclear magnetic resonance signal is measured after each pulse. Provided the repetition time is short in relation to the T_1 relaxation time of the tissue concerned, a proton density image with some T_1 weighting will be produced.

Inversion Recovery. This sequence produces images which contain a greater amount of T_1 information than is provided by saturation recovery, although the principle is very similar. The difference in inversion recovery is that the nuclei are caused to resonate by a 180° pulse prior to the 90° pulse, so that the nuclei are not at equilibrium when the 90° pulse is applied. The resulting image contains approximately twice the amount of T_1 information compared with the saturation recovery mode and is therefore to be preferred when T_1 spin characteristics are to be evaluated.

Spin Echo. The spin echo pulse sequence produces a signal to which both T_1 and T_2 contribute. Although it contains information about T_1 and T_2 it is principally to obtain the latter that it is used in clinical practice. The pulse sequence is the reverse of inversion recovery in that a 90° pulse is followed by a 180° pulse. The time between the initial 90°

pulse and the signal is known as the time to echo; the larger this value the greater the T_2 contribution to the signal.

Paramagnetic Contrast Agents and Proton Relaxation

It is the water content of the tissues which provides the signal for magnetic resonance tomography and it is the behaviour of this water which affects the relaxation times T_1 and T_2. A proportion of the water in the tissues is bound to the surface of proteins, which has the effect of lowering the T_1 value. Unbound (free) water has a much higher T_1 value than the bound water and the T_1 value of any given tissue will depend upon the ratio of bound to free water: the higher the proportion of free water, the higher the T_1 values and vice versa. It is thought that the increase in T_1 in tumour tissue compared with normal tissue is dependent upon the release of free water resulting in a change in the ratio of free to bound water.

The T_1 and T_2 relaxation times can also be affected by the presence of paramagnetic substances. Paramagnetic ions have magnetic moments that are of the order of 1000 times as large as that of protons (Carr and Gadian 1985). These produce large local fields and can enhance the relaxation rates of water protons in the immediate vicinity of the ions. It is found that the increase in the relaxation rate is directly proportional to the concentration of the paramagnetic agent and to the square of its magnetic moment.

The first paramagnetic agent to be introduced into clinical practice as a magnetic resonance contrast medium is the substance gadolinium DTPA. Gadolinium is a very effective paramagnetic agent but as a free ion it is toxic to liver, spleen and bone marrow. However, when chelated to diethylenetriamine penta-acetic acid (DTPA) its toxicity is reduced, permitting it to be used as a safe relaxation enhancing agent. In the soft tissues it is distributed mainly in the extracellular space, and to date no short-term toxicity has been detected (Carr and Gadian 1985).

Technique and Application

Multi-slice Facility. An advantage of magnetic resonance tomography over other methods is the multi-slice facility which is standard on most current machines. This allows multiple sections to be obtained simultaneously using either 1-cm or 0.5-cm contiguous slices to a depth of up to 12 cm. Coupled with the use of a head coil and three-plane imaging it provides total coverage of the head and neck and allows identification of associated disease away from the primary site in the paranasal sinuses: for example neck malignancy (Fig. 14.6, p. 146). This is of great importance to the oncologist in treatment planning and represents a major advance over CT scanning.

Slice Thickness. Thin slices are advantageous in magnetic resonance tomography when trying to visualise small areas. Definition is improved because the amount of overlap between structures lying obliquely through the slice is reduced and their edges become more distinct. However, thin sections suffer from an important disadvantage: the thinner the slice the greater the amount of "noise" on the scan, increasing the signal to "noise" ratio and degrading the image. The effect on the image is less obvious at high static field strengths because the machines with the more powerful magnets possess a higher inherent spatial resolution than low field strength systems.

Choice of Pulse Sequences. For optimum tissue characterisation T_1- and T_2-weighted sequences are required. Spin echo sequences using a long time to echo will give maximum T_2 weighting and signal differentiation between tissues, depending upon the relaxation times of the tissues concerned. T_1 weighting can be achieved using the saturation recovery technique (see above) or a spin echo sequence with a short repetition time, producing images of good anatomical detail and containing some T_2 information. Greater T_1 weighting is achieved, however, by using inversion recovery sequences, and these are to be preferred when available. Multi-slice inversion recovery sequences are not always obtainable on 1.5-tesla scanners, which makes these machines somewhat less versatile than models with medium strength static fields. Inversion recovery sequences are especially important when a paramagnetic contrast agent is employed (see below).

Tissue Characteristics

In general malignant tumours of sinus origin, whether epithelial or mesenchymal, produce signal of medium intensity on T_1-weighted spin echo sequences and a medium to strong signal on T_2 images. In contrast, retained secretion produces high signal on spin echo sequences. Differentiation between tumour and retained secretion is particularly striking on T_2-weighted images, the retained secretion always giving a higher signal than tumour. An additional feature is the heterogeneous signal of tumour in comparison with that shown by retained secretion, which is invariably homogeneous. The vascularity of the tumour is a major contributor to the lack of homogeneous signal seen in juvenile angiofibroma. In these benign tumours large vessels can be identified, both in the tumour itself and in the adjacent musculature. They are shown as areas of negative signal or signal void and when present are totally diagnostic (Lloyd and Phelps 1986; see Chap. 11).

Magnetic resonance scanning can show simple inflammatory or allergic changes in the sinuses. It is possible to show single or multiple polyps, thickened mucosa, or fluid levels in the presence of infection. These conditions are generally well demonstrated by conventional radiography and the need to recognise them is simply to be able to distinguish them from more serious disease. One of the advantages of magnetic resonance scanning is the strong signal which is received on T_2-weighted spin echo images from retained mucus or mucopus in the sinuses. This enables an important distinction to be made between tumour in the sinuses and secondary mucocoele formation, even within the same sinus cavity (Fig. 2.2). In the same way, primary mucocoeles or pyocoeles are optimally demonstrated by this technique (see Fig. 6.20, p. 59). Because of its highly vascular nature, nasal mucosa gives a signal similar to inflamed or oedematous sinus epithelium: that is, a high-intensity signal on T_2-weighted spin echo sequences, allowing discrimination between tumour and normal mucosa. On the other hand, when T_1-weighted sequences are used the signal intensities are very similar and tumour differentiation is difficult. Inflamed or oedematous sinus mucosa also produces a signal of high intensity on T_2-weighted spin echo sequences, and may be distinguished from tumour by this means. This distinction is especially apparent on heavily weighted T_2-weighted sequences, using a long time to echo and a long repetition time (Fig. 18.12, p. 172). This is a valuable feature of the technique, both for post-operative assessment of possible tumour recurrence and for monitoring the effects of radiotherapy or chemotherapy.

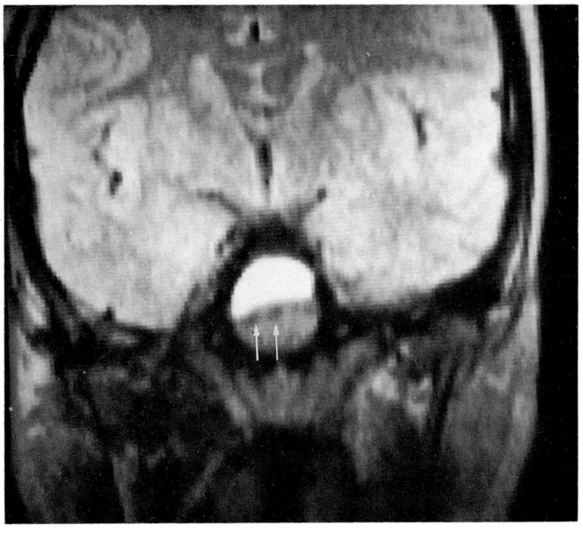

Fig. 2.2. Coronal magnetic resonance scan (T_2-weighted sequence) showing a slightly expanded sphenoid sinus containing tumour in the lower part and retained secretion in the upper. The tumour was an angiofibroma. Areas of signal void are visible (*arrows*) due to large vessels in the tumour mass.

Nasal polyps are oedematous, prolapsed mucosa of the ethmoidal cells, with a histological picture of respiratory epithelium covering a grossly oedematous stroma. The differentiation of retained secretion in the sinuses and nasal polyps is less easily made by magnetic resonance than the distinction between retained secretion and a true neoplasm, either benign or malignant. The close signal characteristics of an oedematous polyp and its associated mucosal secretion make the differentiation difficult, particularly on T_1-weighted images, but differentiation may be possible on T_2-weighted pulse sequences.

The spin characteristics of mucocoeles vary with their surgical history: that is, whether or not the patient has received any previous surgical treatment for the condition. This is important because the presence of haematoma or altered blood affects the spin characteristics. This is thought to be due to the breakdown of red cells, with the release of methaemoglobin – a paramagnetic agent (Gomori et al. 1985). An untouched mucocoele will present a high signal on T_2-weighted spin echo sequences and a very low signal on inversion recovery due to a long T_1 relaxation time (Figs. 6.20 and 6.21, pp. 59 and 60). In contradistinction, a mucocoele in which there has been previous surgical drainage may produce high signal on inversion recovery (Fig. 6.21). This effect is presumed to be due to the release of methaemoglobin from haematoma and shortening of the T_1 relaxation time.

In summary, it can be said that mucocoeles, pyocoeles, retained secretion and inflamed mucosa give a stronger signal on T_2-weighted spin echo sequences than do tumours. In addition, examples of dense fibrous tissue in association with paranasal sinus disease have shown entirely different magnetic resonance features, characterised by a low signal on T_2-weighted spin echo sequences. All these tissue characteristics are represented graphically in Fig. 2.3.

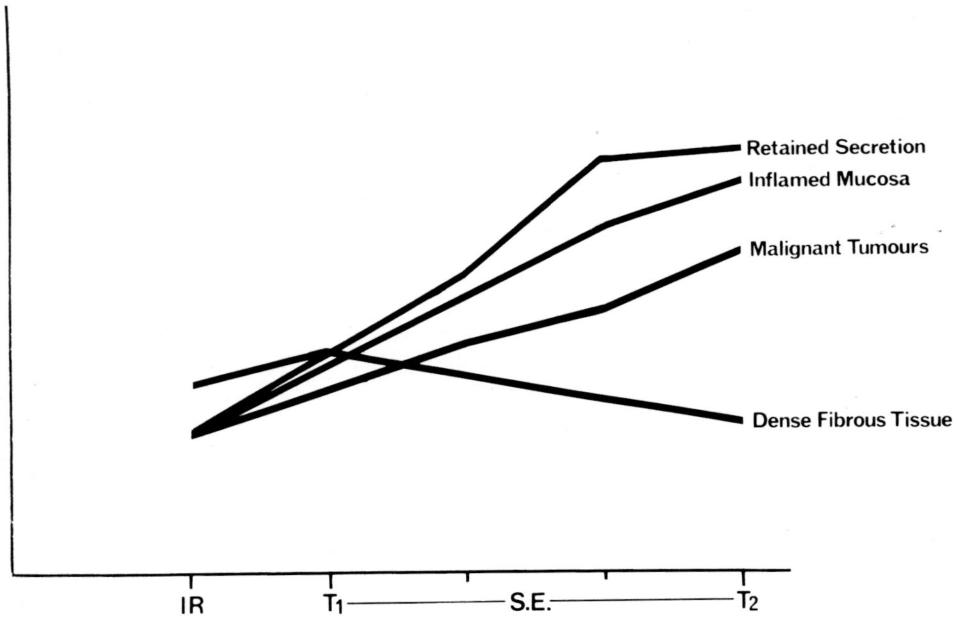

Fig. 2.3. Average signal intensities plotted against inversion recovery (IR) and T_1- and T_2-weighted spin echo (SE) magnetic resonance sequences. Tissue differentiation is possible between retained secretion in the sinuses, inflamed mucosa, malignant tumours and fibrous tissue.

Effects of Gadolinium DTPA

Intravenous injection of the paramagnetic contrast agent gadolinium DTPA has effects on the spin characteristics of both naso-sinus tumours and normal and inflammatory tissue in the nose and sinuses. As described above the majority of tumours show high or moderately high signal on T_2-weighted spin echo sequences. Without the use of contrast, tumours are best shown by this technique and best differentiated from retained secretion or inflammatory mucosa (Lloyd et al. 1987). After intravenous gadolinium DTPA tumour enhancement on the T_1-weighted spin echo and inversion recovery sequences varies from high intensity to zero, the degree of enhancement probably depending upon the vascularity of the individual tumour.

Post-contrast inversion recovery series have been shown to be the best method of tumour demonstration in over 80% of cases and were superior to pre-contrast T_2-weighted spin echo sequences in differentiating tumour both from retained secretion and from inflammatory mucosal thickening in the sinuses. Retained secretion does not enhance after contrast: this helps to make the differentiation between tumour and fluid in the sinuses more obvious than it is on the unenhanced T_2-weighted spin echo scans (Fig. 13.11, p. 137). This applies equally to secondary mucocoele formation and to sinuses in which there is both inflammatory mucosal thickening and fluid. The high-intensity enhancement of the nasal and sinus mucosa is clearly outlined against the non-enhancing fluid (Fig. 13.12, p. 137).

Gadolinium DTPA pools in the capillary bed of the highly vascular nasal mucosa, and this produces a high-intensity signal on magnetic resonance scans. Initially this led to an overestimate of the size of tumour in a patient with an olfactory neuroblastoma, and might be considered a drawback of using paramagnetic enhancement. In practice this is not so: non-enhancing tumour may be outlined against high signal from the nasal mucosa and from inflammatory mucosal thickening in the sinuses (Fig. 2.4); and a totally opaque sinus may be revealed as containing thickened mucosa and fluid rather than tumour. Alternatively all three may be demonstrated in the same sinus cavity (Fig. 2.5); this is clearly important in the selection of the best site for tumour biopsy.

In the nose and sinuses tumour tissue can be recognised by its heterogeneous signal; thickened mucosa, polyps and retained secretion always present a homogeneous signal with or without contrast enhancement. Tumours usually have some degree of signal heterogeneity and this is increased after intravenous gadolinium, especially on T_1 inversion recovery sequences, which are most

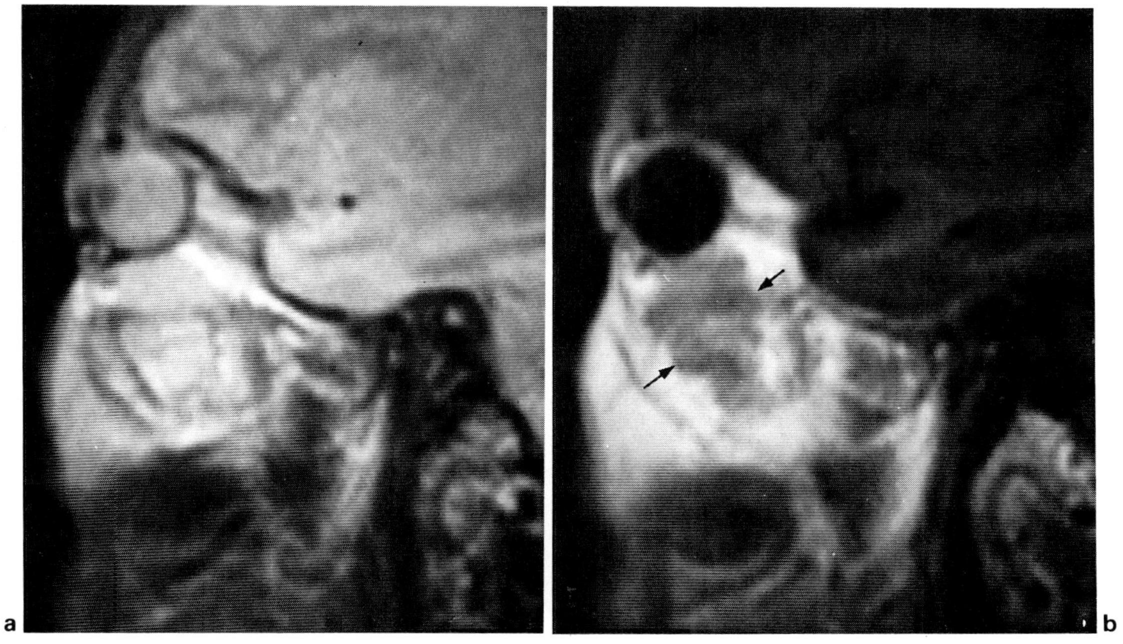

Fig. 2.4. **a** T_2-weighted spin echo sequence showing a lymphoma invading the floor of the orbit from the maxillary antrum. **b** Inversion recovery sequence after intravenous gadolinium DTPA. The non-enhancing lymphoma is outlined against the enhanced mucosa in the antrum and the orbital fat (*arrows*).

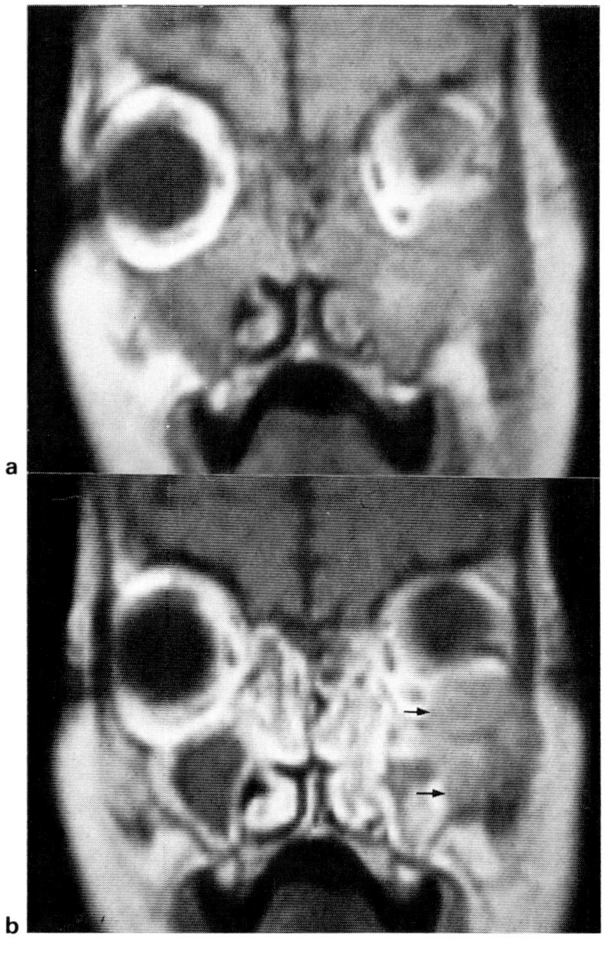

Fig. 2.5a,b. Same patient as Fig. 2.4. T_1-weighted coronal magnetic resonance scans before and after intravenous gadolinium DTPA. **a** The pre-contrast scan shows no discrimination between tumour, thickened mucosa, or fluid in the antrum. **b** The post-contrast scan shows the non-enhanced tumour (*arrow*) outlined against the enhanced mucosa in the left antrum, which contains fluid. The right antrum also contains fluid and thickened mucosa.

affected by the contrast agent (Curati et al. 1986). Differential signal enhancement from tumour is a feature of the post-contrast scans and probably results from areas of varying vascularity in the tumour. This was particularly marked in a chondrosarcoma, where the central chondromatous part of the tumour failed to enhance after intravenous gadolinium while the more peripheral cellular element of the tumour showed fairly intense enhancement (Fig. 17.7, p. 166).

References

Carr DM, Gadian DG (1985) Contrast agents in magnetic resonance. Clin Radiol 36:561–568

Curati WL, Graif M, Kingsley DPE, Niendorf HP, Young IR (1986) Acoustic neuromas: Gd-DTPA enhancement in MR imaging. Radiology 158:447–451

Gomori LM, Grossman RI, Goldberg HI, Zimmerman RA, Bilaniuk LT (1985) Intracranial haematomas: imaging by high field MR. Radiology 157:87–93

Lloyd GAS (1975) Radiology of the orbit. WB Saunders, London

Lloyd GAS (1979) CT scanning in the diagnosis of orbital disease. Comput Tomogr 3:227–239

Lloyd GAS, Phelps PD (1986) Juvenile angiofibroma: imaging by magnetic resonance, CT and conventional techniques. Clin. Otolaryngol 11:247–259

Lloyd GAS, Lund VJ, Phelps PD, Howard DJ (1987) Magnetic resonance imaging in the evaluation of nose and paranasal sinus disease. Br J Radiol 60:957–968

3 Congenital Disease

Choanal Atresia

Posterior choanal atresia is thought to result from a failure of rupture, between the 35th and 38th day of foetal life, of the partition which separates the bucco-nasal and bucco-pharyngeal membranes (Williams 1971). The choanal atresia which results may be unilateral or bilateral, bony or membranous, complete or incomplete. The bony atresia is commonly located 1–2 mm anterior to the posterior margin of the hard palate; the membranous form usually occurs more posteriorly. In approximately one third of patients the atresia is bilateral, and some patients have associated deformities including a high arched palate and increased thickness of the vomer and nasal septum. A familial incidence has been described by some authors (Wilkerson and Coyce 1948).

Ronaldson (1881) gave the first clinical description of this condition, which may present as an acute respiratory emergency in the neonate. In the newborn oropharyngeal breathing is an acquired function (Williams 1971) and takes several days or weeks to learn. Initially the baby is obliged to breath through the nose, so that bilateral choanal atresia usually presents with respiratory distress soon after birth. A problem may also arise during attempts at feeding, when the infant is unable to breathe and suck simultaneously. Unilateral choanal atresia is seen in later childhood or in adult life, the chief symptom being nasal obstruction and nasal discharge.

The diagnosis of choanal atresia is established by the inability to pass a probe or catheter through the posterior nares, and may be confirmed radiologically. The bony septum can be demonstrated by conventional or computerised tomography (CT). Positive confirmation is made by instilling contrast medium into the nasal cavities and taking radiographs with a horizontal beam so that the contrast can be seen to pool posteriorly (Fig. 3.1). Alternatively, a submentovertical projection can be used.

The above techniques clearly demonstrate the presence of choanal atresia, but do not provide information concerning the nature of the stenosis: whether it is bony or membranous. This, however, can be established by CT scan. Hasegawa et al. (1983) investigated a 6-year-old female with bilateral choanal atresia and were able to demonstrate an osseous stenosis on the right side and a membranous stenosis of the left choana. These authors point out that once an infant is suspected, by rhinoscopic examination or nasal catheterisation, of having congenital choanal atresia, CT is the preferred method of examination because it is a non-invasive technique. It should therefore be used prior to any contrast radiography.

Congenital Nasal Masses

The nasal encephalocoele, the glioma and the dermoid are the most common of the congenital mid-line nasal masses. All three may be associated with leakage of cerebrospinal fluid (CSF) and the possibility of fatal meningitis. A variety of theories have been advanced to explain the development of

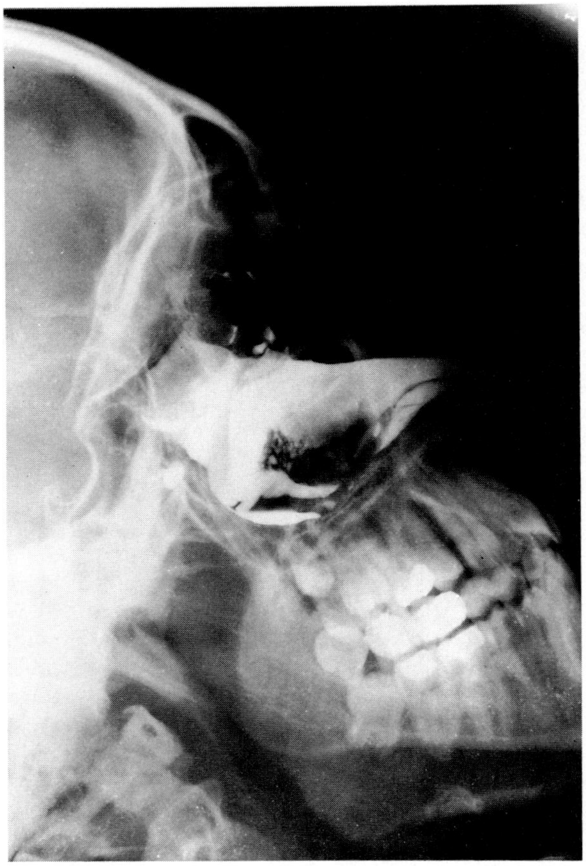

Fig. 3.1. Choanal atresia. Lateral view with horizontal beam and chin elevated showing obstruction of the contrast posteriorly.

the nasal glioma and encephalocoele. Embryologically, the latter is thought to result from faulty closure of the neuropore leading to herniation of the intracranial contents into the nasal area. A nasal glioma is regarded as a nasal encephalocoele which has lost its connection with the sub-arachnoid space (Bradley and Singh 1982).

Nasal dermoids are either simple or complex. Complex dermoids are thought to be due to faulty closure of the bones in the nasal area with entrapment of dermal elements. Other theories have been advanced to explain the simple dermoid, including entrapment of facial cleft, simple inclusion cyst, and aberrant development of skin appendages (Hughes et al. 1980).

Nasal Dermoids

A classical description of a dermoid was given by Matson and Ingraham (1951). They described the lesion as a cutaneous defect lined by stratified squamous epithelium in a congenital dermal sinus. At any point along a congenital dermal sinus there may be expansion into a cyst. Such a cyst is termed an epidermoid if it contains only epithelial debris and is lined by stratified squamous epithelium; it is called a dermoid if in addition it contains sebaceous material and hair, and the lining includes elements of the deeper layer of the skin.

The clinical and imaging features of nasal dermoids depend upon the extent and location of the sinus and its cystic component. The sinus presents as a dimple or pit over the mid-dorsum of the nose or at the junction of the bony and cartilaginous tip, and characteristically contains a hair (Johnson and Weisman 1964). These changes are often associated with hypertelorism and a soft tissue swelling over the bridge of the nose. Dermoids may have a communication with the dura through a patent foramen caecum (Hughes et al. 1980) and in some the cystic component of the sinus is intracranial, carrying the threat of meningitis and CSF leakage.

On plain radiograph, conventional tomography or CT it may be possible to show a well-demarcated defect in the mid-line of the nasal bones or fronto-

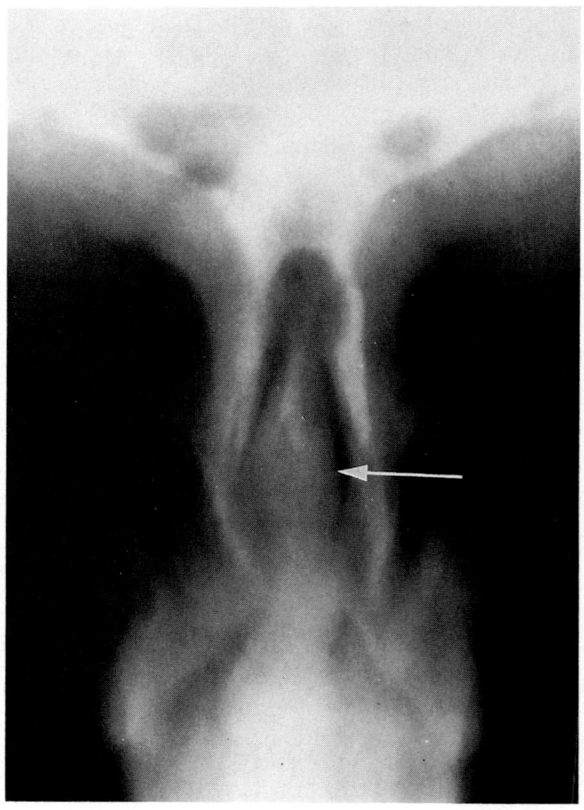

Fig. 3.2. Nasal dermoid. Well-demarcated defect in the mid-line of the nasal bones shown on coronal tomography. Note the fusiform enlargement of the septum (*arrow*).

nasal suture (Fig. 3.2). The cyst may involve the nasal septum producing a fusiform soft tissue mass within the septum or the latter may be bifid. When large the cyst may erode the glabellar region or extend into the ethmoid cells or frontal sinus; in those which have an intracranial communication it may be possible to show a defect in the cribriform plate area.

Nasal Glioma

The first comprehensive account of nasal glioma was given by Schmidt (1900), since when several extensive reviews have appeared (Black and Smith 1950; Walker and Resler 1963; Enfers and Herngren 1975).

The term nasal glioma is really a misnomer since the lesion is not a true tumour but results from sequestration of primitive brain tissue and consists histologically of glial cells (astrocytes) interlaced with vascular fibrous tissue septa. Gliomata may be entirely extranasal, intranasal, or both components may be present, and they may or may not have intracranial connections. When intracranial communications do exist the connection passes through a defect in the cribriform plate or in the region of the nasal attachment of the frontal bone (Walker and Resler 1963). Gliomata most frequently present at birth or shortly afterwards as a mass on the bridge of the nose. The lesion tends to be located to one side of the nasal bridge; this distinguishes them from dermoids, which are mid-line. A small number of cases are encountered in an older age group occasionally extending into middle life (Smith et al. 1963; Blumenfeld and Skolnik 1965).

The extranasal glioma is the commonest type and may show radiologically as a bone defect in the locality of the soft tissue mass. In the intranasal variety the mass tends to occur high in the nasal fossa, to one side of the nasal septum (Fig. 3.3), the latter being displaced to one side. When both intranasal and extranasal components are present there is a communication between the two, usually through a defect in the nasal bone or at its lateral margin (Dodd and Bao-Shan Jing 1977). A cranial defect may be present in the intranasal variety but difficult to identify. Routine computerised tomography (Hughes et al. 1980) in the management of patients with suspected nasal glioma is advised, in addition to magnetic resonance studies.

Encephalo-meningocoeles

Encephalo-meningocoeles may be grouped, according to their anatomical site, as occipital, sincipital and basal; the latter two are the only varieties of interest to the otolaryngologist.

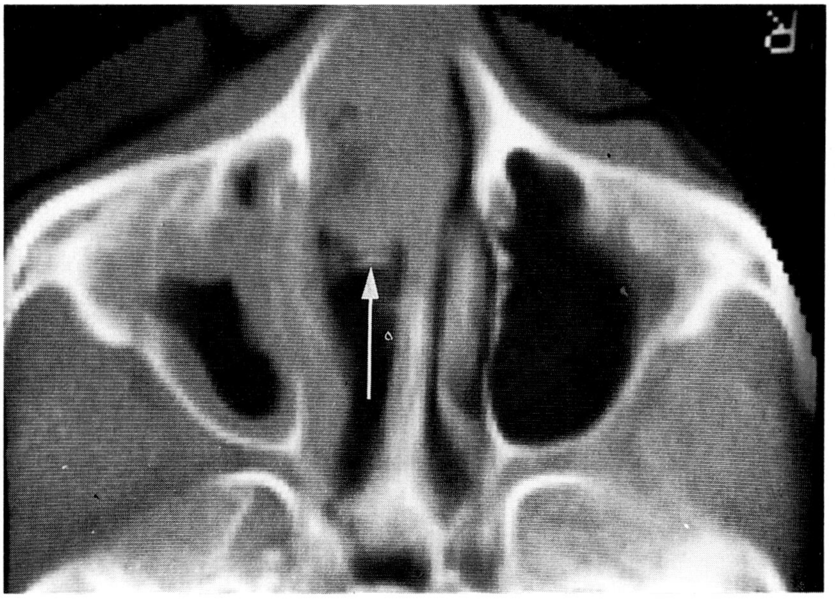

Fig. 3.3. Nasal glioma shown on axial CT scan (*arrow*).

Sincipital Encephalo-meningocoeles

The sincipital (pertaining to the anterior and upper part of the head) encephalo-meningocoeles are sub-classified as follows:

1. *Naso-frontal*, in which the herniated cerebral or meningeal tissue passes between the nasal and frontal bone, giving rise to a mid-line protuberance at the nasal root.
2. *Naso-ethmoid*, in which the defect lies between the nasal, ethmoid and frontal bones, the swelling presenting at the junction of the bony and cartilaginous part of the nose.
3. *Naso-orbital*, in which there is a defect in the suture line between the frontal, lacrimal and ethmoid bones with herniation into the orbit, causing a protuberance at the inner canthus of the eye.

Haverson et al. (1974) have recorded the radiological changes associated with naso-frontal and naso-ethmoid encephalocoeles. Naso-frontal lesions present with a V-shaped defect in the frontal bone, lateral displacement of the medial orbital wall, depression of the nasal bones, which are attached to the cribriform plate below the hernia, a low cribriform plate and a mid-line soft tissue mass. The naso-ethmoidal variety present with a circular defect between the orbits, increased inter-orbital distance, elevation of the nasal bones, which are attached to the frontal bones above the hernia, a normal position of the cribriform plate, and a soft tissue mass to one side of the mid-line.

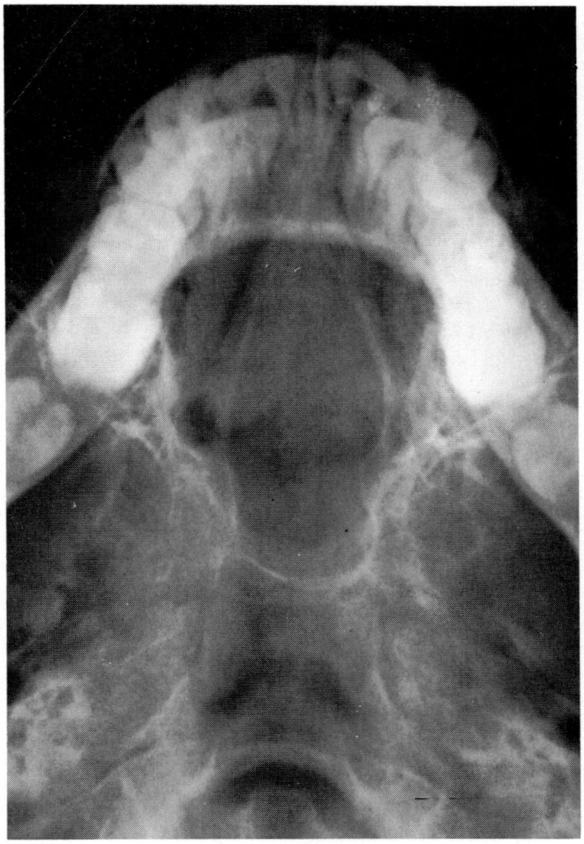

Fig. 3.5. Same patient as Fig. 3.4. The submentovertical view shows the defect in the sphenoid and posterior ethmoids.

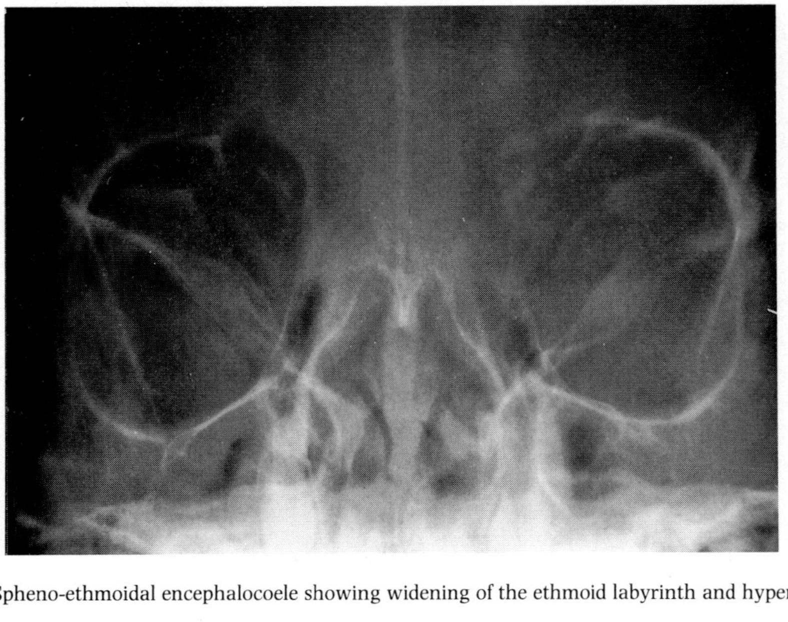

Fig. 3.4. Spheno-ethmoidal encephalocoele showing widening of the ethmoid labyrinth and hypertelorism.

Basal Encephalo-meningocoeles

The basal cerebral hernias are those which occur through the cribriform plate and through the sphenoid bone, the herniation appearing in the nasal cavity, nasopharynx, sphenoid sinus, posterior orbit, or pterygo-palatine fossa. The different varieties of basal encephalocoeles may be described as:

1. Transsphenoidal
1. Transethmoidal
3. Spheno-ethmoidal
4. Spheno-orbital

Pollock et al. (1968) have described the clinical and radiological features of eight patients with basal encephalocoeles (five transsphenoidal, three transethmoidal). Two clinical findings suggest a basal encephalocoele: a facial abnormality with hypertelorism (Fig. 3.4), and a mid-line soft tissue mass in the nose or epipharyngeal space. In the transsphenoidal variety a defect in the base of the skull can be shown in plain axial views (Fig. 3.5), but in the transethmoidal variety some form of tomography is needed for their proper demonstration.

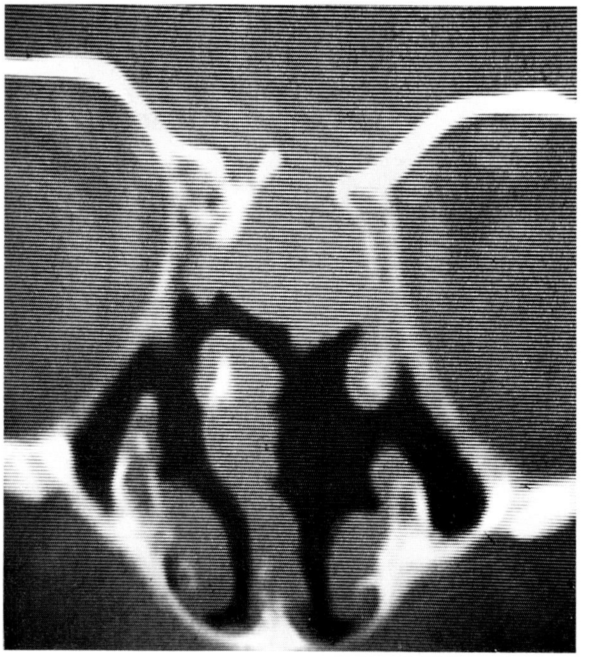

Fig. 3.6. Coronal CT scan showing a bone defect at the cribriform plate and a transethmoidal encephalocoele.

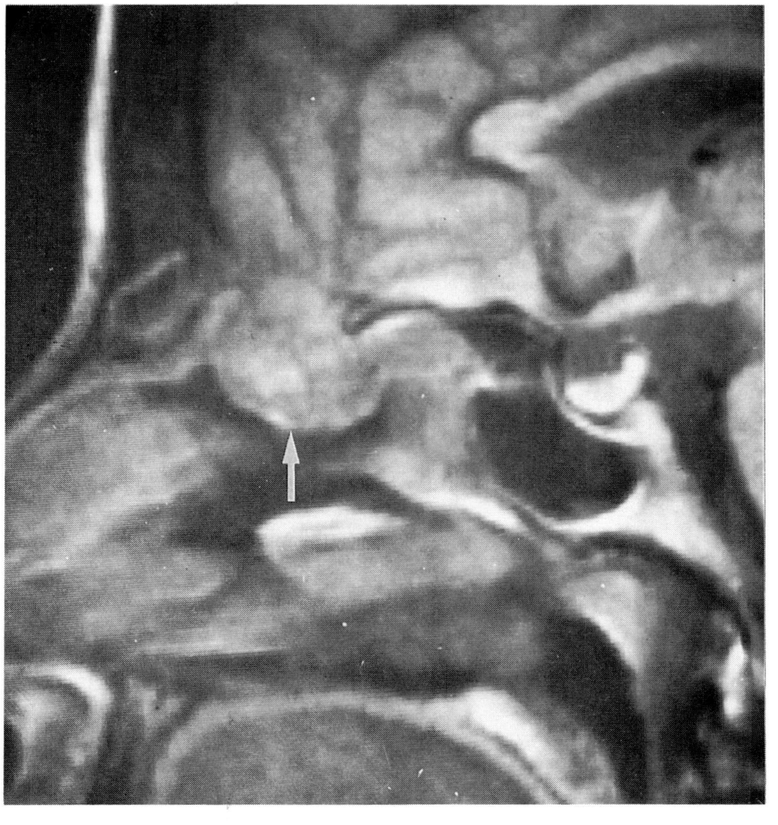

Fig. 3.7. Same patient as Fig. 3.6. The encephalocoele (*arrow*) is demonstrated by a sagittal magnetic resonance scan.

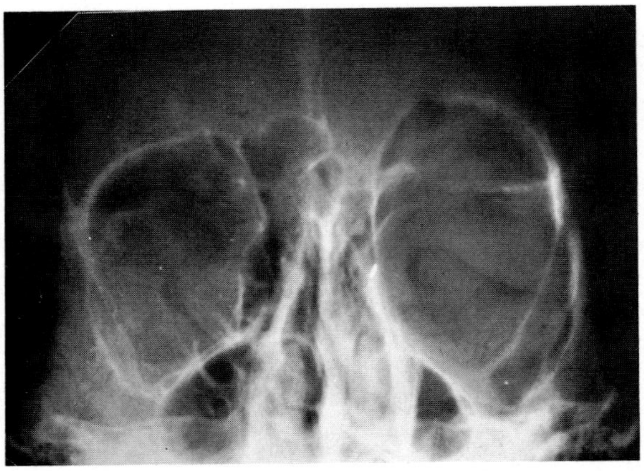

Fig. 3.8. Posterior encephalocoele (spheno-orbital) of the left orbit in neurofibromatosis.

Lusk and Dunn (1986) have recorded the findings of magnetic resonance studies in a 19-month-old child who was shown at surgery to have an anterior cranial fossa encephalocoele. In this case magnetic resonance proved superior to CT in its ability to distinguish brain from normal tissues in the nose. Examples of an encephalocoele demonstrated by CT and magnetic resonance are shown in Figs. 3.6 and 3.7.

The spheno-orbital type of encephalocoele involves the posterior orbit and superior orbital fissure and is commonly associated with neurofibromatosis. It produces the characteristic empty orbit sign of this condition (Fig. 3.8).

Progressive Hemifacial Atrophy (Parry–Romberg Disease)

Hemifacial atrophy is a condition of unknown aetiology originally described by Parry (1825) and by Romberg (1846). Eulenberg (1871) gave the condition the name progressive facial hemiatrophy, but the disease now goes under the more appropriate title of progressive hemifacial atrophy (PHA) (Goldhammer et al. 1981).

The disease is usually first noticed in the first or second decade of life and is more common in females. Over several years there is a slow, progressive unilateral atrophy of the face with involvement of skin and subcutaneous fat, and more rarely of muscles and bone (Bramley and Forbes 1960; Gorlin and Pindborg 1964). The atrophy begins in the skin or subcutaneous tissues and spreads to involve the whole of one side of the face, with atrophy of orbital fat and enophthalmos. Neurological complications including migraine, trigeminal neuralgia, epilepsy and hemiplegia may ensue (Wolf and Verity 1974; Asher and Berg 1982).

In rare cases hemifacial atrophy affects the bone more than the soft tissues (Goldhammer et al. 1981), resulting in a general atrophic collapse of the walls of the maxillary antrum (Figs. 3.9, 3.10). In this respect hemifacial atrophy needs to be distinguished from other causes of inward collapse of the antral boundaries – most notably the orbital

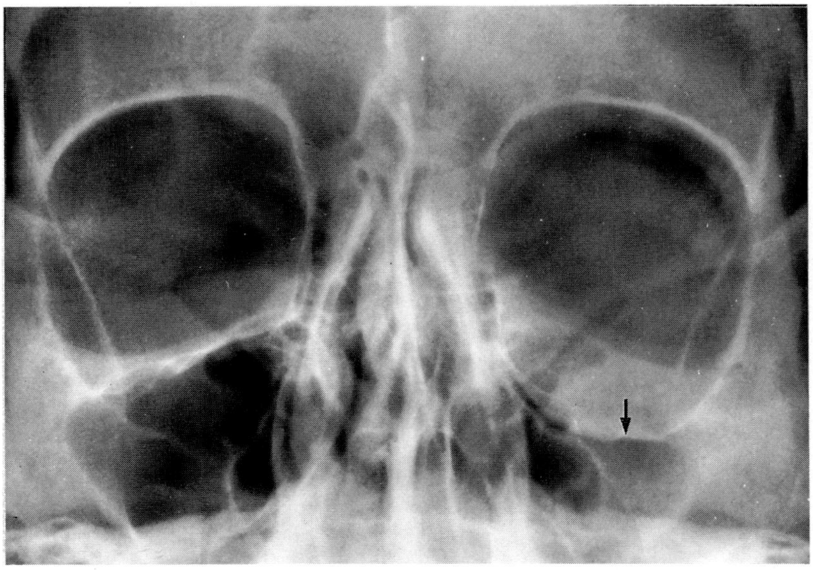

Fig. 3.9. Progressive hemifacial atrophy. There is collapse of the orbital floor (*arrow*) resembling a severe blow-out fracture, but without a history of trauma.

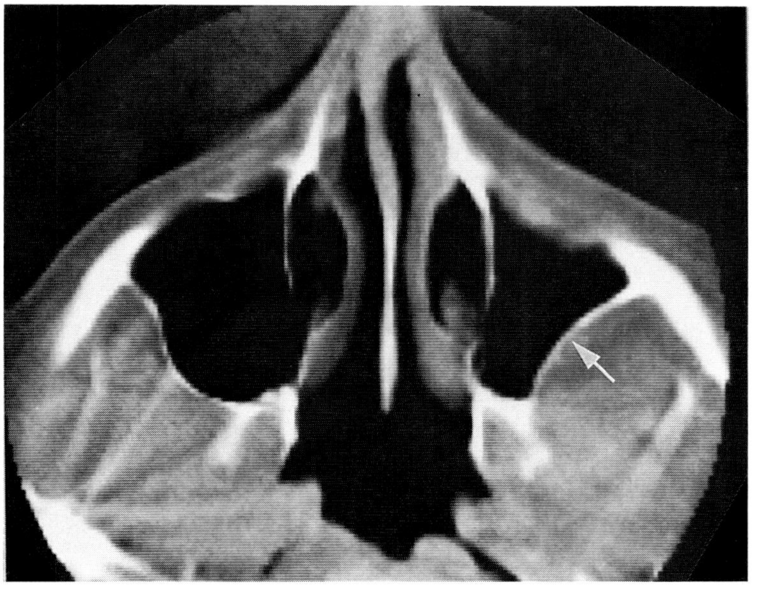

Fig. 3.10. Same patient as Fig. 3.9. The axial CT scan shows that there is also an incurvation of the posterior wall of the maxillary antrum (*arrow*).

floor – such as blow-out fractures and chronic osteomyelitis of the maxilla (see Chap. 5). The diagnosis is made on the clinical findings and the radiological features shown on plain radiographs and CT scan; the striking change is the collapse and incurvation of all the sinus walls of the maxillary antrum on the side involved (Figs. 3.9, 3.10).

References

Asher SW, Berg BO (1982) Progressive hemifacial atrophy. Arch Neurol 39:44–46

Black BK, Smith DE (1950) Nasal glioma. Arch Neurol Psychiatr 64:614–630

Blumenfeld R, Skolnik EM (1965) Intranasal encephalocoeles. Arch Otolaryngol 82:527–531

Bradley PJ, Singh SD (1982) Congenital nasal masses: diagnosis and management. Clin Otolaryngol 7:89–97

Bramley P, Forbes A (1960) A case of progressive hemiatrophy presenting with spontaneous fractures of the lower jaw. Br Med J i: 1476–1478

Dodd GD, Bao-Shan Jing (1977) Radiology of the nose, paranasal sinuses and nasopharynx. Williams and Wilkins, Baltimore

Enfers B, Herngren L (1975) Nasal glioma. J Laryngol Otol 89:863–868

Eulenberg A (1871) Lehrbuch des functionellen Nervenkrankheiten. Berlin

Goldhammer Y, Kronenberg J, Tadmor R, Braham J, Leventon G (1981) Progressive hemifacial atrophy (Parry–Romberg's disease), principally involving bone. J Laryngol Otol 95:643–647

Gorlin RJ, Pindborg JJ (1964) Syndromes of the head and neck. McGraw-Hill Book Company, New York, pp 341–344

Hasegawa M, Oku T, Tanaka H et al. (1983) Evaluation of CT in the diagnosis of congenital choanal atresia. J Laryngol Otol 97:1013–1015

Haverson G, Bailey IC, Kiryabwire JWM (1974) The radiological diagnosis of anterior encephalocoeles. Clin Radiol 25:317–322

Hughes GH, Sharpino G, Hunt W, Tucker HN (1980) Management of the congenital mid-line mass: a review. Head Neck Surg 2:222–233

Johnson GF, Weisman PA (1964) Radiological features of dermoid cysts of the nose. Radiology 82:1016–1023

Lusk RP, Dunn VD (1986) Magnetic resonance imaging in encephalocoeles. Ann Otol Rhinol Laryngol 95:432–433

Matson DD, Ingraham FD (1951) Intracranial complications of congenital dermal sinuses. Pediatrics 8:463–474

Parry CH (1825) Collections from the unpublished medical writings of the late Caleb Hillier Parry. Underwood, London

Pollock JA, Newton TH, Hoyt WF (1968) Transsphenoidal and transethmoidal encephalocoeles. Radiology 90:442–453

Romberg HM (1846) Klinische Ergebnisse. A Forstrer, Berlin

Ronaldson TR (1881) Note on a case of congenital closure of the posterior nares. Edinburgh Med J 26:1035–1037

Schmidt MB (1900) Über seltene Spaltbildungen in Beireiche des mittleren Stimfortsatzes. Virchows Arch [A] 162:340–370

Smith KR, Schwartz HG, Luse SA, Ogura JH (1963) Nasal gliomas; a report of 5 cases, with electron microscopy of one. J Neurosurg 20:968–982

Walker EA, Resler DR (1963) Nasal glioma. Laryngoscope 73:93–107

Wilkerson WW, Coyce LE (1948) Congenital choanal occlusion. Trans Am Acad Ophthalmol 52:234–236

Williams HJ (1971) Posterior choanal atresia. Am J Roentgenol 112:1–11

Wolf SM, Verity MA (1974) Neurological complications of progressive facial hemiatrophy. J Neurol Neurosurg Psychiatry 37:997–1004

4 Trauma

Fractures of the Nasal Bones

Fractures of the nasal bones are common and are seen either in isolation or, not infrequently, associated with other injuries to the facial bones. They have been classified into linear fractures, lateral fractures, frontal fractures (partial or complete) and fronto-lateral fractures. There is also the very severe "smash fracture", with flattening and comminution of the nasal bones. This involves the frontal processes of the maxilla, the lacrimal bones and nasal septum, and may extend to the anterior ethmoid cells and cribriform plate.

Lateral fractures are the commonest variety and all may be demonstrated using the techniques described in Chap. 1. However, radiology has little part to play in the management of nasal bone fractures unless there is extension to other facial bones. De Lacey et al. (1977) concluded from a study of 100 patients with nasal bone trauma that radiography of the nasal bones is an unnecessary procedure and should only be carried out at the specific request of an ENT surgeon. Reduction of a nasal bone fracture is indicated because of the clinical deformity rather than the radiological appearance. When the blow to the nose may have involved other facial bones clinically appropriate views of facial bone should be obtained; but views of the nasal bone itself are often unnecessary. However, most ENT surgeons would justify nasal radiographs after trauma on medico-legal grounds as a permanent record of the trauma, and to give some idea of the severity of the injury, particularly with regard to displacement of the fragments and degree of comminution (T.R. Bull, 1987, personal communication).

Fractures of the Maxillary Antrum

Isolated fractures of the maxillary antrum occur most commonly in the antero-lateral wall. Usually they are due to direct impact and are often comminuted with depressed fragments. They are best demonstrated by CT (Fig. 4.1). A less common type of isolated fracture involves the orbital rim at the infraorbital margin and may or may not be associated with an antral blow-out fracture. In addition fracture of the alveolus and floor of the maxillary antrum may occur during dental extraction or oral surgery.

Malar Fractures

Malar fractures are common fractures of the middle third of the facial skeleton and occur as the result of laterally disposed violence to the facial bones. The force of the blow causes the malar bone to be displaced downwards, inwards and posteriorly, with a hinge movement and slight separation or fracture at the fronto-malar suture. This is accompanied by a fracture of the zygomatic arch with little or no displacement. At the same time separation occurs at the junction of the medial third

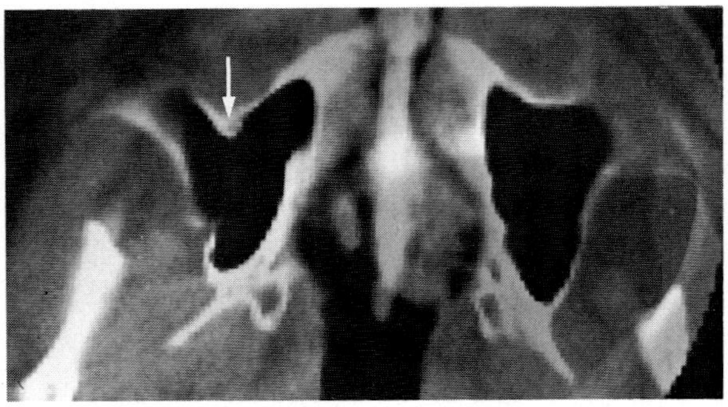

Fig. 4.1. Depressed fracture (*arrow*) of the anterior wall of the maxillary antrum shown on axial CT.

and lateral two thirds of the inferior orbital rim, usually with depression of the outer fragment. Sometimes this is also associated with disruption of the orbital floor. If the force is moderate the body of the malar bone is left intact, but in a more severe injury this bone is comminuted and may become impacted into the maxillary antrum. With severe degrees of impaction there is nearly always a comminution of the orbital floor, with prolapse of the orbital contents into the maxillary antrum.

The patient may present with facial anaesthesia or paraesthesia due to damage to the infraorbital nerve, and if the zygomatic arch is depressed it may interfere with the action of the mandible by pressure on the coronoid process. A fracture through the floor of the orbit may cause trapping of the inferior rectus muscle or the inferior oblique in a similar manner to an antral blow-out fracture.

The best plain radiographic projection for the demonstration of these fractures is an overtilted occipito-mental view or 30° occipito-mental projection. In this the radiographic base line is adjusted as in the standard occipito-mental view but with the tube centred through the vertex of the skull and angled 30° towards the feet. In this view the lower orbital margin and the zygoma form a continuous arch; any break in this arch will indicate the site of fracture and can also show displacement and impaction of the main fragment (Fig. 4.2). The injury may be accompanied by a generalised opacity of the antrum due to haemorrhage into the sinus cavity, or more commonly there is a localised thickening of the mucosa of the sinus due to mucosal

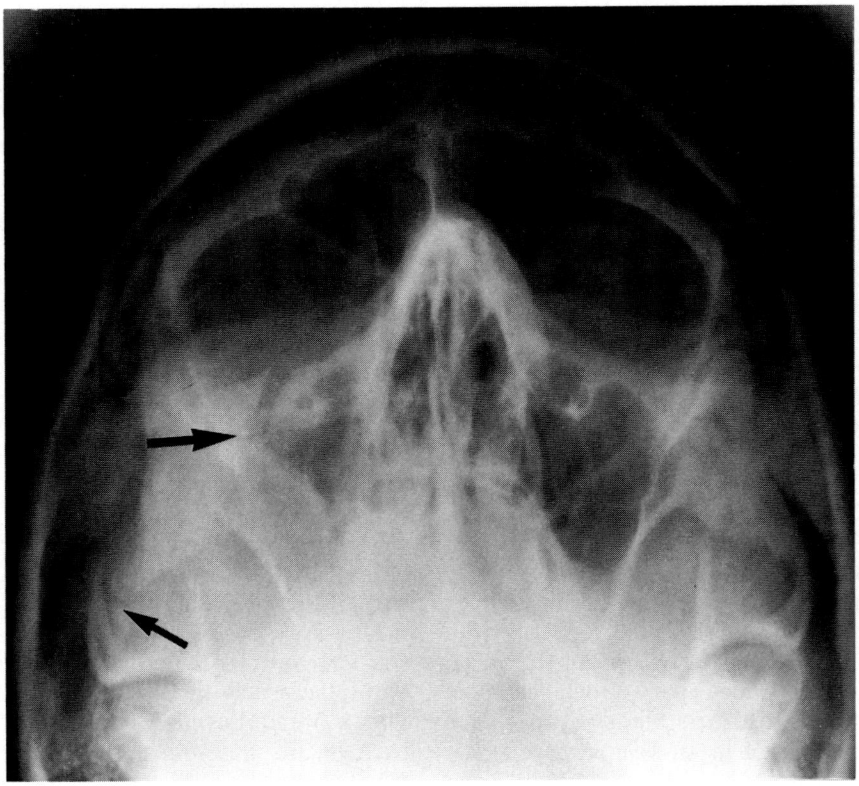

Fig. 4.2. Overtilted occipito-mental projection showing a malar fracture with fractures of the zygoma and infraorbital margin (*arrows*). Note the thickening of the mucosal shadow along the lateral wall of the sinus due to submucosal haemorrhage or oedema.

oedema or submucosal haemorrhage (Fig. 4.2). Occasionally a blood fluid level may be present.

Of all the imaging methods available, axial CT is the best for showing displacements of the malar eminence in the antero-posterior plane and may also show the lateral wall fractures. In addition coronal CT will show inferior displacement of the outer fragment with a characteristic step deformity of the inferior orbital rim (Figs. 4.3, 4.4).

LeFort Injuries

The sinuses may be involved in severe fractures of the facial bones, usually sustained during a road accident. Le Fort (1901) described anatomical lines of weakness in the facial skeleton, and his name has been given to the fractures which follow these lines. LeFort I fractures are usually sustained by force from in front and in an upward direction. Minor LeFort I injuries may simply consist of a fracture of the upper alveolar margin, but in more severe injuries to the lower third of the facial bones the fracture may extend through the dense bone of the maxilla to the pterygoid plates, and separate the hard palate and alveolus (Fig. 4.5).

In a LeFort II injury the line of fracture runs across the nasal bones, extending on both sides across the frontal process of the maxilla, and across the naso-lacrimal canal. From this point the line of weakness (or fracture) traverses the floor of the orbit, crossing the lower orbital margin in the region of the maxillo-zygomatic suture, and eventually involving the lateral wall of the maxillary antrum and pterygoid laminae (Fig. 4.5). This type of fracture is produced by force applied to the middle level of the facial skeleton from an antero-posterior direction, or sometimes from a lateral direction.

Force applied to the facial bones from a superior direction will result in a local fracture of the nasal and ethmoid bones if the momentum is small, or a high-level LeFort III fracture when the impact is severe. The line of fracture starts at the upper part of the nasal bones close to the fronto-nasal suture. In some cases the dislocation of the nasal bones at this site results in disruption of the cribriform plate of the ethmoid, and fracture of the floor of the anterior fossa. From its mid-line point of origin the

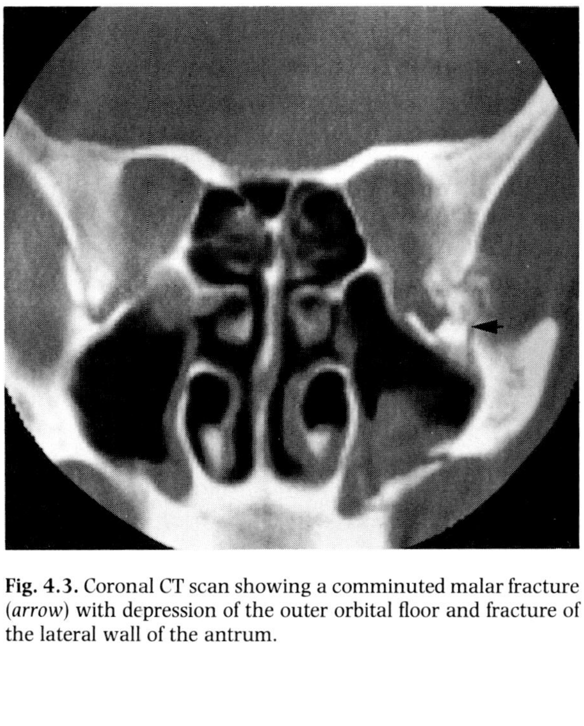

Fig. 4.3. Coronal CT scan showing a comminuted malar fracture (*arrow*) with depression of the outer orbital floor and fracture of the lateral wall of the antrum.

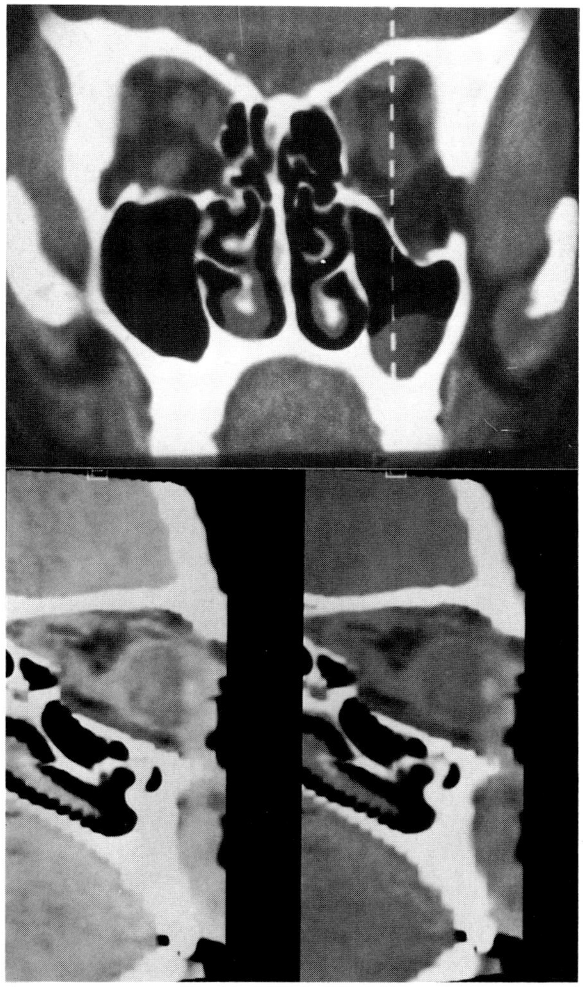

a

b

Fig. 4.4. a Coronal CT scan of a malar fracture with characteristic step deformity of the outer orbital rim. **b** The parasagittal reformatted views also show a step deformity of the orbital floor anteriorly.

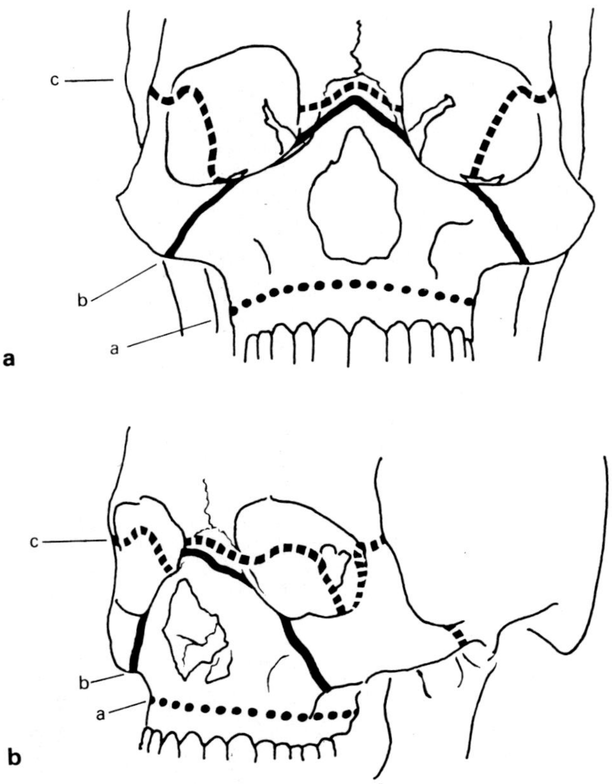

Fig. 4.5. The lines of fracture in LeFort injuries to the mid-face: **a** LeFort I, **b** LeFort II and **c** LeFort III injuries.

fracture runs through the medial wall of the orbit in the neighbourhood of the fronto-maxillary suture, crossing the lacrimal bone and medial orbital wall and extending backwards to the inferior orbital fissure. At this point the line of fracture bifurcates, one component running upwards and forwards to terminate in the lateral wall of the orbit immediately below the fronto-malar suture, while the lower component of the fracture runs downwards and backwards from the pterygo-maxillary fissure to involve the roots of the pterygoid laminae (Fig. 4.5). In this injury, therefore, there is a virtual separation or partial separation of the facial bones from the cranium above the zygomatic arch, the force of the trauma causing the entire middle third of the facial skeleton to hinge backwards through the ethmoid labyrinth.

Blow-Out Fractures

Blow-out fractures are classically caused by violent contact with a round hard object such as a cricket ball, hurling ball or baseball, but the vast majority are the result of a blow in the eye from a closed fist.

According to Smith and Regan (1957) the mechanism of injury is as follows: Part of the impact of the blow is absorbed by the orbital rim, which remains intact, but the force of the blow causes a backward displacement of the eye and an increase in intraorbital pressure, with a resultant fracture of the orbital floor or medial wall, decompressing the orbit and causing a herniation of orbital soft tissues into the maxillary antrum or ethmoid cells. The adjacent extraocular muscles may be impeded as a result of this herniation, producing an impairment of ocular mobility and diplopia. The inferior rectus, inferior oblique, or more rarely the medial rectus muscle may be involved.

It is the impairment of muscle function which constitutes the particular gravity of a blow-out fracture. Early surgical intervention may be required to release the trapped muscle and restore free movement of the globe. Unless this is done the diplopia will persist. Clinically, vertical diplopia is the most important feature and may be accompanied by paraesthesia or anaesthesia of the cheek and upper lip, in the distribution of the infraorbital nerve, which may be involved in the fracture.

The characteristic radiological appearance is that of a small, soft tissue shadow in the roof of the maxillary antrum, due to the herniated soft tissues from the orbit; this is associated with a depressed bony fragment, which is nearly always visible on plain radiography at the fracture site. Sometimes oblique views are necessary to show a depressed fragment convincingly or it may be best shown on a lateral film. But the projection which consistently shows these fractures, even when only minimal changes are present, is the nose–chin or occipito-oral view of the orbits (Fig. 4.6). Tomography in the coronal plane may confirm the fracture but is not usually needed to make the diagnosis.

CT can be used as a substitute for conventional tomography in the demonstration of blow-out fractures. CT has the advantage of showing not only the site of fracture, but also the state of the rectus muscles; and it is possible to show herniation of the extraocular muscle. However, herniation of the medial or inferior rectus muscle does not imply that the muscle action is impeded; conversely the absence of herniation does not necessarily mean that there is no entrapment. The explanation for this has been given by Koorneef (1982) (Fig. 4.7). Anatomical studies have shown that in the normal orbit there is a series of fibrous connective tissue septa which join the periorbita to the rectus muscle sheaths and the tendinous ring connecting them. Frequently it is the entrapment of these connective

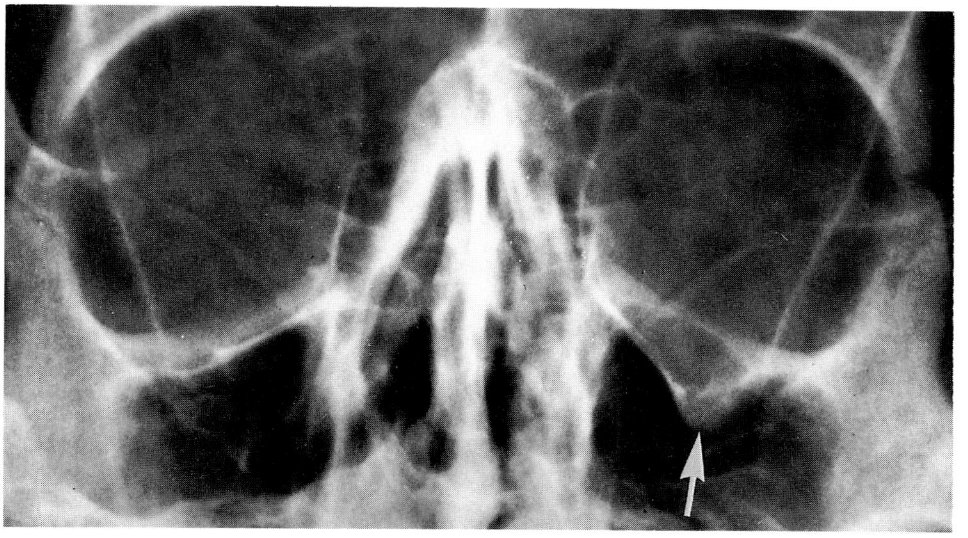

Fig. 4.6. Undertilted occipito-mental projection showing a typical antral blow-out fracture (*arrow*).

tissue elements which causes the muscle dysfunction without the muscle being displaced. Entrapment of the connective tissue septa in an orbital blow-out fracture may cause a downward drag on the medial or lateral rectus muscle via the tendinous ring which joins them to the inferior rectus muscle (Fig. 4.7). This change can be detected on CT scan as a downward displacement of the medial rectus muscle (Figs. 4.8, 4.9) and,

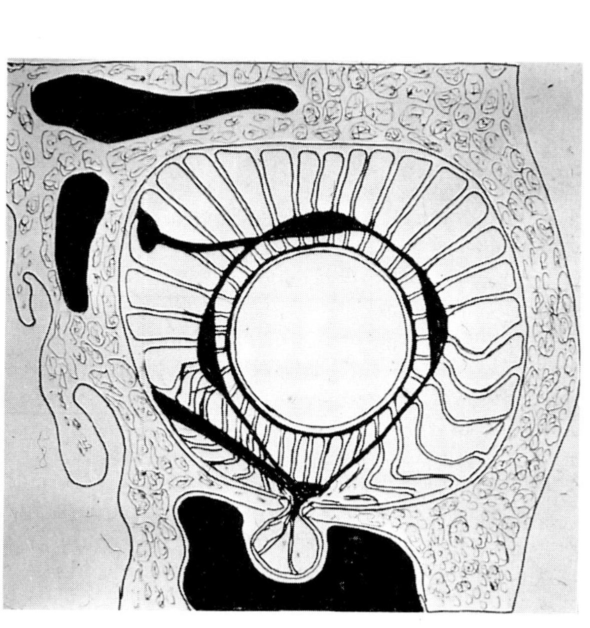

Fig. 4.7. Line drawing showing entrapment of the fibrous septa at the site of an antral blow-out fracture, with downward drag on the medial rectus muscle. (Reproduced with permission from Koorneef 1982.)

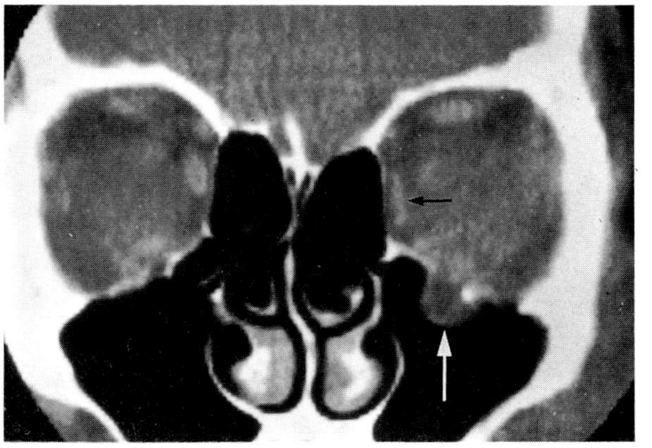

Fig. 4.8. Blow-out fracture of the floor of the orbit (*white arrow*). Medial rectus drag is shown on coronal CT scan (*black arrow*), indicating entrapment of the inferior rectus muscle.

when present, may indicate the need for surgical intervention to release the trapped muscle. Another feature which can be shown on CT is the presence of fibrous adhesions within the antrum (Fig. 4.10), which clearly need to be dealt with if restoration of a normal orbital floor is to be achieved surgically.

Fractures of the Frontal Sinus

Fractures of the frontal sinus are usually caused by direct violence, but the sinuses may be involved indirectly by a fracture of the skull vault and frontal

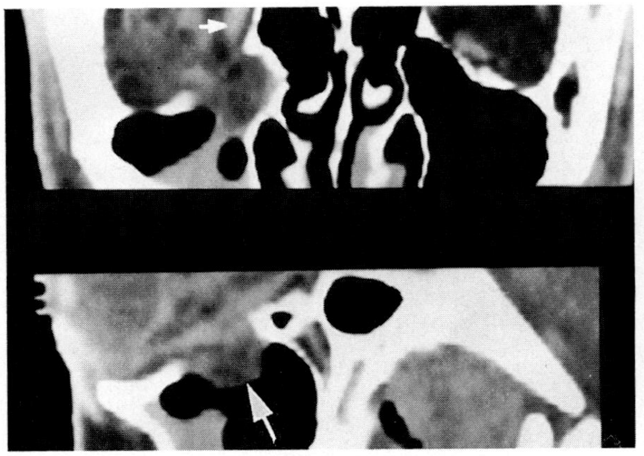

Fig. 4.9. Reformatted coronal and sagittal CT sections of the orbital floor. The coronal section (*top*) shows a downward displacement of the medial rectus muscle, or medial rectus drag (*small arrow*). The sagittal section (*bottom*) shows the herniation of soft tissues (*large arrow*) and confirms muscle entrapment.

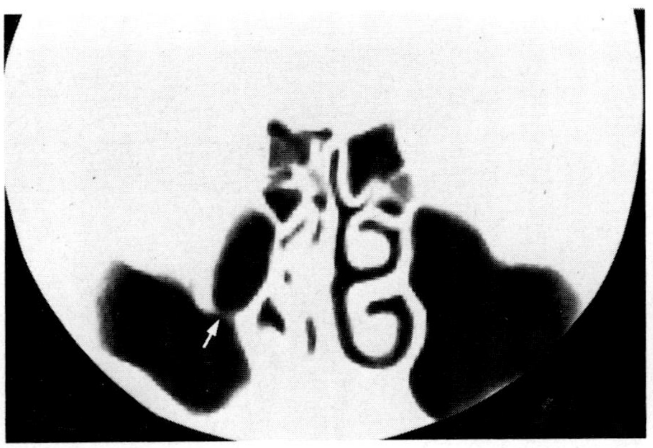

Fig. 4.10. In orbital floor fractures adhesions (*arrow*) may form between the herniated soft tissues or bony fragments and the opposite walls of the maxillary antrum. These can only be appreciated if the CT section is imaged with the window level set in the negative range of the Hounsfield scale (i.e. −400 to −500 Hounsfield units).

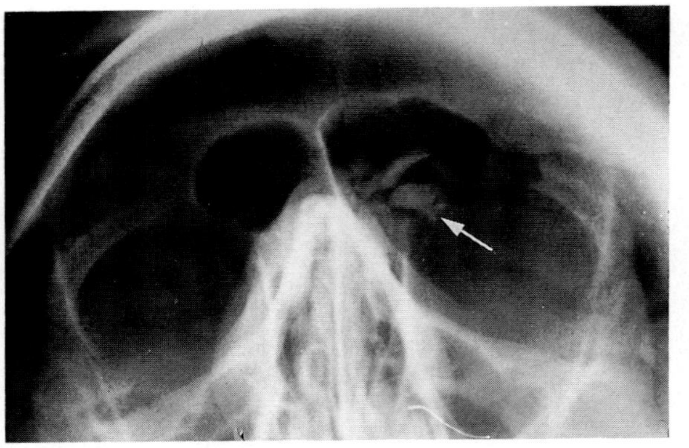

Fig. 4.11. Fragments of windscreen glass in the frontal sinus (*arrow*).

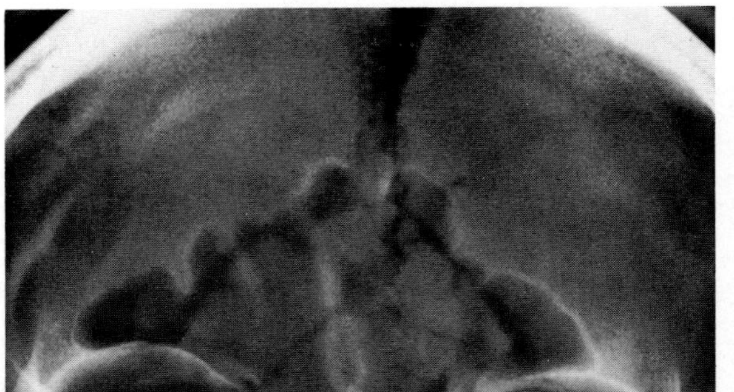

Fig. 4.12. Severe comminuted fracture of the frontal bone and frontal sinuses. The patient received a direct blow to the forehead by the blunt end of a wood-chopper.

bones. Fractures may involve one or both walls. Fracture of the posterior wall of the frontal sinus is rare and is usually associated with fractures of the floor of the anterior fossa. These fractures may be complicated by traumatic rhinorrhoea, pneumocephalus or meningitis. Fractures involving the anterior wall are far more common. They are usually due to direct trauma in road accidents and not infrequently fragments of windscreen glass are present within the sinus cavity (Fig. 4.11). the radiological demonstration of these injuries is by plain radiograph and CT scan (Figs. 4.12, 4.13 and 4.14).

Fractures of the Sphenoid Sinus

Fractures of the sphenoid sinus usually occur as part of a severe fracture of the skull base, frequently associated with a longitudinal fracture of the petrous bones. The sphenoid may also be involved in a LeFort III fracture of the facial bones or a very severe direct injury to the frontal bones (Fig. 4.15). On plain radiographs direct evidence of a fracture line may be difficult to see unless the floor of the pituitary fossa is involved (Fig. 4.15), but indirect

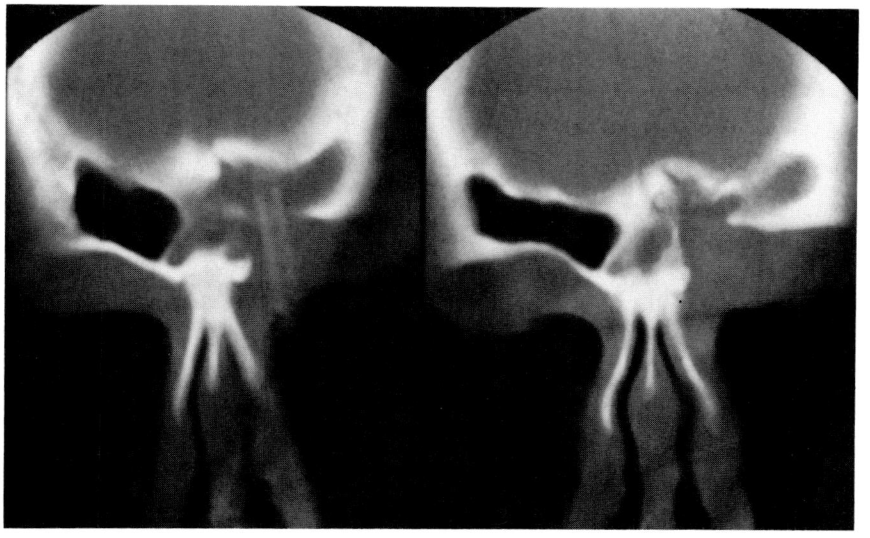

Fig. 4.13. Comminuted fracture of the frontal sinus shown on coronal CT scan, with drainage tube in situ.

evidence may be given by the presence of an effusion into the sinus cavity with a fluid level (Fig. 4.16). The only way to show the full extent of these bony injuries is by CT scan, and the axial projections are particularly helpful in this respect (Fig. 4.17). Pneumocephalus or CSF rhinorrhoea is a not infrequent complication of these fractures.

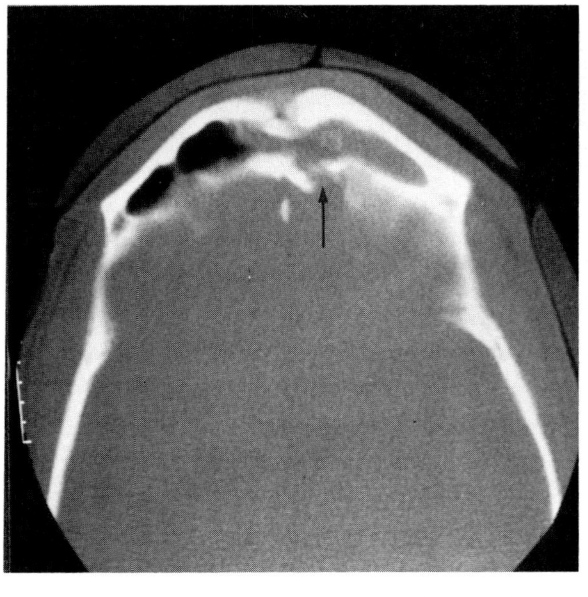

Fig. 4.14. Same patient as Fig. 4.13. Axial CT section showing fractures of both anterior and posterior walls of the sinus cavity (*arrow*).

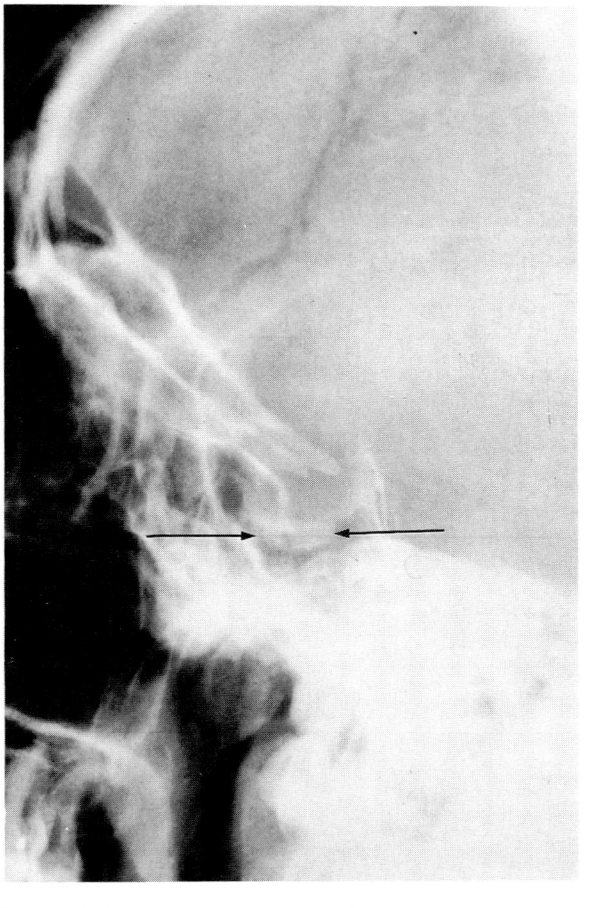

Fig. 4.15. Same patient as Fig. 4.12. Force transmitted from a blow on the forehead has fractured the sphenoid and disrupted the floor of the pituitary fossa (*arrows*).

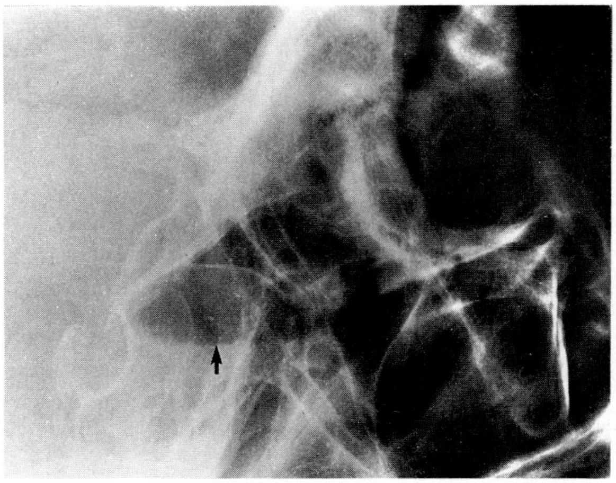

Fig. 4.16. Fluid level in the sphenoid sinus (*arrow*) due to a fracture with CSF leak.

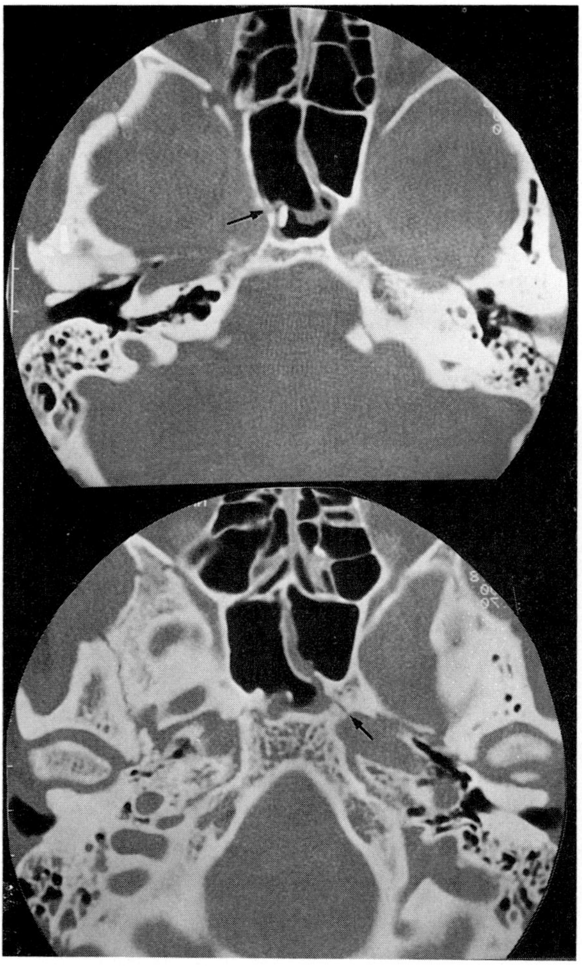

Fig. 4.17. Axial CT sections of a fracture of the skull base involving both petro-mastoids and the sphenoid sinus (*arrows*).

Fractures of the Ethmoid Cells

Fractures of the ethmoid cells occur as part of the more extensive LeFort III injuries (see above), but they may also be found in isolation. They may occur in the roof of the ethmoid labyrinth in the region of the cribriform plate or they may involve the medial orbital wall. In the latter fractures the evidence from a series collected by the author (Lloyd 1966) would suggest that the mechanism of injury can be of two types:

1. A direct injury to the medial wall of the orbit and base of the nose. Sometimes the injury is associated with a fracture of the nasal bones.
2. A medial wall blow-out fracture through the lamina papyracea of the ethmoids. This may be associated with a blow-out fracture into the antrum; many patients with blow-out injuries have a combined antro-ethmoidal fracture.

Logically the lamina papyracea would seem the part of the orbit most likely to fracture by the blow-out mechanism, since it is the most fragile part of the orbital wall. This type of injury presents radiologically in many patients as an orbital emphysema, associated with clouding of the ethmoid cells on the injured side (Fig. 4.18). The actual site of the fracture is seldom directly visible on plain radiographs, but the injury is not usually of any significance and only rarely is there any interference with muscle function, so that more detailed radiological investigation is seldom required. If it does

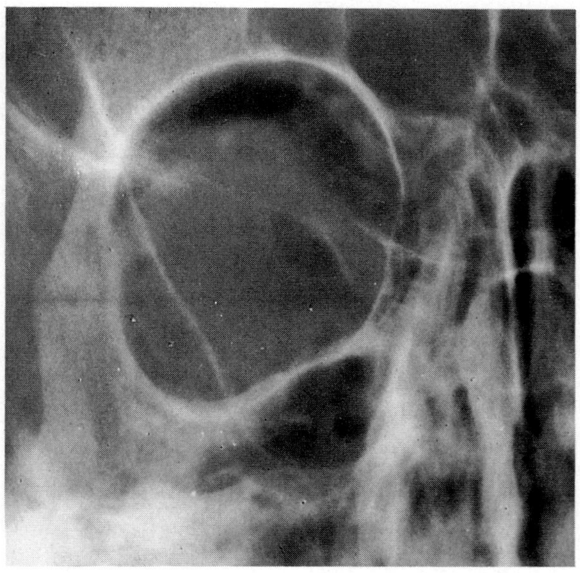

Fig. 4.18. Ethmoid blow-out fracture. Clouding of the ethmoid cells and orbital emphysema can be seen.

become necessary to show the exact site of the fracture and its extent then CT scan in the axial plane is the procedure of choice. These fractures can also be demonstrated by coronal CT and magnetic resonance scans (Figs. 4.19, 4.20).

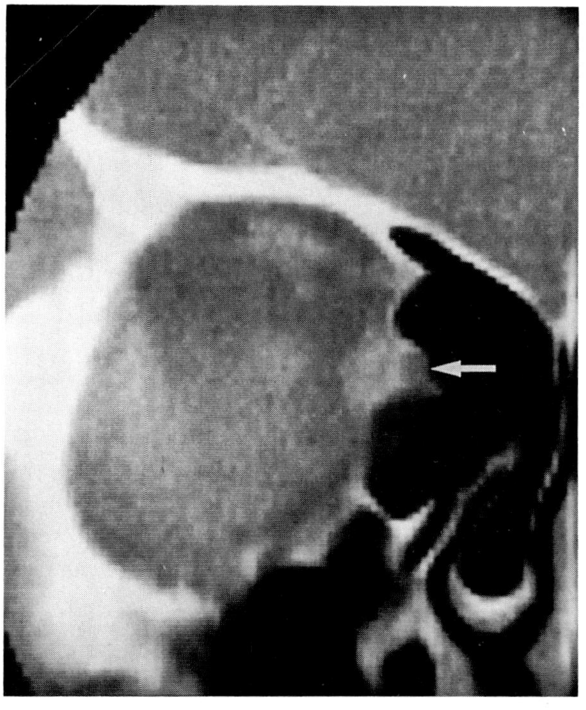

Fig. 4.19. Coronal CT scan of a small blow-out fracture of the medial wall (*arrow*) caused by an elbow in the face during basketball. At surgery the medial rectus muscle was shown to be tethered.

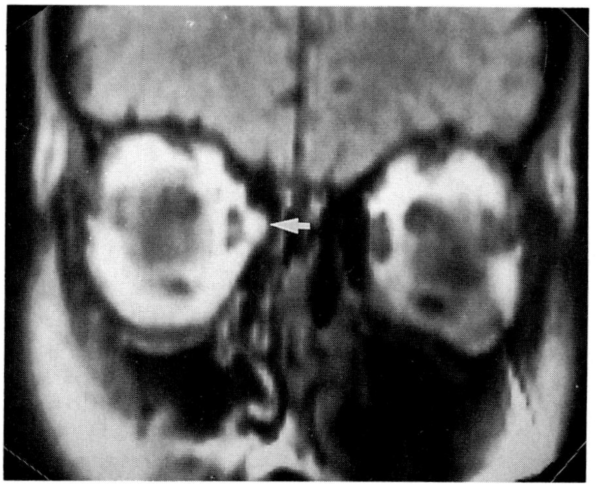

Fig. 4.20. Coronal magnetic resonance scan showing an antro-ethmoidal blow-out fracture. The medial component of the fracture is *arrowed*.

Cerebrospinal Fluid Leaks

CSF rhinorrhoea is caused by an abnormal connection between the sub-arachnoid space and the nasal cavity. This may be directly from the anterior cranial fossa via the cribriform plate or ethmoid cells; from the middle fossa via the sphenoid sinus; or indirectly from the posterior fossa via the eustachian tube. The majority of leaks are either traumatic or spontaneous in origin, some occurring after surgery (Fig. 4.21). Intracranial and extracranial tumours, encephalocoeles and infection are all rare causes.

Fractures with CSF leaks are most common in the ethmoid cells and cribriform plate area. CSF leaks also occur in fractures of the frontal sinuses, sphenoid sinuses and temporal bones. CSF fistulae arising spontaneously without a history of trauma occur at several sites: (1) from the cribriform plate area in which the CSF can reach the nasal cavity via a prolongation of the sub-arachnoid space along the olfactory nerve filaments; (2) via a dehiscence in the basi-sphenoid (Fig. 4.22); (3) via the petromastoid in certain congenital anomalies of the inner ear; and (4) into the sphenoid sinus from the middle cranial fossa. Fistulae of the last type may occur in cases of non-tumorous enlargement of the pituitary fossa – the so-called empty sella (Kaufman et al. 1977). These authors also reported five patients in whom the aetiology was thought to be herniation of the meninges through natural pits or holes in the floor of the middle fossa into a well-pneumatised sphenoid sinus. With fixation of these prolongations the dura becomes thinned and eventually allows herniation of the arachnoid with the formation of an arachnoid diverticulum; rupture then results in a CSF fistula.

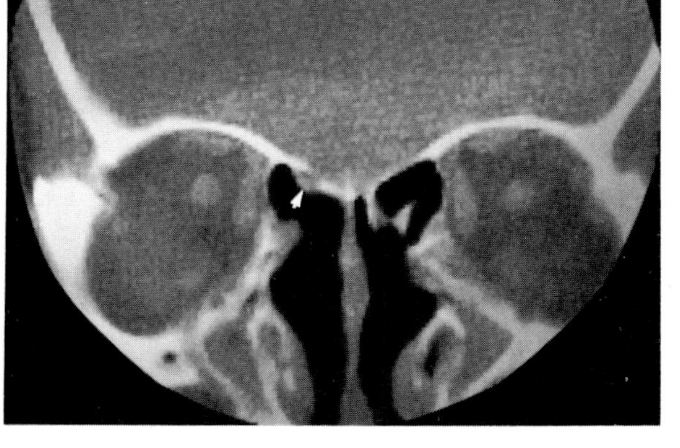

Fig. 4.21. Coronal CT scan showing a CSF leak resulting from fracture of the roof of the ethmoid labyrinth (*arrow*) following nasal polypectomy.

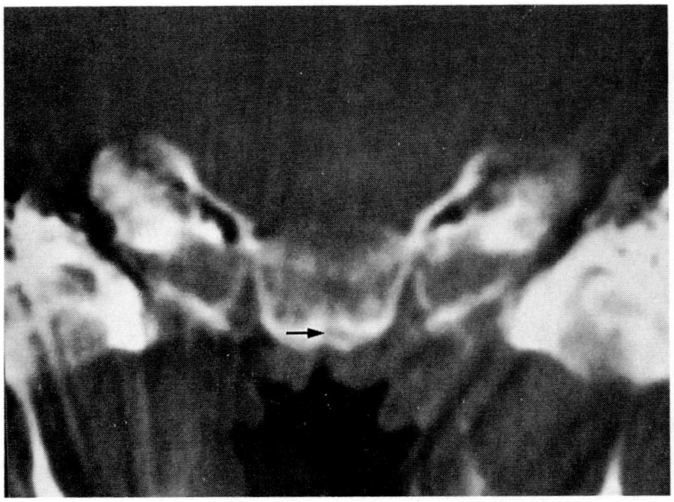

Fig. 4.22. Congenital dehiscence (*arrow*) shown in the basi-sphenoid on coronal CT scan. At surgery CSF was seen to be leaking into the nasopharynx from this site.

Recognition of spontaneous rhinorrhoea is dependent upon a good clinical history to differentiate it from chronic rhinitis. The discharge is usually unilateral and the flow changes with alterations in posture. Confirmation of the presence of CSF is obtained by demonstrating a high concentration of glucose in the fluid.

Radiological Investigation

Initially enquiry should be made into the state of the patient's hearing: if there is unilateral sensorineural hearing loss then the petrous bones need to be examined tomographically to exclude a congenital abnormality of the inner ear, or fracture if there is a history of trauma. The common deformity associated with recurrent meningitis and a spontaneous CSF leak has been described by Phelps and Lloyd (1978, 1983) and consists of a severely dysplastic labyrinth with a wide communication between the vestibule and cochlea and a tapering internal auditory meatus. This deformity allows CSF to pass into the vestibule and thence through the oval window into the middle ear.

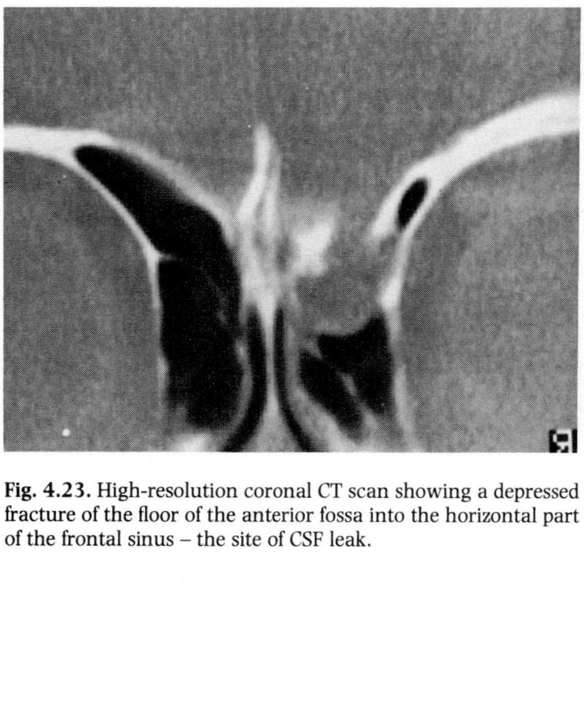

Fig. 4.23. High-resolution coronal CT scan showing a depressed fracture of the floor of the anterior fossa into the horizontal part of the frontal sinus – the site of CSF leak.

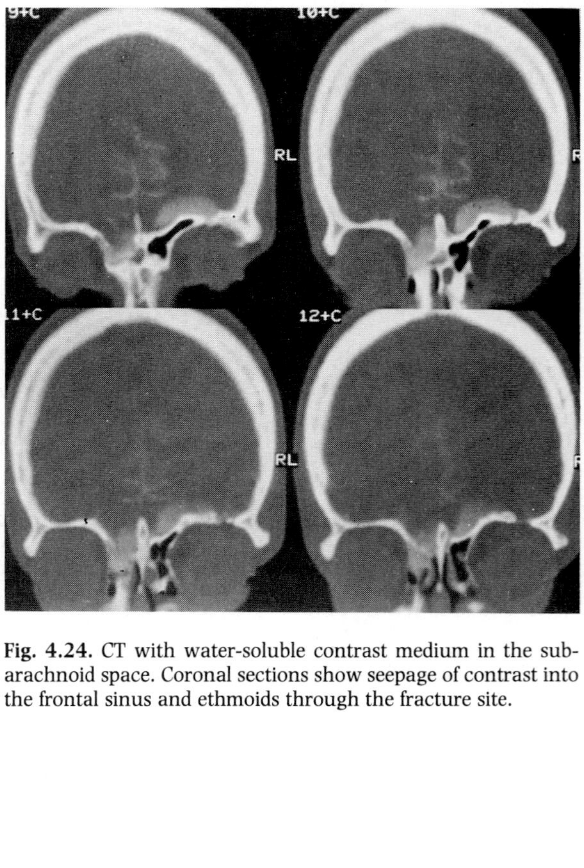

Fig. 4.24. CT with water-soluble contrast medium in the subarachnoid space. Coronal sections show seepage of contrast into the frontal sinus and ethmoids through the fracture site.

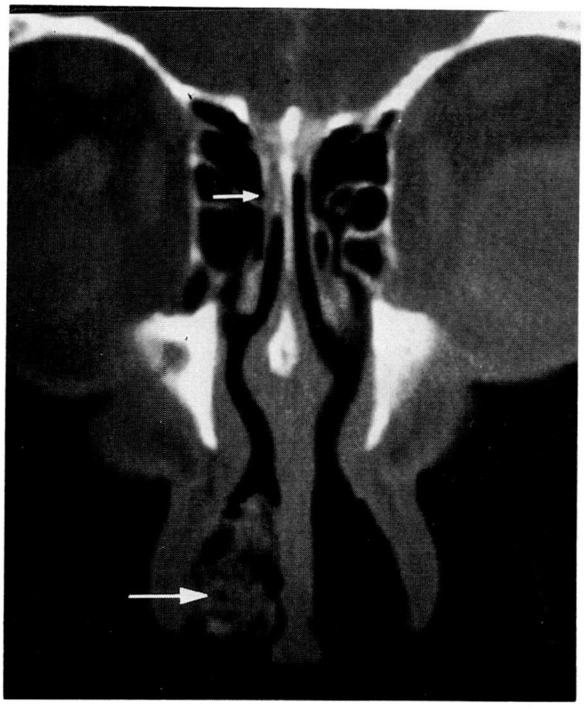

Fig. 4.25. Spontaneous CSF leak shown by positive contrast cisternography. A small quantity of the contrast medium has leaked into the upper part of the nasal cavity via the cribriform plate (*small arrow*). Note the contrast medium absorbed by the nasal plug (*large arrow*).

On plain radiographs and CT the presence of a fluid level will give a clue to the probable site of a CSF fistula into the sinuses (Fig. 4.16); or there may be simple opacification of an air cell in the neighbourhood of the leak. In post-traumatic fistulae direct evidence of bone interruption should be sought if a fracture is not obvious on the plain radiograph. This is best achieved by a combination of CT (Fig. 4.23) and conventional tomography, the latter being used to provide sagittal sections of the anterior fossa.

If these methods fail to show the site of a fistula then CT with an intrathecal contrast agent (Iohexol) may identify the leak (Figs. 4.24, 4.25). The contrast medium is injected into the sub-arachnoid space by the translumbar or transcervical route. The contrast is run into the cranial cavity with the patient prone and made to gravitate over the suspected area (i.e. the cribriform plate and planum sphenoidale); the scans are made with the patient in this position. Delayed scans after several minutes may be helpful to allow time for the contrast to seep into the appropriate air cell.

References

De Lacey GJ, Wignall BK, Hussain S, Reidy JR (1977) The radiology of nasal injuries: problems of interpretation and clinical relevance. Br J Radiol 50:412–414

Kaufman B, Neilson FE, Weiss MH, Brodkey JS, White RJ, Sykora GF (1977) Acute spontaneous, non-traumatic normal-pressure cerebro-spinal fluid fistulas originating from the middle fossa. Radiology 122:379–387

Koorneef L (1982) Current concepts in the management of orbital blow-out fractures. Ann Plast Surg 9(3):185–200

LeFort R (1901) Etude experimentale sur les fractures de la mouchoire supérieuse. Rev Chir 23:208–220

Lloyd GAS (1966) Orbital emphysema. Br J Radiol 39:933–938

Phelps PD, Lloyd GAS (1978) Congenital deformity of the internal auditory meatus and labyrinth associated with CSF fistula. Adv Otorhinolaryngol 24:51–55

Phelps PD, Lloyd GAS (1983) Radiology of the ear. Blackwell Scientific Publications, Oxford

Smith B, Regan WF (1957) Blow-out fracture of the orbit. Am J Ophthalmol 44:733–740

5 Inflammatory and Allergic Sinus Disease

Acute Sinusitis

Acute sinusitis is most often due to secondary bacterial infection following an upper respiratory tract infection of viral origin. Infection can also occur in the maxillary antrum by secondary extension from an infected tooth in the upper jaw. The primary site of infection is the lining mucosa of the sinus. The accompanying oedematous swelling of the mucosa will show on the radiograph as an opaque rim around the periphery of the sinus. It is said that in infective sinusitis the rim of mucous membrane follows the contour of and is parallel to the walls of the sinus – in contradistinction to allergic sinus disease in which the mucosa assumes a polypoid aspect with a convex inner border. This distinction is by no means clear-cut, however, and does not always co-relate well with the findings at antroscopy and mucosal biopsy. Acute sinusitis is accompanied by an outpouring of fluid into the sinus cavity causing a total loss of translucence in the affected sinus on the radiograph. This is a non-specific sign and, although in the vast majority of patients it denotes infection, a sinus filled with blood or new growth could give a similar appearance. More certain evidence of infection is provided radiologically when there is a fluid level (always assuming that there has been no trauma or recent antral washout). It is important for the demonstration of fluid levels that all sinus radiographs are taken with the patient in an upright position with a horizontal X-ray beam; if there is any doubt about the presence of fluid a tilted view should

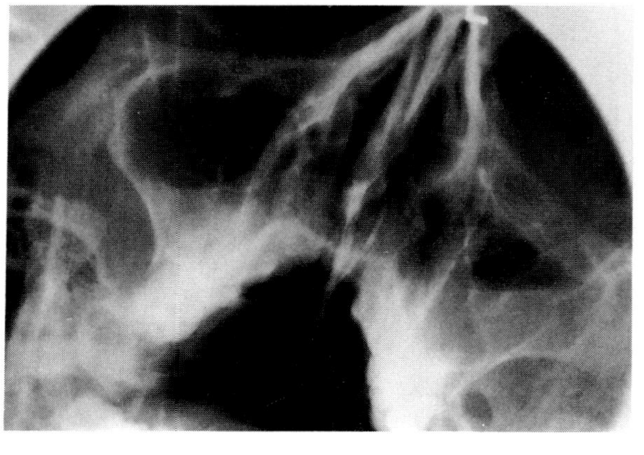

Fig. 5.1. Bilateral fluid levels in the antra shown on a tilted occipito-mental view.

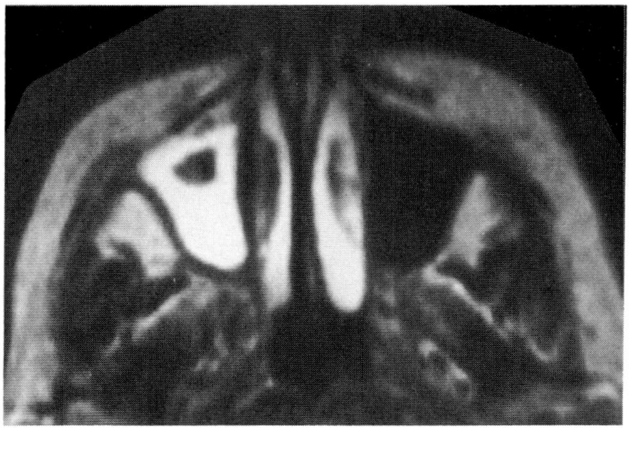

Fig. 5.2. Magnetic resonance scan. T_2-weighted spin echo sequence showing high signal and fluid level in the antrum.

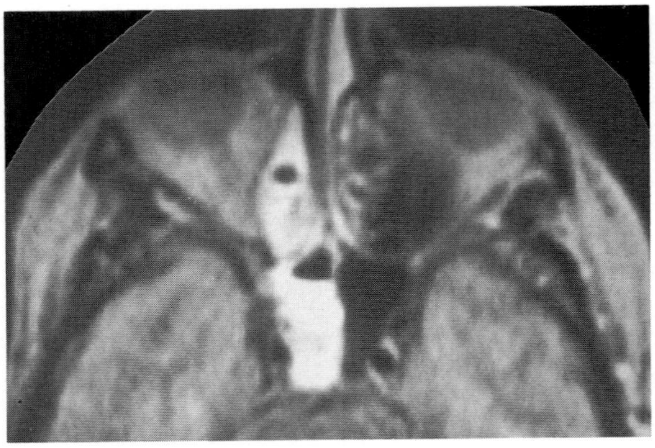

Fig. 5.3. Same patient as Fig. 5.2. Fluid levels are also present in the ethmoids and sphenoid sinus.

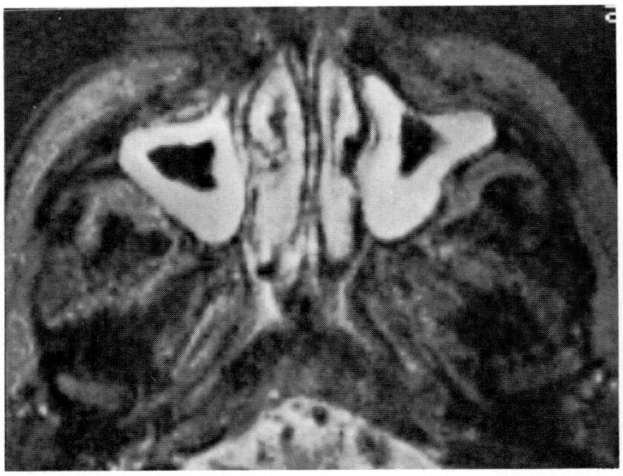

Fig. 5.4. Inflammatory mucosal thickening shown on a T_2-weighted spin echo sequence using a long time to echo.

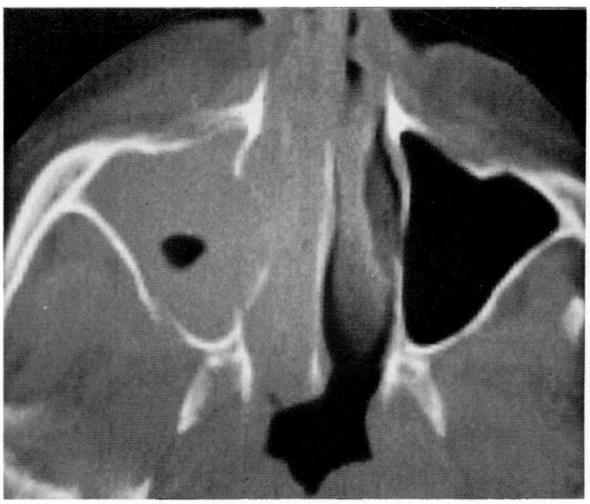

Fig. 5.5. Infection in the maxillary antrum with early osteolysis and fragmentation of the sinus walls. There is a fluid level centrally in the antrum. (The CT scan was made in the prone position.)

be obtained (Fig. 5.1), which will lead to a new horizontal level. It should be remembered that thick viscid pus or mucus may require a few moments to assume a new level when the head is tilted. In many patients the upright lateral view of the sinuses shows a fluid level most obviously. Although fluid levels can be adequately demonstrated in all the sinuses on plain radiographs, they are often more obvious on CT scan. They may also be shown by magnetic resonance studies, when the fluid in the sinus cavity and the thickened mucosa give a characteristically strong signal on T_2-weighted spin echo sequences (Figs. 5.2, 5.3 and 5.4).

Complications

Osteomyelitis

An empyema is found most commonly in the maxillary antrum and is a complication of acute sinusitis.

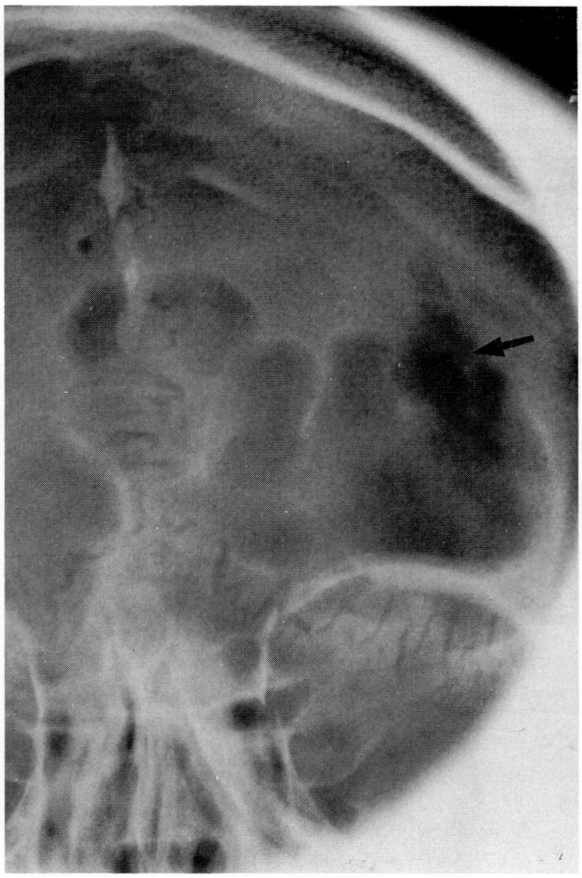

Fig. 5.6. Osteomyelitis in the left frontal bone following frontal sinus infection. An osteolytic area in the wall of the sinus with a sequestrum (*arrow*) can be seen.

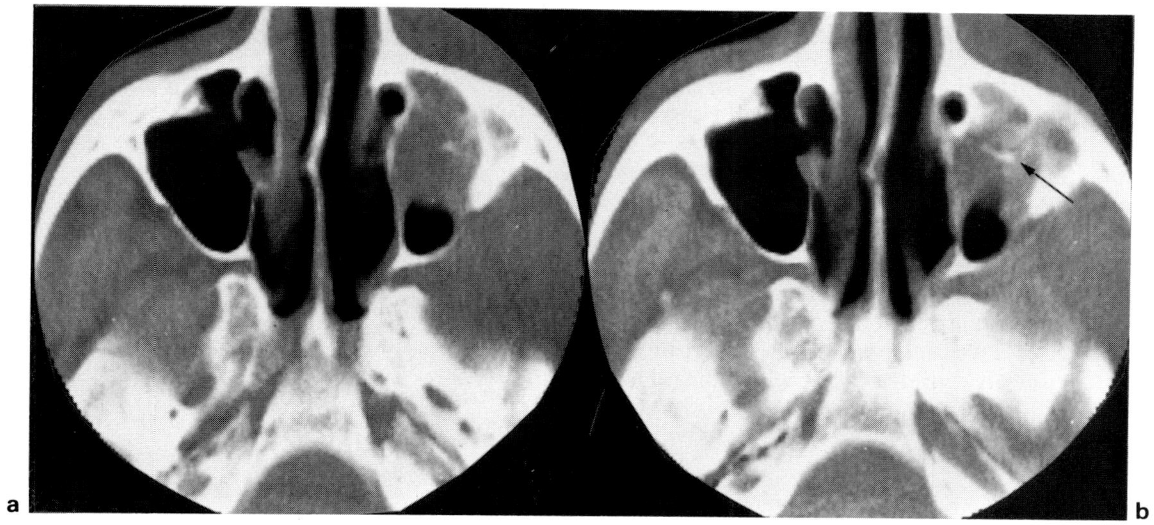

Fig. 5.7a,b. Two axial CT sections showing sequestration of the orbital floor in a patient with maxillary antritis (*arrow*). Note the associated loss of muscle outline in the infratemporal fossa.

It may rarely lead to bone involvement and osteomyelitis in the surrounding bony walls. The radiological changes of osteomyelitis lag behind the clinical signs. Loss of outline of the sinus wall followed by frank osteolysis (Fig. 5.5) and sequestration is the usual sequence. The latter change is most often seen in the frontal bone in association with frontal sinus infection (Fig. 5.6), but may also occur in the maxilla. In this situation the bone destruction may be difficult to differentiate on a plain radiograph from that caused by malignant sinus disease, and the identification of a sequestrum is important in this respect (Fig. 5.7a,b). Other points of differentiation from tumour malignancy may be observed. For example, bone destruction in a malignant tumour is almost always accompanied by a large tumour mass in the sinus concerned and in the adjacent nasal cavity, with outward expansion and destruction of the sinus walls. Conversely infections accompanied by osteomyelitis may produce an inward collapse of the sinus (Fig. 5.8), frequently with a central air shadow in the sinus cavity and sometimes with a fluid level. Inflammatory disease in the antrum may also be associated on CT with loss of muscle definition in the infratemporal fossa due to oedematous changes (Fig. 5.7b).

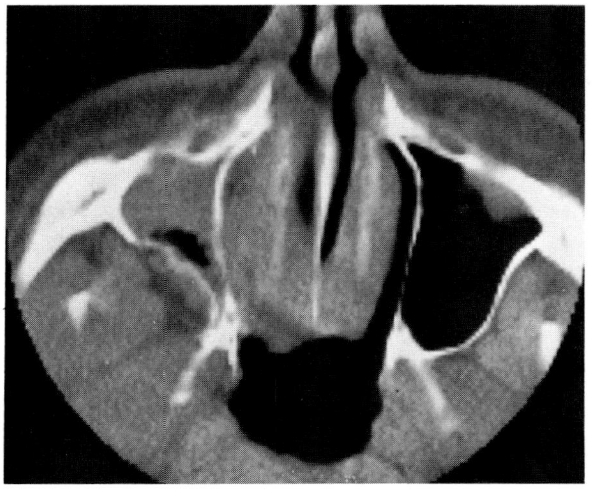

Fig. 5.8. Chronic osteomyelitis. Axial CT section showing inward collapse of the antral walls (*right*). This serves to distinguish osteomyelitis from neoplastic bone destruction, where there is inevitably some bone expansion present and a soft tissue mass.

Frontal Lobe Abscess

Spread of infection posteriorly through the wall of the frontal sinus may result in an intracranial abscess either within the brain substance, subdurally or extradurally. Plain radiographs give no indication of the development of this complication unless gas is present in the cavity of the abscess, but CT scan after intravenous contrast will show the characteristic ring enhancement of an abscess (Fig. 5.9). An abscess may also be demonstrated by magnetic resonance (Fig. 5.10).

Orbital Cellulitis

Orbital cellulitis may result from acute bacterial infection in the ethmoids (Fig. 5.11), frontal sinus or maxillary antrum. In the author's series of patients with this condition, two thirds showed obvious sinus infection on plain radiographs, and very occasionally gas or air with a fluid level was visible within the soft tissues at the site of an abscess (Fig. 5.12). The abscess is better demonstrated by CT as a soft tissue mass frequently extraconal and outside the periorbita adjacent to the infected ethmoid cells (Fig. 5.13a,b). In this situation the abscess may assume a crescentic shape around the periphery of the supero-medial orbit. It is important to localise abscess formation when it is present. Unequivocal evidence is provided when there is soft tissue gas with or without a fluid level. Ring enhancement on CT after intravenous contrast will also indicate abscess formation.

Chronic Sinusitis

Chronic sinusitis usually follows an acute infection, but infections such as tuberculosis or actinomycosis are chronic from inception. A chronically infected sinus shows mucosal thickening, which as a rule alters little with time or treatment unless there is an acute exacerbation. In some patients a reactive change may occur in the sinus walls. Skillern (1936) described the conversion of granulation tissue in the sinus to osteoid by osteoblast invasion, followed by calcium deposition and the conversion of the osteoid into true bone. The result is a proliferative osteitis causing an increased bone density in the sinus walls and leading to partial or complete obliteration of the sinus cavity.

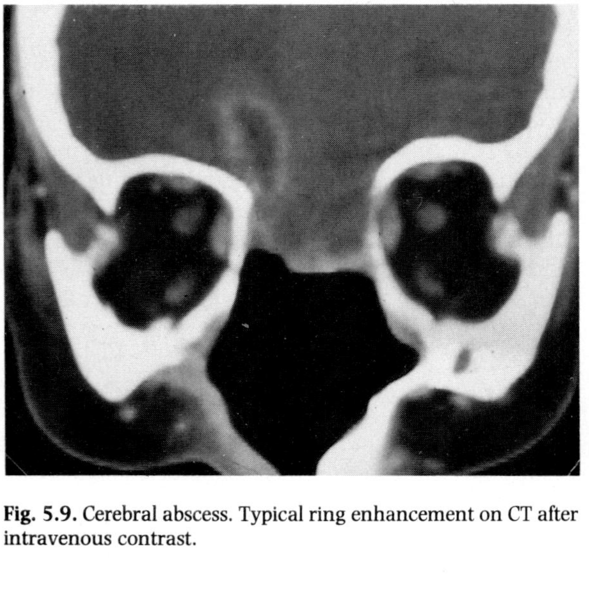

Fig. 5.9. Cerebral abscess. Typical ring enhancement on CT after intravenous contrast.

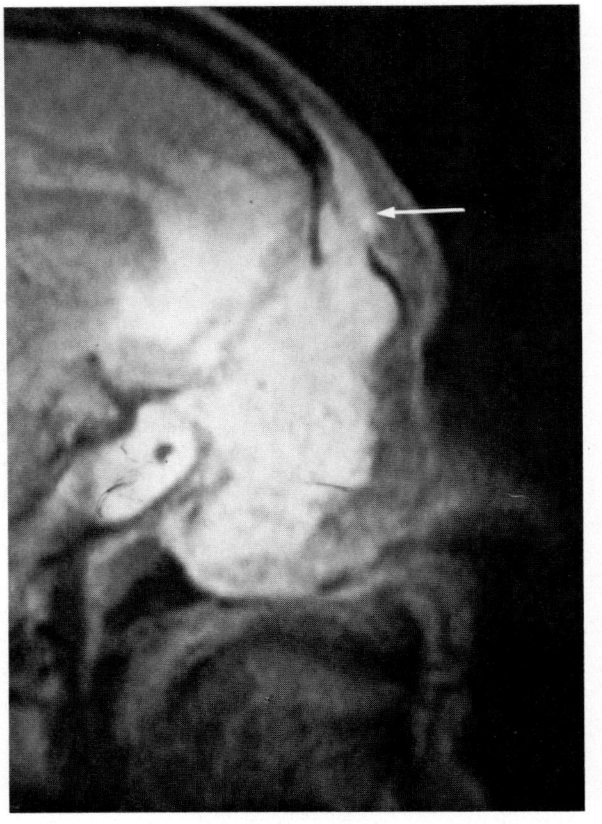

Fig. 5.10. Sagittal magnetic resonance scan of a large squamous cell carcinoma associated with frontal sinus infection, osteomyelitis and an abscess in the scalp (Pott's "puffy tumour": *arrow*).

The frontal sinuses are most commonly affected, but in some patients there may be multiple sinus involvement, and the condition may be associated with granulomatous changes in the orbit or recurrent orbital cellulitis (Fig. 5.14).

A similar bony thickening of the sinus walls may be seen in the maxillary antrum after a Caldwell–Luc operation, producing an overall opacity of the sinus on the plain radiograph (Cable et al. 1981).

Allergic Sinusitis

Allergic sinusitis is caused by an allergic reaction in the mucosa of the upper respiratory tract giving rise to oedema of the lining membranes of the sinuses and nose accompanied by hypersecretion. The radiological changes may be difficult to distinguish from those of infective sinusitis, especially as the two conditions may coexist: for example total

a

b

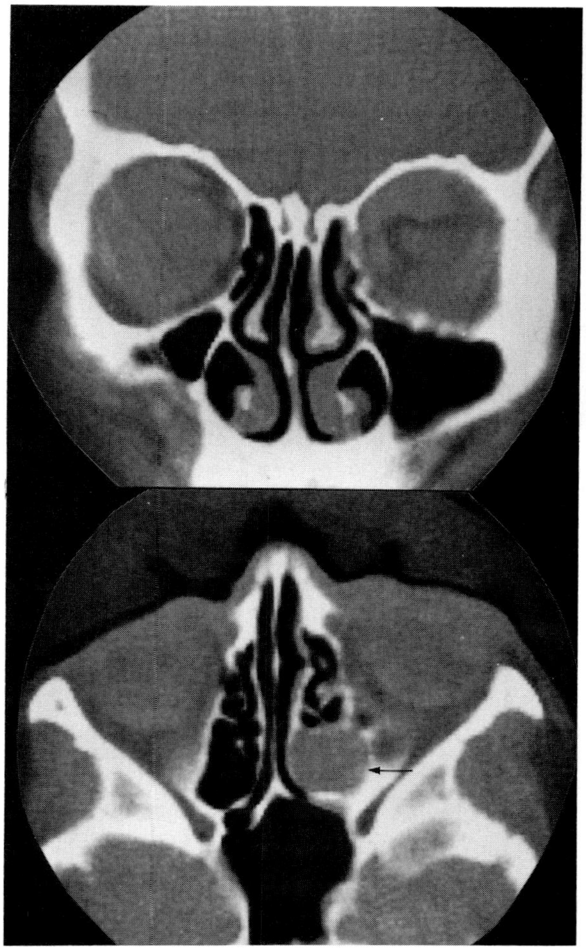

Fig. 5.11a,b. Orbital cellulitis. **a** The coronal CT section shows an abscess in the supero-medial quadrant of the left orbit. **b** The underlying cause is shown in the axial section: a pyocoele in the posterior ethmoid cells (*arrow*).

opacity of a sinus may occur in both conditions. Thickening of the nasal turbinates is characteristic of allergic sinusitis and is usually accompanied by the polypoid type of mucosal thickening in the adjacent sinuses, the thickened mucosa producing a convex indentation into the cavity of the sinus. Bilateral changes in the sinuses are also typical of allergic sinus disease and the condition is frequently complicated by nasal polyposis. Sinus fluid levels are uncommon in allergic sinusitis and when present usually denote superadded infection.

Nasal Polyposis

Simple nasal polyps are pedunculated sections of oedematous upper respiratory mucosa. They can

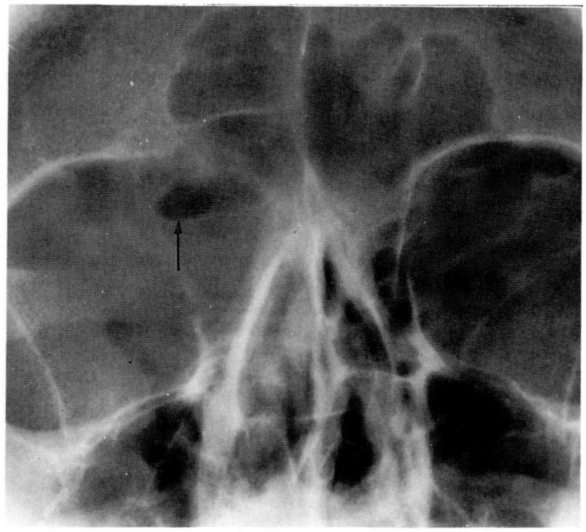

Fig. 5.12. Orbital cellulitis. The plain radiograph shows an abscess in the medial orbit and right ethmoid cells with fluid level (*arrow*).

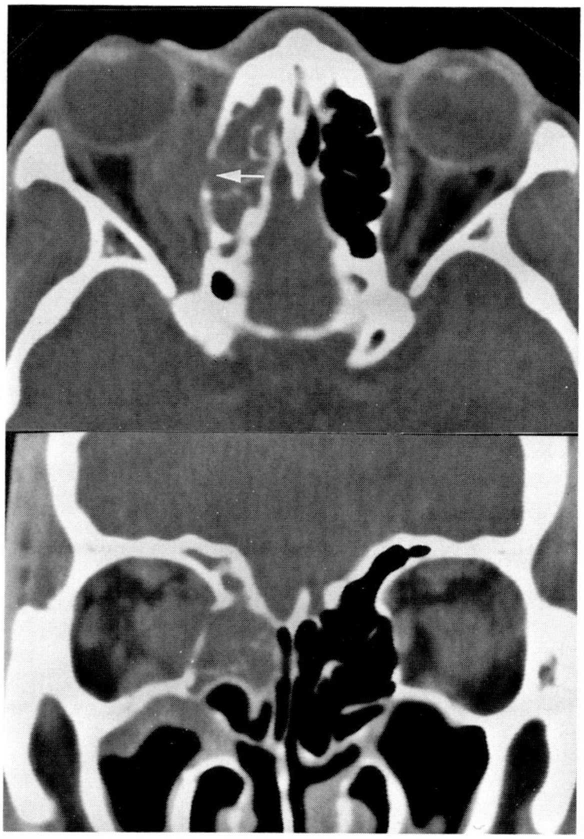

a

b

Fig. 5.13a,b. Axial and coronal CT scans showing infection in the ethmoid cells with orbital cellulitis and abscess formation. Note the bone destruction in the medial orbital wall (*arrow*) at the site of the abscess.

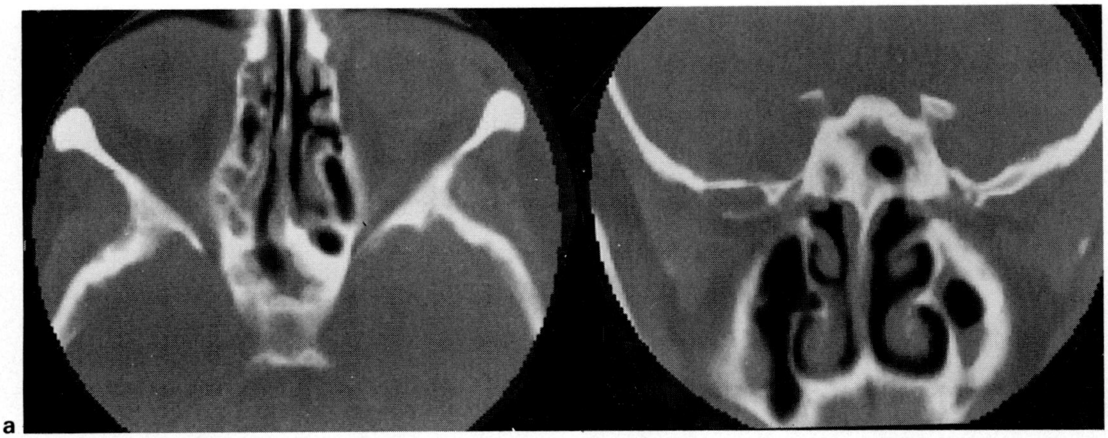

Fig. 5.14a,b. Chronic infection in the sinuses with sclerosing osteitis. An increase in the thickness and density of the walls of the sphenoid sinus is shown on axial (**a**) and coronal (**b**) CT scans. Note that the maxillary antrum is also affected. These sinuses were chronically infected, and the patient had had six attacks of orbital cellulitis in the preceding year.

arise from any part of the nasal and sinus mucosa and are often multiple and bilateral. Histologically they consist of a grossly oedematous stroma covered by respiratory epithelium, which in some places may undergo squamous metaplasia (Fig. 5.26). Their pathogenesis is poorly understood, but it is likely that the majority of polyps are related to hypersensitivity mediated through mast cells (Michaels 1987). Non-allergic asthma is a common accompanying condition and polyps are also associated with aspirin hypersensitivity. Their commonest site of origin is the mucosa of the ethmoid cells; much less commonly they arise in one or other maxillary antrum, entering the nasal cavity through its ostium and passing backwards through the posterior nares to form an antro-choanal polyp. In either case the mechanism of formation is the same.

Multiple nasal polyps may completely occlude the nasal cavity and prolapse through the posterior choana into the nasopharynx, where they may be

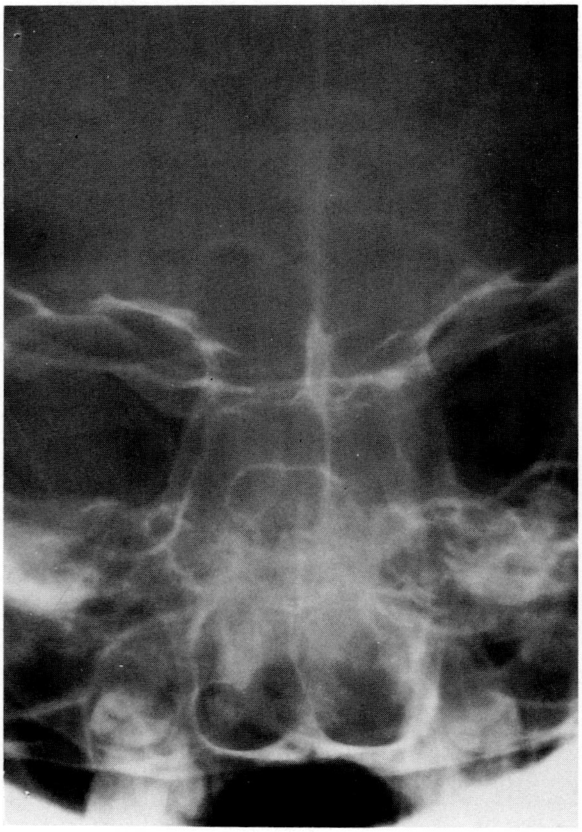

Fig. 5.15. Nasal polyposis. Plain radiograph showing loss of translucence in the ethmoid cells, frontal sinuses and the upper part of the nasal cavity.

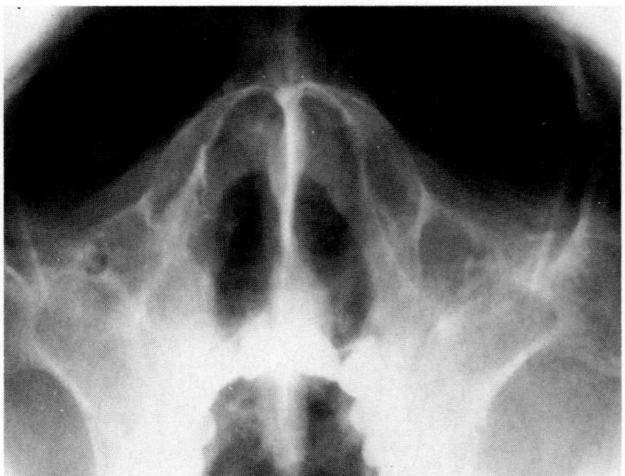

Fig. 5.16. Nasal polyposis. Widening of the nasal bones shown on an occipito-mental projection.

visible on lateral radiographs. However they are seldom recognised individually on plain radiographs; more often the nasal cavity presents a homogeneous loss of translucence, especially in its upper part (Fig. 5.15). Long-standing polyps can enlarge to such a degree that they produce expansion of the walls of the nasal cavity (Fig. 5.16) and ethmoid labyrinth (Lloyd 1971, 1975). This affects the whole labyrinth, but its general form is largely retained (Fig. 5.17). The expansion may produce an increase in the interpupillary distance of the eyes and hypertelorism. Typically the changes are accompanied by a generalised loss of translucence in the sinuses on plain radiographs, decalcification of their walls and erosion of the ethmoid septa (Fig. 5.18). The essential feature of nasal polyposis with or without ethmoid expansion is the bilateral nature of the changes. This distinguishes it from antro-ethmoidal malignancy, where the changes often remain unilateral even when the disease is far advanced.

The changes described above are, in a small minority of patients, associated with a local expansion of a sinus or bone erosion (Wilson 1976; Winestock et al. 1978). In some patients there may be mucocoele formation – either single or multiple. When these changes are found in children under 10 they are usually associated with mucoviscidosis (Fig. 5.19), and nasal polyps are sometimes the presenting symptom of this disease (Schwachman et al. 1962). They are estimated to occur in 6.7% of children with cystic fibrosis. Microscopically they resemble the adult polyp, produce similar radiographic features and exhibit the same tendency to recur (Toma and Stein 1968; Fonsman 1970).

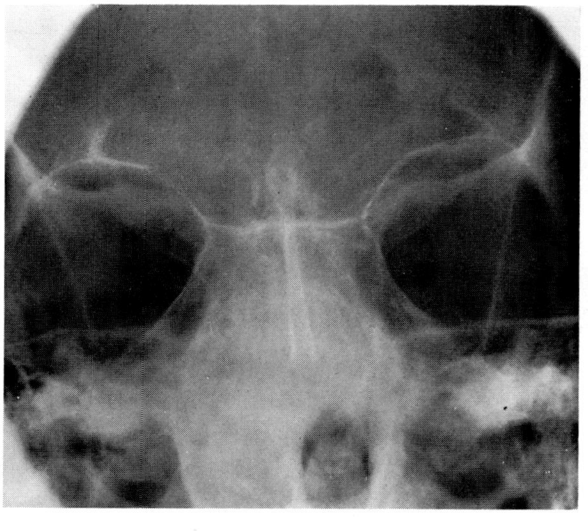

Fig. 5.17. Nasal polyposis. Lateral expansion of the ethmoids shown on a postero-anterior radiograph.

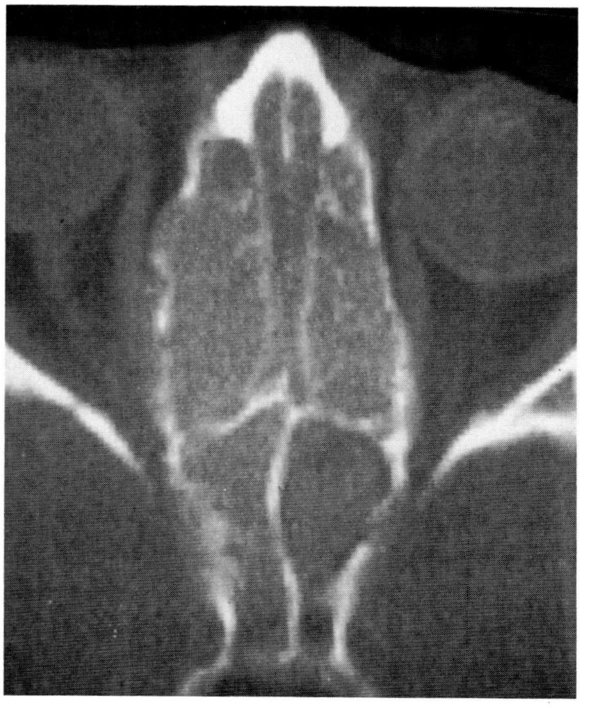

Fig. 5.18. Nasal polyposis. Axial CT scan showing total loss of translucence in the ethmoid cells with erosion of the ethmoid septa. Note the slight expansion of the lamina papyracea on the right side.

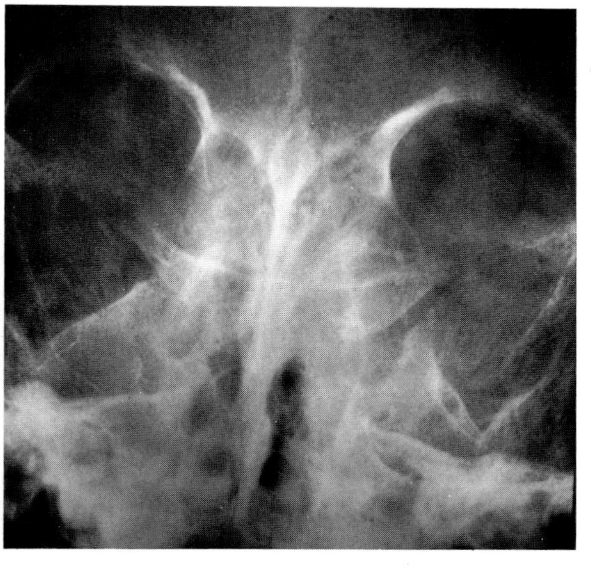

Fig. 5.19. Plain radiograph showing changes due to benign nasal polyps associated with mucoviscidosis.

Bone destruction in the paranasal sinus associated with benign nasal polyps was described in 1885 by Edward Woakes, although it may have been recognised as early as 1628 (Wentges 1972).

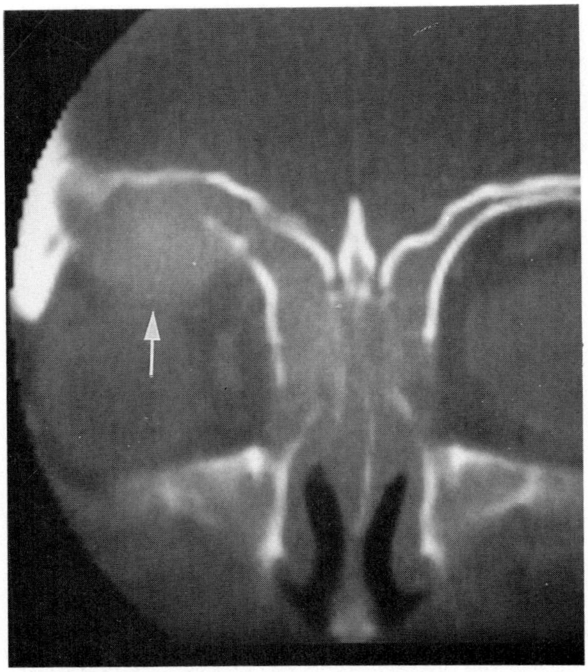

Fig. 5.20. Coronal CT scan showing nasal polyposis and mucocoele formation in the right frontal sinus (*arrow*).

The bone erosion is usually completely asymptomatic without evidence of infection and is therefore unlikely to be due to an associated osteomyelitis. The mechanism is probably one of passive bone erosion in a similar fashion to that produced by mucocoeles. Once the nose or sinus cavity has filled with polypoid tissue any further increase in tissue bulk is likely to lead to bone expansion, with thinning of bony septa, widening of the bridge of the nose and hypertelorism.

In a series of 100 patients with nasal polyposis reviewed by Lund and Lloyd (1983) there was evidence of expansion of the ethmoids on plain radiographs in 20 cases (19 bilateral and 1 unilateral). The changes were seen most often in the younger patients in whom the onset of symptoms occurred below the age of 30 years. In addition in the same group, 11 patients were observed in whom there was mucocoele or pyocoele formation in association with polyps. The majority of these patients had developed the mucocoele following surgery to the nose and sinuses, and this would appear to be the predominant cause of mucocoele formation in these patients rather than direct obstruction of sinus drainage by the polyps.

CT and Magnetic Resonance

On CT nasal polyposis shows as a mass of homogeneous density in the ethmoid cells with erosion of the ethmoid septa (Fig. 5.18). In some patients this is associated with expansion of the ethmoid cells and mucocoele formation (Figs. 5.20, 5.21). Polyps produce a strong signal on T_2-weighted spin echo magnetic resonance sequences, the signal intensity approaching that of retained secretion in the sinuses. This makes differentiation of fluid and polyp difficult, which is not surprising because the fluid secretions are almost identical in composition to the oedema fluid in the polyp. The polyps can be identified by both CT and magnetic resonance as a soft tissue mass occupying the upper part of the nasal cavity and ethmoids (Fig. 5.22).

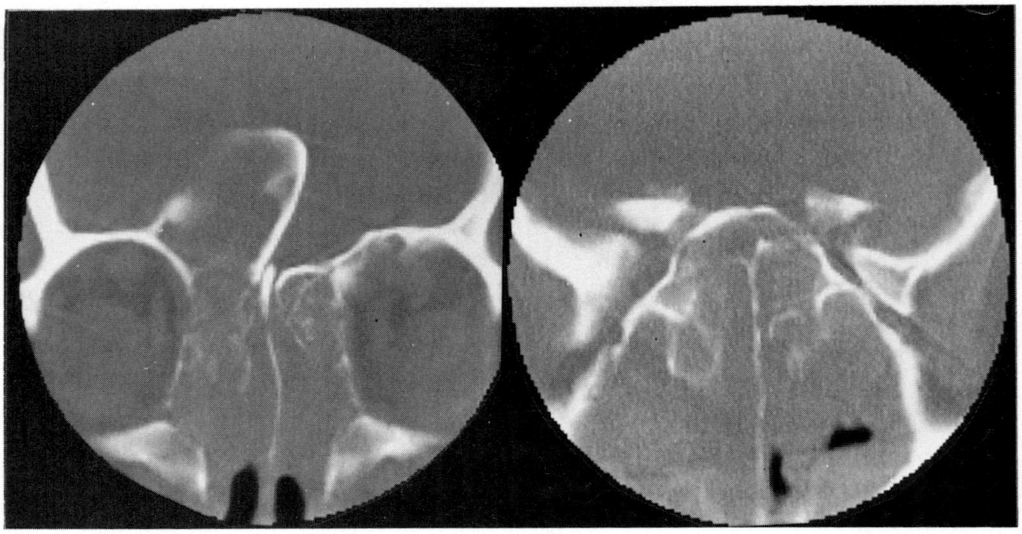

Fig. 5.21a,b. Coronal CT sections showing nasal polyposis with upward expansion of the ethmoids (**a**) due to mucocoele formation.

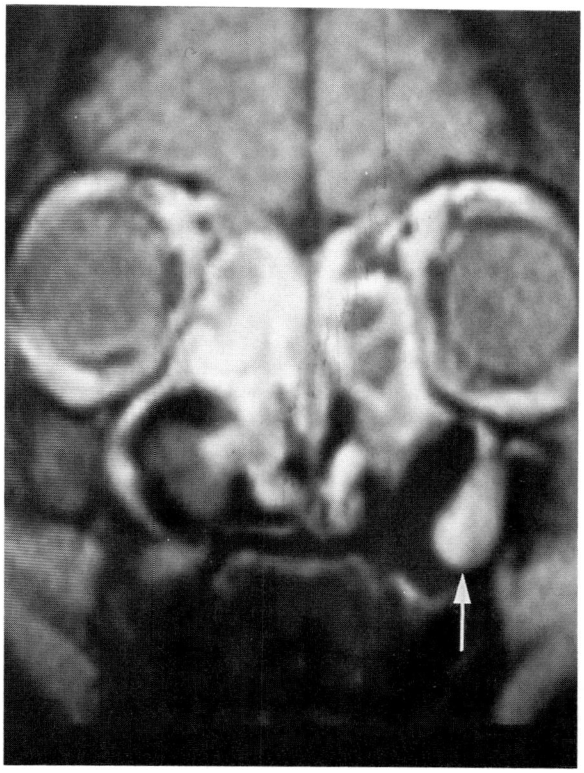

Fig. 5.22. Coronal magnetic resonance scan showing nasal polyps in the ethmoids and upper part of the nasal cavity. Note the typical appearance of a polyp in the cavity of the left antrum (*arrow*).

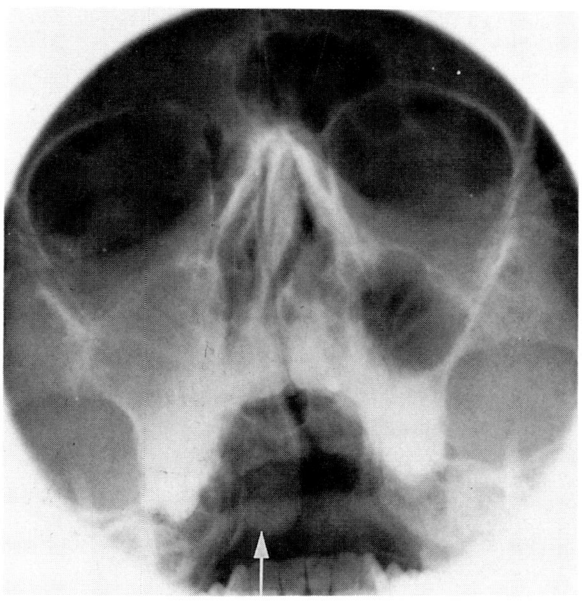

Fig. 5.23. Occipito-mental projection of a child with an antro-choanal polyp. Note the opaque antrum with the outline of a polyp in the nasopharynx (*arrow*).

Antro-choanal Polyps

Antro-choanal polyps are readily seen on plain radiographs when they project into the air-filled nasopharynx. They then appear as a soft tissue shadow with a well-defined smooth outline. They may be seen on lateral views of the nasopharynx or through the open mouth on an occipito-mental projection (Fig. 5.23). Usually the maxillary antrum and the nasal cavity on the side of the polyp are opaque.

The diagnosis of antro-choanal polyp is important in adolescent boys, when the condition needs to be differentiated from juvenile angiofibroma. Surgeons

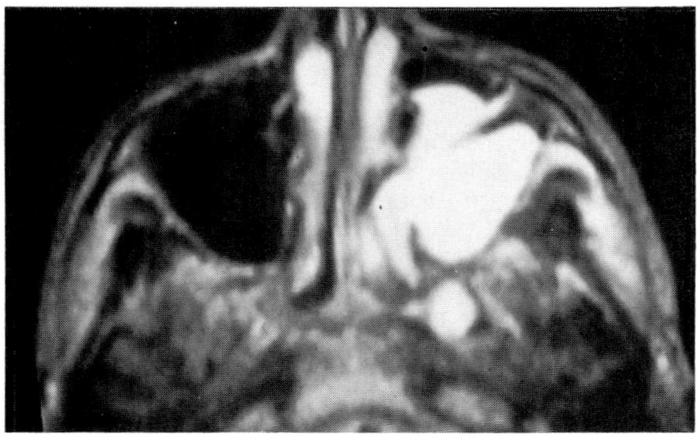

Fig. 5.24. Axial magnetic resonance section showing an antro-choanal polyp within the antrum and emerging from the ostium.

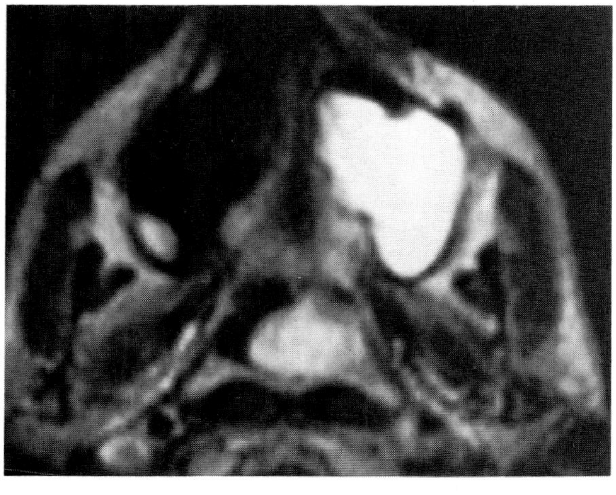

Fig. 5.25. Same patient as Fig. 5.24. T_2-weighted spin echo sequences showing the nasopharyngeal component of the polyp.

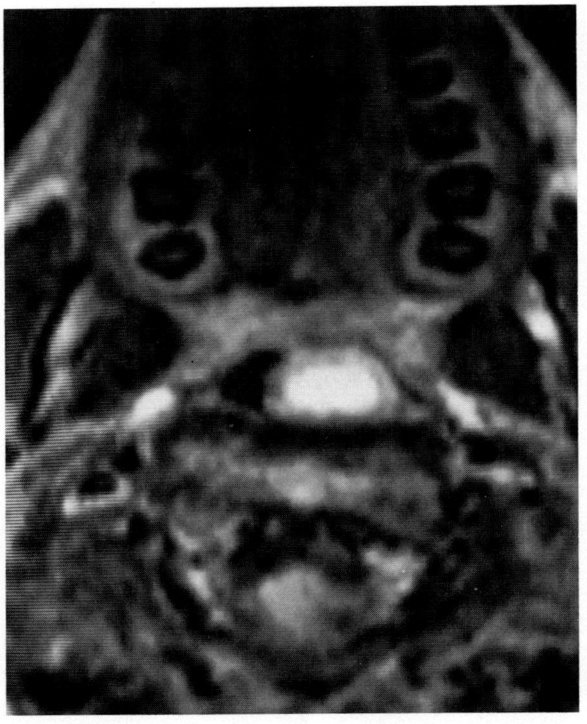

Fig. 5.26. Same patient as Fig. 5.24. The diminished signal from the peripheral part of the mass is shown. This is due to squamous metaplasia of the epithelium, forming a hard outer capsule.

are naturally reluctant to biopsy a nasopharyngeal mass in this age group because of the danger of profuse haemorrhage from the highly vascular angiofibroma. Diagnosis can be achieved in the first instance by demonstrating the typical bone changes associated with angiofibroma on conventional tomography or CT scan (Chap. 11). Magnetic resonance can also show diagnostic features of this tumour (Lloyd and Phelps 1986); alternatively it can demonstrate an antro-choanal polyp directly, showing the site of origin of the polyp in the antrum and the characteristically high homogeneous signal on T_2-weighted spin-echo sequences (Figs. 5.24, 5.25). Figure 5.26 shows the thickened capsule of the nasopharyngeal component of the polyp – the result of squamous metaplasia.

References

Cable HR, Jeans WD, Cullen FJ, Bull PD, Maw AR (1981) Computerized tomography of the Caldwell–Luc cavity. J Laryngol Otol 95:775–783

Fonsman J (1970) Mucoviscidosis and nasal polyps. Acta Otolaryngol 69:152–154

Lloyd GAS (1971) Axial tomography of the orbits and paranasal sinuses. Br J Radiol 44:373–381

Lloyd GAS (1975) Radiology of the orbit. WB Saunders, Philadelphia

Lloyd GAS, Phelps PD (1986) Juvenile angiofibroma: imaging by magnetic resonance, CT and conventional techniques. Clin Otolaryngol 11:247–259

Lund VJ, Lloyd GAS (1983) Radiological changes associated with benign nasal polyps. J Laryngol Otol 97:503–510

Michaels L (1987) Ear, nose and throat histopathology. Springer, Berlin Heidelberg New York

Schwachman H, Kulczycki LL, Mueller HL, Flake CG (1962) Nasal polyposis in patients with cystic fibrosis. Paediatrics 30:389–401

Skillern SR (1936) Obliterative frontal sinusitis. Arch Otolaryngol 23:267–276

Toma GA, Stein GE (1968) Nasal polyposis in cystic fibrosis. J Laryngol Otol 82:265–268

Wentges RTR (1972) Edward Woakes: the history of an eponym. J Laryngol Otol 86:501–512

Wilson M (1976) Chronic hypertrophic polypoid rhinosinusitis. Radiology 120:609–616

Winestock DP, Bartlett PC, Sondheimer FK (1978) Benign nasal polyps causing bone destruction in the nasal cavity and paranasal sinuses. Laryngoscope 88:675–679

Woakes E (1885) Necrosing ethmoiditis and mucous polypi. Lancet I:619–620

6 Cysts and Mucocoeles

Cysts of the Paranasal Sinuses

Cysts of the paranasal sinuses may be classified into two groups: intrinsic cysts arising within the sinuses and extrinsic cysts which take origin in adjacent structures such as the orbit or dental tissues.

Intrinsic Cysts

Mucous Retention Cysts

Mucous retention cysts are lined by columnar epithelium and are the result of obstruction of the ducts of a mucous gland, usually as an aftermath of infection. They commonly occur in the maxillary antrum and appear on the radiograph as a dome-shaped, clearly demarcated opacity (Fig. 6.1) which may sometimes grow to fill the sinus cavity. These cysts are opaque to both transillumination and X-rays; in contrast non-secreting cysts transilluminate normally and are opaque only to X-rays.

Non-secreting Cysts

Non-secreting cysts are formed from oedematous dehiscence of a connective tissue plane underlying the sinus mucosa. They therefore have no epithelial lining and are regarded as post-inflammatory in origin; they contain fluid with a high cholesterol content, which allows normal light transillumination (Samuel and Lloyd 1978). Radiologically they present as a smooth, dome-shaped opacity, resembling a mucous cyst. They are often associated with apparently normal mucosa in the rest of the sinus and do not attain sufficient size to expand the bony sinus cavity.

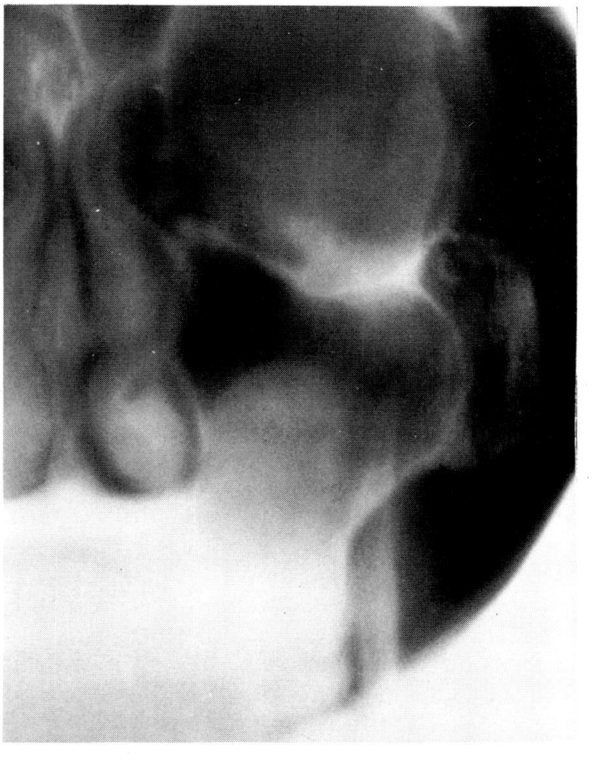

Fig 6.1. Mucous retention cyst shown as a dome-shaped opacity in the floor of the maxillary antrum on conventional tomography.

Cholesteatoma

In the past there has been some confusion between cholesteatoma and cholesterol granuloma, and a proportion of cases reported under the former diagnosis were clearly cholesterol granulomata (Osborn and Wallace 1967). True cholesteatoma is a rarity in the sinuses, and several theories have been advanced to explain its presence. The most likely is a metaplasia of the normal sinus mucosa to keratinising stratified squamous epithelium as a result of chronic sinus infection. Another theory is that cholesteatoma occurs by invasion of buccal epithelium via an oro-antral fistula. This would account for the majority of these lesions being reported in the maxillary antrum. Clinically the patient may have pain in the cheek, deformity of the face and displacement of the eye. Radiologically they present with expansion of the sinus very like that caused by a mucocoele (Fig. 6.2).

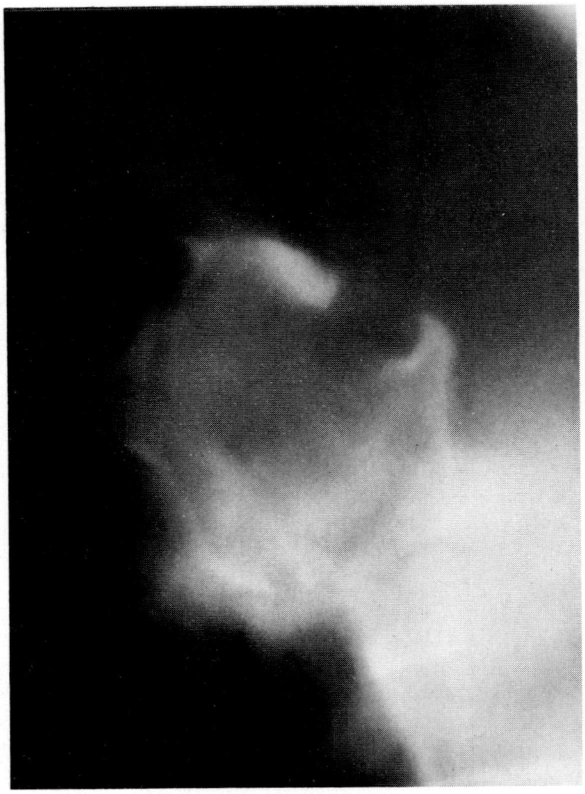

Fig. 6.2. Expansion of the sphenoid sinus and erosion of the floor of the pituitary fossa by a cholesteatoma.

Cysts of Dental Origin

There are three common types of dental cyst which may affect the maxillary antrum:

1. *Odontogenic keratocysts.* These are formed by cystic degeneration of the enamel organ before the formation of a tooth.
2. *Dentigerous cysts.* These are due to cystic degeneration of the enamel organ after it has been formed but before it has erupted.
3. *Radicular cysts.* These are formed after eruption of a tooth and are associated with apical infection.

Odontogenic Keratocysts

Odontogenic keratocysts arise from cystic degeneration of the enamel organ as described above, or they may sometimes take origin from ectopic odontogenic epithelium. The term odontogenic keratocyst was first used by Philipsen (1956); previously these lesions were generally called primordial cysts. The majority occur in the mandible but a minority arise in the maxilla and in this situation they may expand into the maxillary antrum and buccal soft tissues (McIvor 1972). Radiologically most of them present as a well-defined monolocular translucence in the bone with a well-corticated margin, but others may be multilocular.

The most striking feature of the odontogenic keratocyst is its tendency to recur after surgical treatment. The recurrence rate varies from 12% to 62.5% depending on the series (Lund 1985). The high recurrence rate and aggressive behaviour are thought to derive from the highly proliferative keratocyst epithelium. An example of late recurrence of this lesion has been recorded by Lund (1985). The patient was a 64-year-old female who gave a 44-year history of problems associated with the left upper jaw. Plain radiography and CT showed a large multiloculated area occupying the lateral wall of the left maxilla and adjacent orbit (Fig. 6.3). This was associated with osteolysis and new bone formation in the areas affected. Tissue removed at surgery from the antral floor and lateral wall of the orbit showed numerous cysts in both soft tissue and bone, containing large amounts of keratin.

Dentigerous Cysts

Dentigerous cysts are associated with maldevelopment of a tooth bud, resulting in a unilocular cyst related to the crown of an unerupted tooth. The cyst may extend upwards to invade the maxillary antrum and nasal cavity (Fig. 6.4).

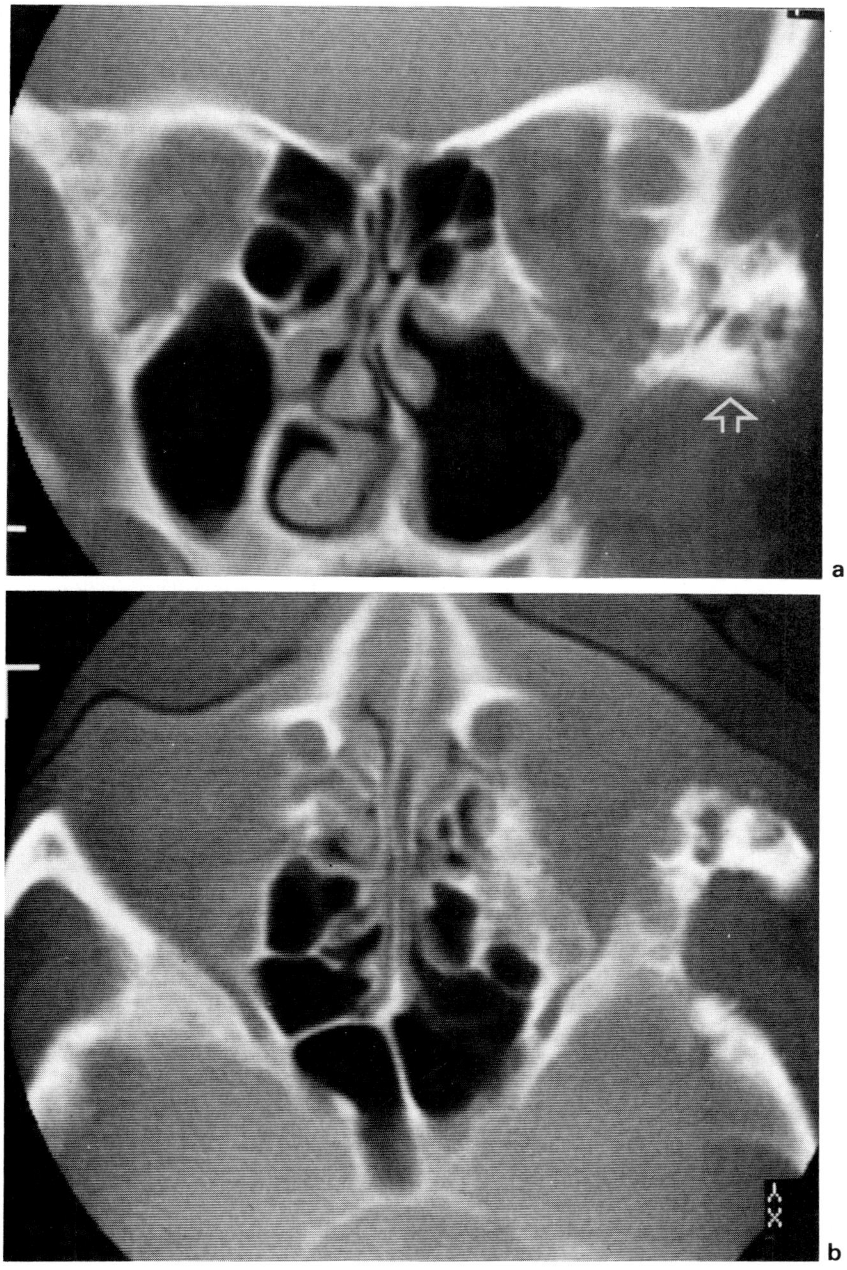

Fig. 6.3a,b. Coronal (**a**) and axial (**b**) CT scans showing erosion of the antrum and lateral orbital wall by a recurrent odontogenic kerotocyst. There is much reactive new bone formation on the lateral wall of the orbit (*arrow*).

Radicular Cysts

Radicular cysts are the commonest variety of dental cyst. They lie directly on the apex of a tooth and follow apical infection. The cyst has a well-defined margin and the adjacent tooth root shows loss of lamina dura. These cysts may discharge into the maxillary sinus, and an antro-oral fistula may develop when the tooth is extracted.

Calcifying Odontogenic Cysts

A calcifying odontogenic cyst is not a true cyst but is in fact a tumour with cystic change. Like other odontogenic tumours it may invade the maxilla, producing the appearance of a multiloculated cyst on CT (Fig. 6.5).

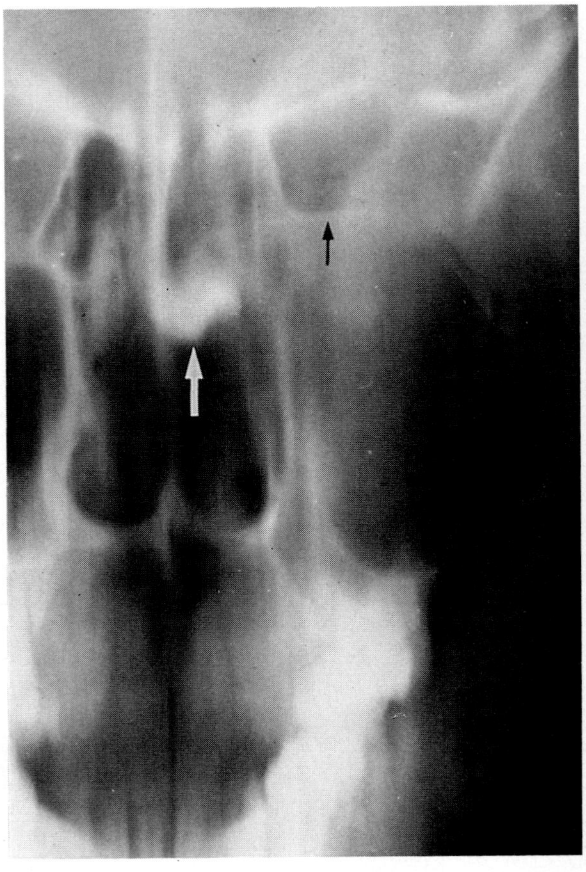

Fig. 6.4. Dentigerous cyst shown on conventional tomography. The cyst has expanded upwards into the maxillary antrum (*black arrow*). An unerupted tooth (*white arrow*) is seen associated with the cyst and lying within the nasal cavity.

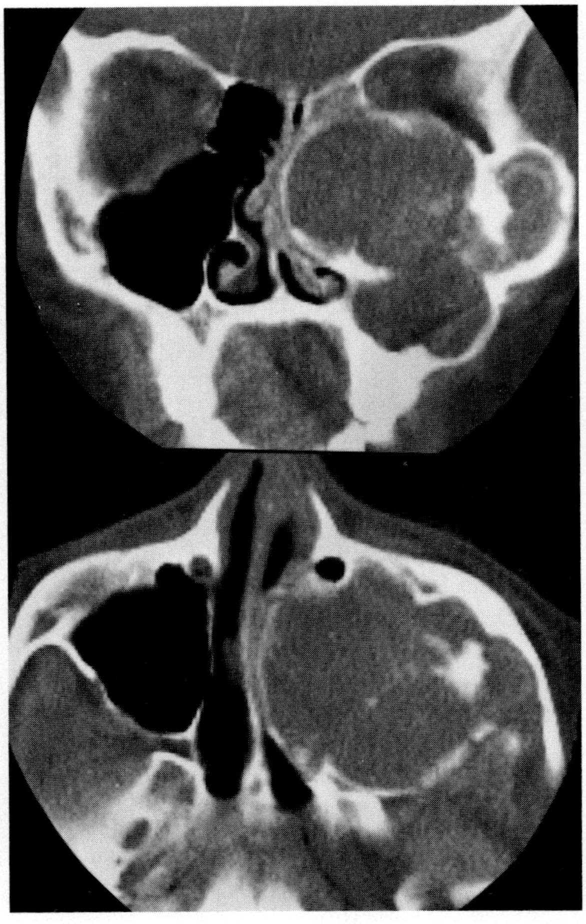

a

b

Fig. 6.5a,b. Coronal (**a**) and axial (**b**) CT scans of a calcifying odontogenic cyst. The cyst appears as a huge multilocular expansion of the maxillary antrum.

Miscellaneous Cysts

Globulo-Maxillary Cysts

Globulo-maxillary cysts are the most common type of fissural cyst. They are of developmental origin and represent cystic inclusions from a failure of complete fusion between the globular and maxillary processes of the maxilla. They appear in the canine fossa and as they grow they erode the hard palate and invade the maxillary antrum. Their characteristic feature on the radiograph is a separation of the roots of the canine and incisor teeth.

Incisor Canal Cysts

Incisor canal cysts are also designated as fissural cysts and result from failure of the two halves of the premaxilla to fuse. They lie in the mid-line and cause enlargement of the incisive canal. On plain radiographs they are best demonstrated on occlusal films, but may be better shown by conventional tomography or coronal CT.

Dermoid Cysts

Dermoid cysts are rare in the paranasal sinuses and generally occur as a secondary extension from a cyst of the nose or orbit (Figs. 6.6, 6.7 and 6.8). The radiological features are variable depending upon the location of the cyst, but generally they show an expanding cystic lesion which, in the bones forming the orbit, is characterised by a well-defined, slightly sclerotic edge. From the orbit the cyst may displace and erode the bony wall of the adjacent sinus. Dermoids have a diagnostic feature on CT: approximately two thirds of them will show areas of negative attenuation due to the presence of a loculus of fat or oil within the cyst.

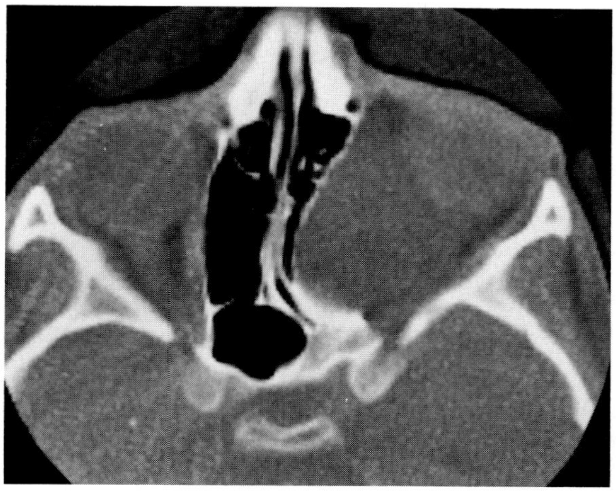

Fig. 6.6. Dermoid cyst. Axial CT scan showing a huge indentation into the ethmoid labyrinth.

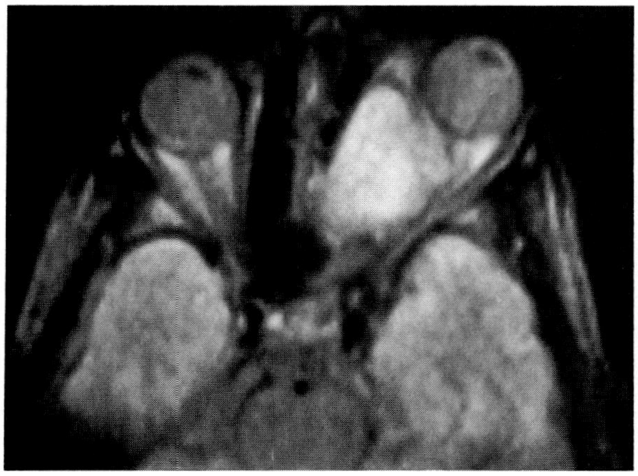

Fig. 6.7. Same patient as Fig. 6.7. On magnetic resonance scan there is strong signal from the tumour on T_2-weighted spin echo sequences.

Hydatid Cysts

One example of a hydatid cyst has been seen in the nasal cavity (Fig. 6.9).

Mucocoeles

Cyst-like expansions of the paranasal sinuses have been recognised for over 160 years (Evans 1981), the term mucocoele first being used to describe them by Rollet (1896). Mucocoeles are defined as expansile lesions occurring within the cavity of a sinus, containing mucoid secretions and limited by the

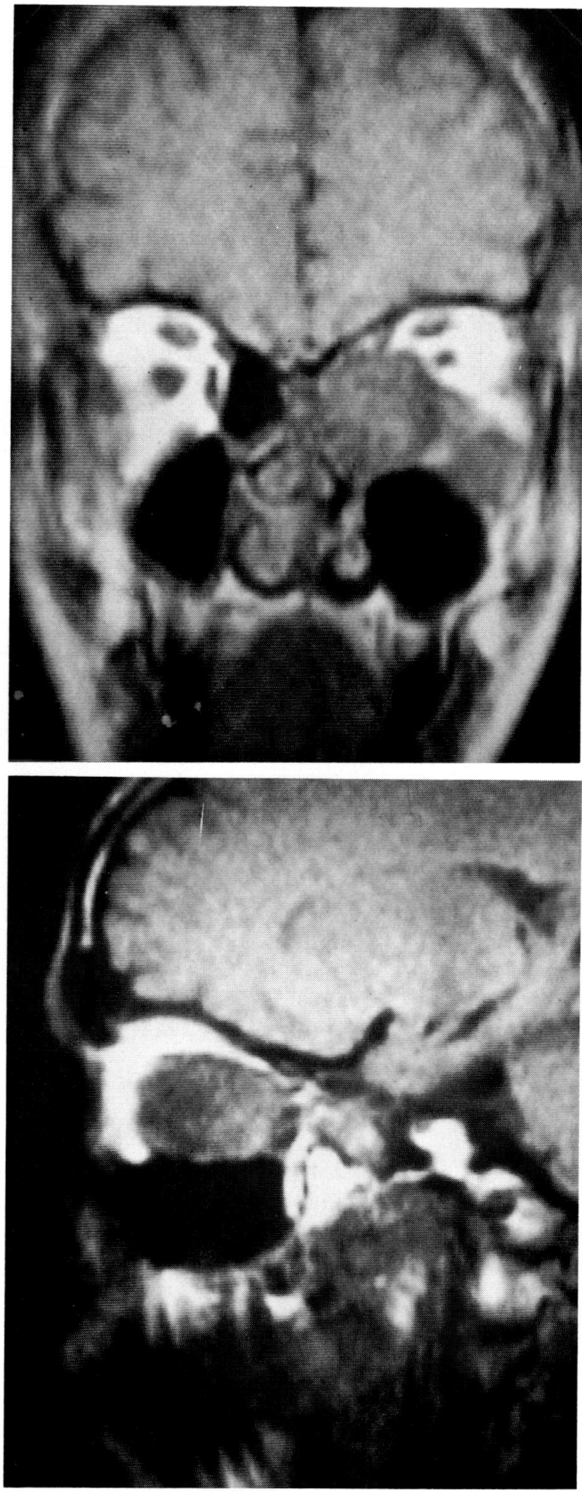

Fig. 6.8a,b. Dermoid cyst. T_1-weighted spin echo sequence (**a**) and sagittal inversion recovery sequence (**b**) both show moderate to low signal. In this case the dermoid did not show evidence of fat on magnetic resonance scan.

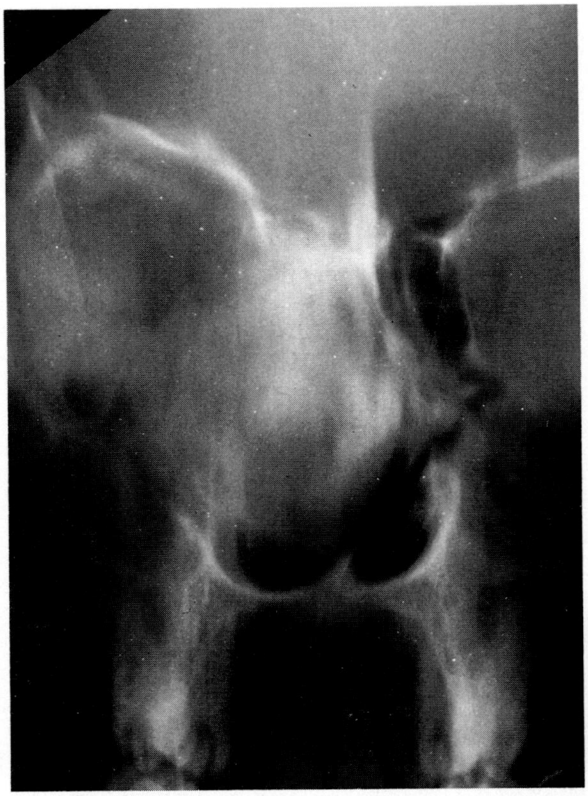

Fig. 6.9. Conventional tomogram showing a hydatid cyst in the nasal cavity.

lining membrane (Canalis et al. 1978). They are locally destructive and cause pressure erosion and expansion of the bony walls, encroaching upon and displacing adjacent structures. They occur most frequently in the frontal sinus but may arise in the ethmoids and rarely in the sphenoid sinus or maxillary antrum. The relative frequency of mucocoele formation is illustrated by the author's series seen over a 20-year period: there were 91 frontal sinus mucocoeles, 27 ethmoid, 11 sphenoid, and 5 in the maxillary antrum.

Mucocoeles may result from infection; trauma, including bullet wounds and previous surgery; polyps (Lund and Lloyd 1983); or tumours, benign and malignant. All of these may affect the normal drainage of the sinus concerned. In some cases there is histological evidence of an increase in the number of secretory cells in the lining epithelium, and hypersecretion of mucus may then be a contributory causative factor. Lund (1987), in reviewing 80 cases of mucocoeles treated surgically, identified seven groups according to their aetiology. These included direct sinus trauma involving fracture of the facial bone; nasal polyps untreated or removed surgically; a miscellaneous group which included osteomata and maxillary sinus surgery for cysts; and those in which no aetiological factor could be determined. This last group was the largest, representing 29% of the total.

Frontal Sinus Mucocoeles

The classical radiographic appearance of a frontal sinus mucocoele is that of an expanded sinus showing loss of the scalloped margin and an overall loss of translucence on the affected side (Fig. 6.10). Other features include depression or erosion of the supraorbital ridge and extension across the mid-line through the septum to the opposite frontal sinus (Fig. 6.11). Loss of translucence is not, however, an invariable feature. While it is true that in most patients the affected sinus is more opaque than normal, due to displacement of the air content and

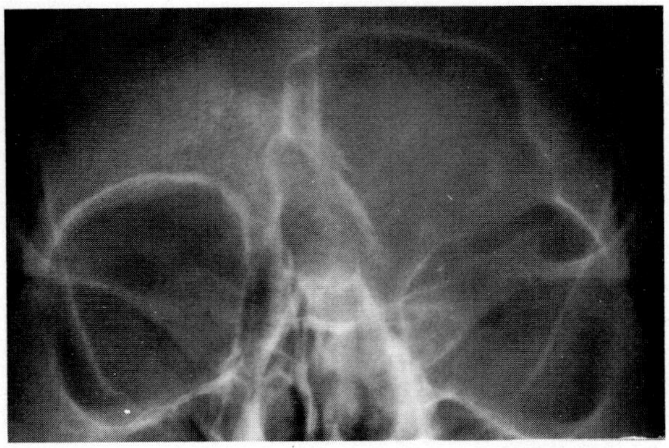

Fig. 6.10. Frontal sinus mucocoele. Expansion with loss of scalloped edge, loss of translucence and erosion of the supraorbital ridge, can be seen on the radiograph.

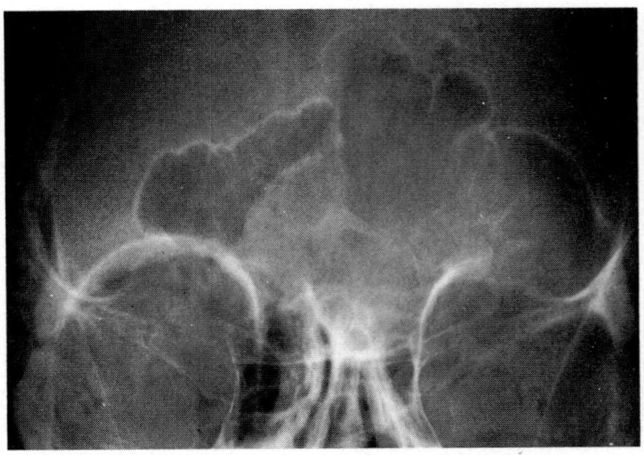

Fig. 6.11. Frontal sinus mucocoele. There is depression of the supraorbital ridge and extension of the mucocoele across the septum to the opposite frontal sinus.

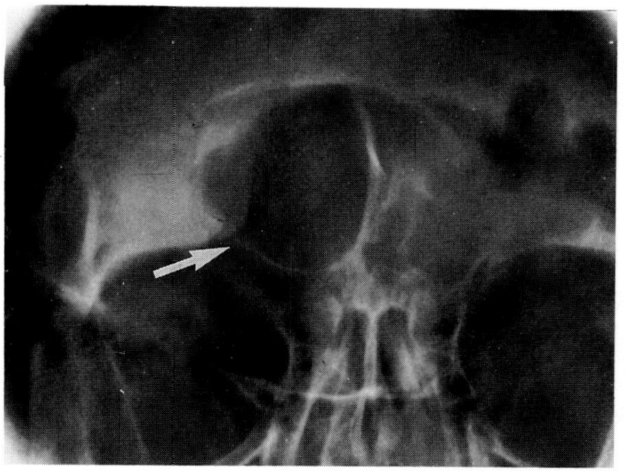

Fig. 6.12. Frontal sinus mucocoele. Mucocoele formation (*arrow*) has produced hypertranslucence.

fluid accumulation, in others it may appear more radiolucent in some projections if local bone erosion is sufficient to produce an image that is less dense overall (Fig. 6.12).

Another inconstant feature is loss of the scalloped margin of the sinus walls. This occurs when the vertical section of the sinus is involved in the expansion, but will not be seen when only the horizontal part is expanded – as it was in nearly 25% of the patients in the author's series (Fig. 6.13). This group of patients present with proptosis and downward displacement of the eye, and are usually seen initially by the ophthalmologist. The mucocoele can be missed on plain radiographs when the full classical picture of a frontal sinus mucocoele is not present. An important observation is the downward displacement of the supraorbital ridge. CT scan may be necessary to make the diagnosis (Price and Danziger 1980), as this allows both the bone changes and the soft tissue expansion to be appreciated (Figs. 6.14, 6.15); but these expansions affecting the roof of the orbit are now optimally demonstrated by magnetic resonance studies (Fig. 6.16). The importance of accurate pre-operative assessment of this type of mucocoele cannot be overemphasised; it is essential that all loculi of the expansion are recognised by the surgeon and drained adequately, since this variety of mucocoele is very prone to recur.

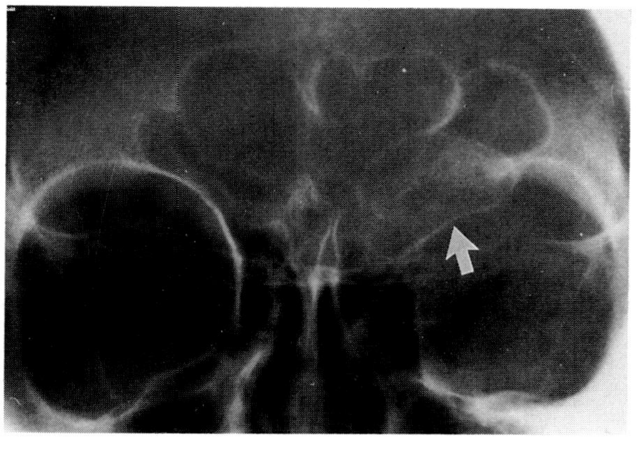

Fig. 6.13. Frontal sinus mucocoele. There is expansion of the horizontal section of the sinus cavity (*arrow*) but the scalloped outline of the vertical part of the sinus is preserved.

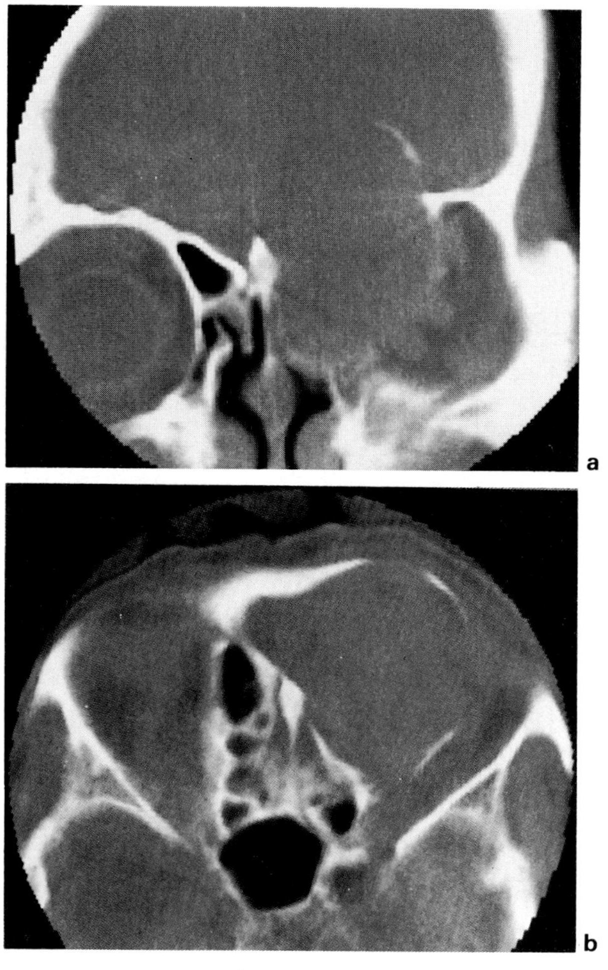

Fig. 6.14a,b. Large fronto-ethmoidal mucocoele demonstrated on coronal (**a**) and axial (**b**) CT scans.

Ethmoid Mucocoeles

Ethmoid mucocoeles are predominantly anterior in location: in the author's series 25 out of 27 were anterior and only 2 posterior. Most posterior mucocoeles are in fact part of a generalised expansion of both the sphenoid sinus and posterior ethmoid cells – a result of their common drainage into the superior meatus of the nose, and presumably a common obstructing agent.

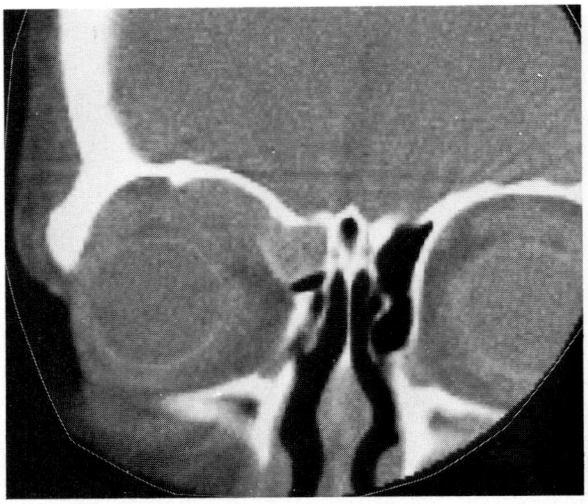

Fig. 6.15. Frontal sinus mucocoele. Coronal CT scan showing expansion of the horizontal part of the frontal sinus with downward displacement of the globe.

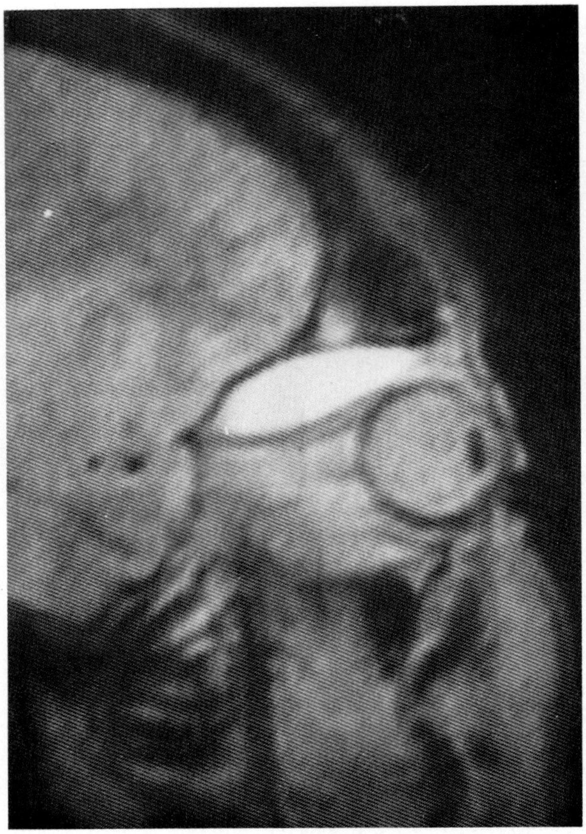

Fig. 6.16. Sagittal magnetic resonance scan showing a mucocoele in the roof of the orbit derived from the horizontal part of the frontal sinus.

Mucocoeles arising from the frontal sinus are not as a rule difficult to recognise radiologically, but those derived from the anterior ethmoid cells may be surprisingly difficult to detect on standard sinus views. They are usually more obvious clinically than radiologically, over 90% presenting with a mass at the medial canthus in addition to proptosis and lateral displacement of the eyeball. Some patients may present to the ophthalmologist with

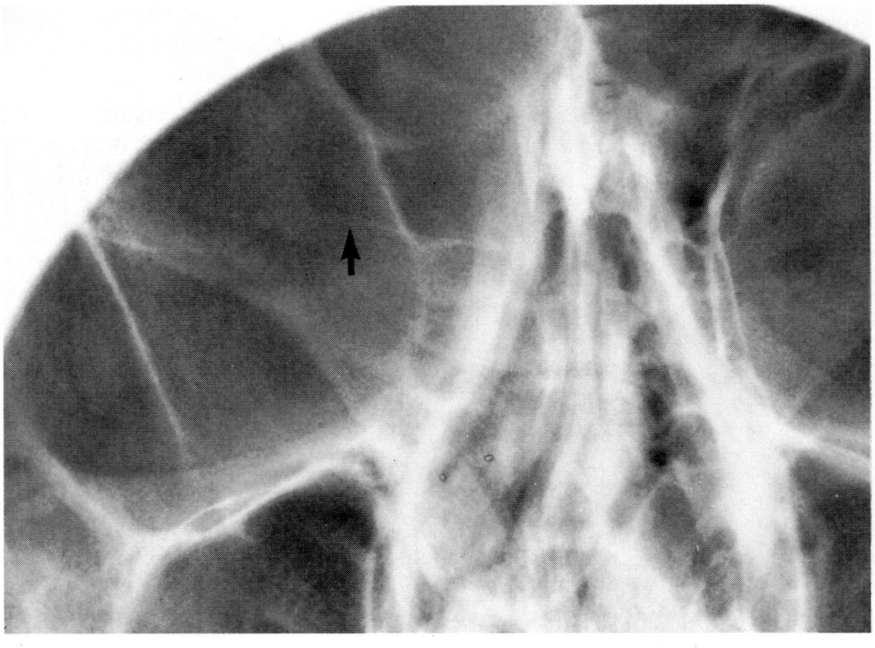

Fig. 6.17. Ethmoid mucocoele. The edge of the ethmoid expansion is shown as a fine bony rim (*arrow*).

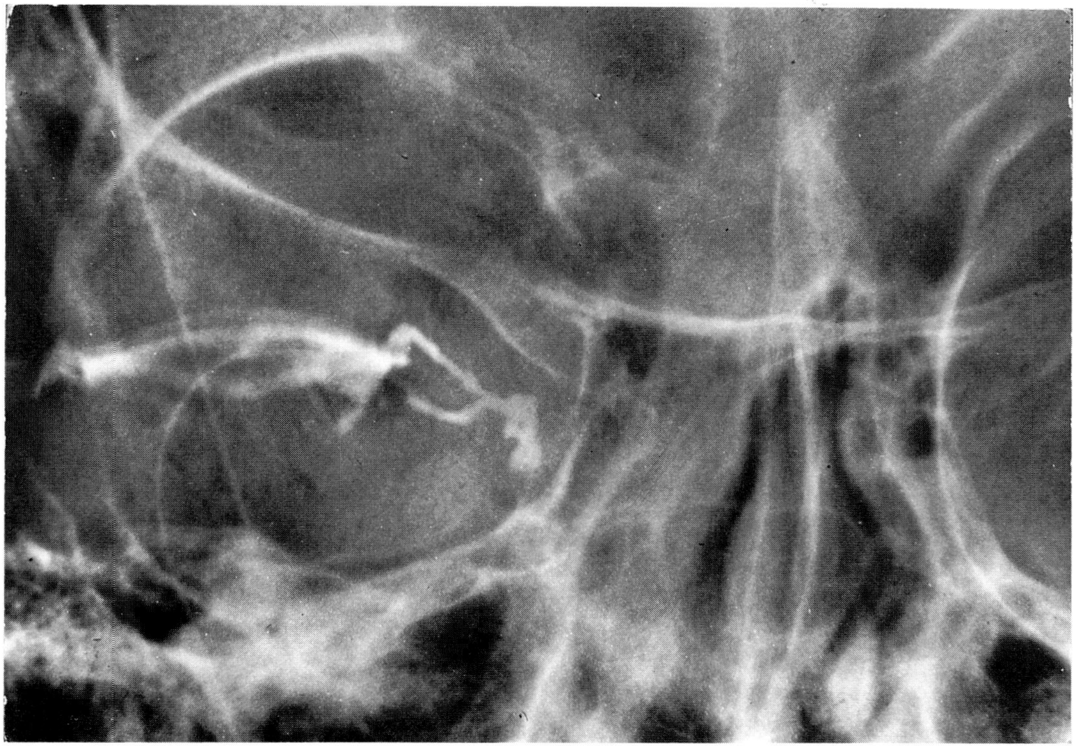

Fig. 6.18. Dacryocystogram of a right ethmoid mucocoele causing extrinsic obstruction and displacement of the lacrimal sac.

epiphora if the expansion impinges on the lacrimal drainage apparatus at an early stage. Loss of translucence in the affected ethmoid cells is the change most often observed on plain radiographs but is entirely non-specific. Other patients may show loss of the vertical line forming the anterior component of the medial orbital wall (Lloyd et al. 1974), and in some a bony rim may be demonstrated within the orbit partly covering the expansion (Fig. 6.17). Obstruction with forward and lateral displacement of the lacrimal sac and canaliculi may be demonstrated on dacryocystography (Fig. 6.18), but axial CT (Fig. 6.19) and axial magnetic resonance scans show the lesion optimally (Figs. 6.20, 6,21).

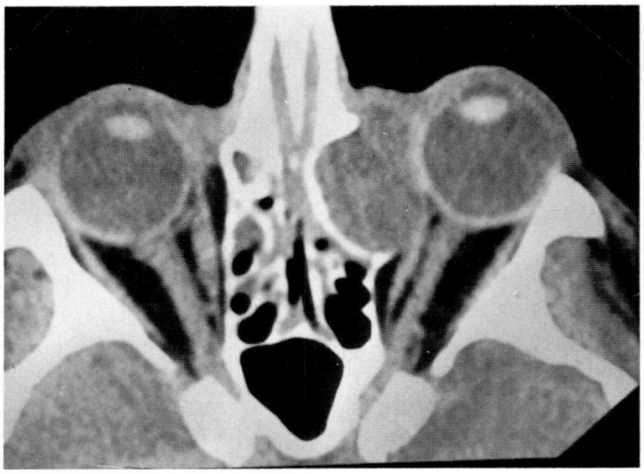

Fig. 6.19. Anterior ethmoid mucocoele shown on axial CT scan.

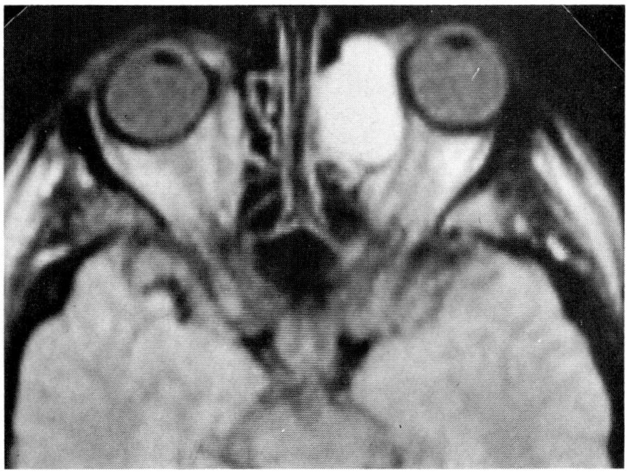

Fig. 6.20. Same patient as Fig. 6.19. On magnetic resonance scan the mucocoele gives a strong signal on a T_2-weighted spin echo sequence.

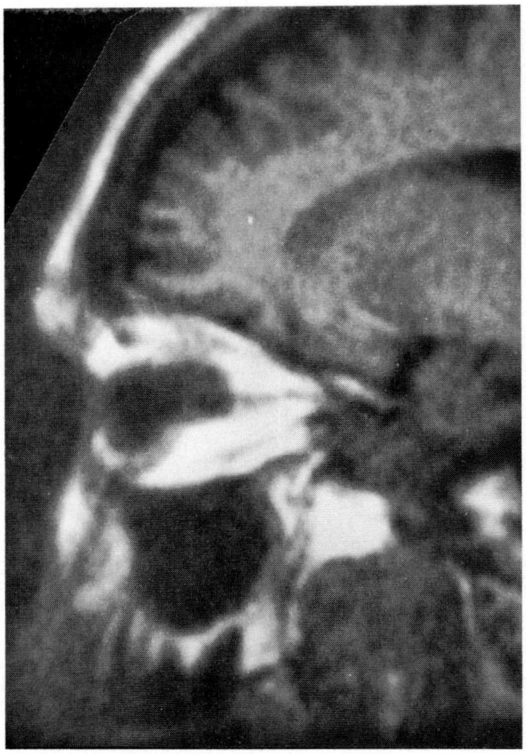

Fig. 6.21. Same patient as Figs. 6.19 and 6.20. On sagittal magnetic resonance scan there is a characteristic low signal on the T_1-weighted inversion recovery sequence.

Sphenoid Mucocoeles

Under the heading of sphenoid mucocoeles are included expansions of the posterior ethmoid cells in addition to sphenoid expansions; they are probably better designated as spheno-ethmoidal mucocoeles. Imaging techniques play a key role in the diagnosis of this condition and it is important that it be recognised by the radiologist at an early stage and dealt with surgically before vision is seriously compromised. Because of the proximity of the sphenoid sinus to the optic nerve, cavernous sinus and ocular motor nerves, mucocoele expansion of the sinus results in symptoms due to involvement of these structures. The patient commonly presents with headache combined with eye symptoms such as blurred vision or diplopia. On plain radiographs sphenoid mucocoeles are liable to be misdiagnosed as pituitary tumours, or as nasopharyngeal carcinoma invading the sphenoid; in either case this may result in inappropriate treatment.

In general the radiological features become more emphatic as the lesion expands the sphenoid sinus. Early in the course of the disease the changes may be limited to the sinus itself (Minagi et al. 1972) and may consist only of opacification of one or both sphenoid sinuses. At this stage certain radiological diagnosis is not always possible, but as pressure within the sinus continues, an expansion of the

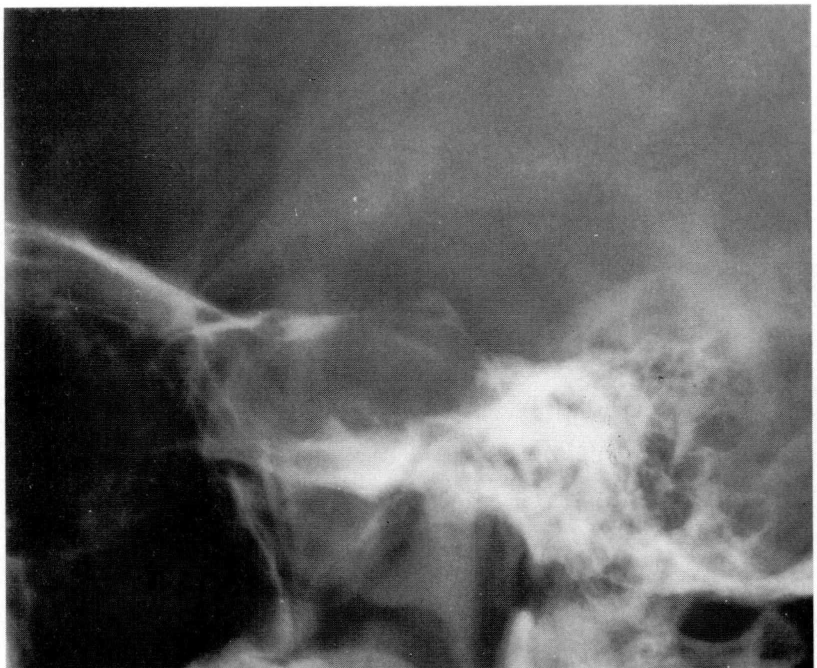

Fig. 6.22. Sphenoid mucocoele. Expansion of the sphenoid sinus with elevation of the floor of the pituitary fossa.

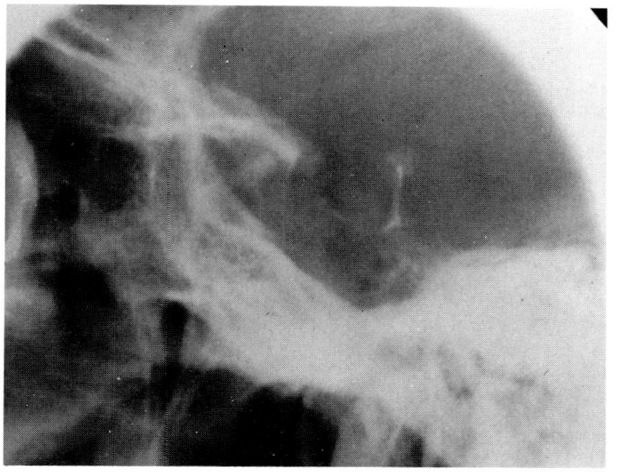

Fig. 6.23. Erosion of the floor of the pituitary fossa caused by a sphenoid mucocoele.

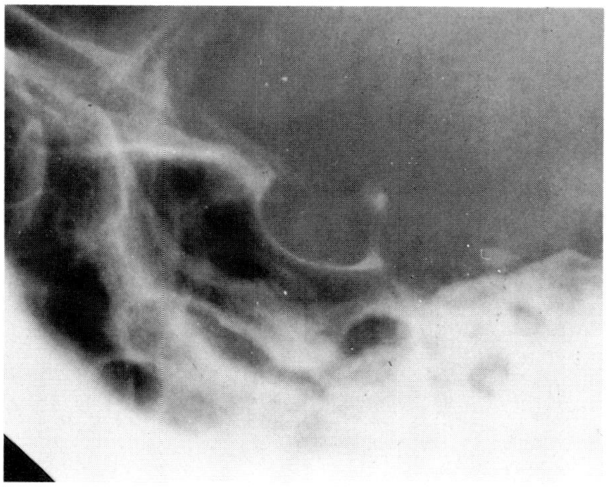

Fig. 6.24. Re-formation of the sella floor following drainage of the sphenoid mucocoele.

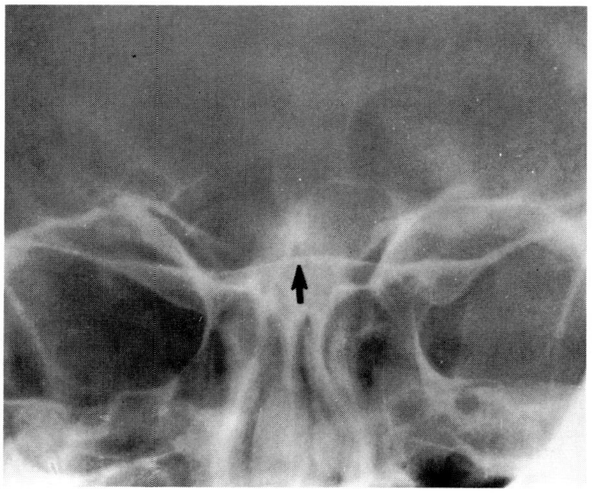

Fig. 6.25. Sphenoid mucocoele showing elevation of the planum sphenoidale (*arrow*).

cavity ensues with elevation (Fig. 6.22) or bone erosion of the floor of the pituitary fossa on the lateral skull film. This change may be reversible after drainage of the mucocoele (Figs. 6.23, 6.24). Progressive expansion of the sinus will also cause erosion of the medial wall of the optic canal, and elevation of the planum sphenoidale (Fig. 6.25). Another sign may be present if the posterior ethmoid cells are involved in the expansion: the oblique line which forms the lateral boundary of the posterior ethmoid cells on the occipito-frontal projection may be lost as the posterior ethmoid cells expand with the sphenoid sinus.

Once the diagnosis is suspected then CT or magnetic resonance studies should be undertaken (Fig. 6.26). Takahashi et al. (1973) described a diagnostic feature on conventional tomography which can also be shown on CT (Fig. 6.27a–c). They reported the presence of multiple cyst-like expansions, which appeared to be intercommunicating and connected to the expanded sinus. These result from expansion of compartments of the sinus or aberrant pneumatisations participating in the general expansion of the main sinus cavity. Magnetic resonance has now replaced CT as the method of choice for identifying sphenoid mucocoeles. The recognition of retained secretions in the sinus is more easily demonstrated and discrimination between mucocoele and any associated tumour or polyps is more obvious (see below) (Figs. 6.28, 6.29).

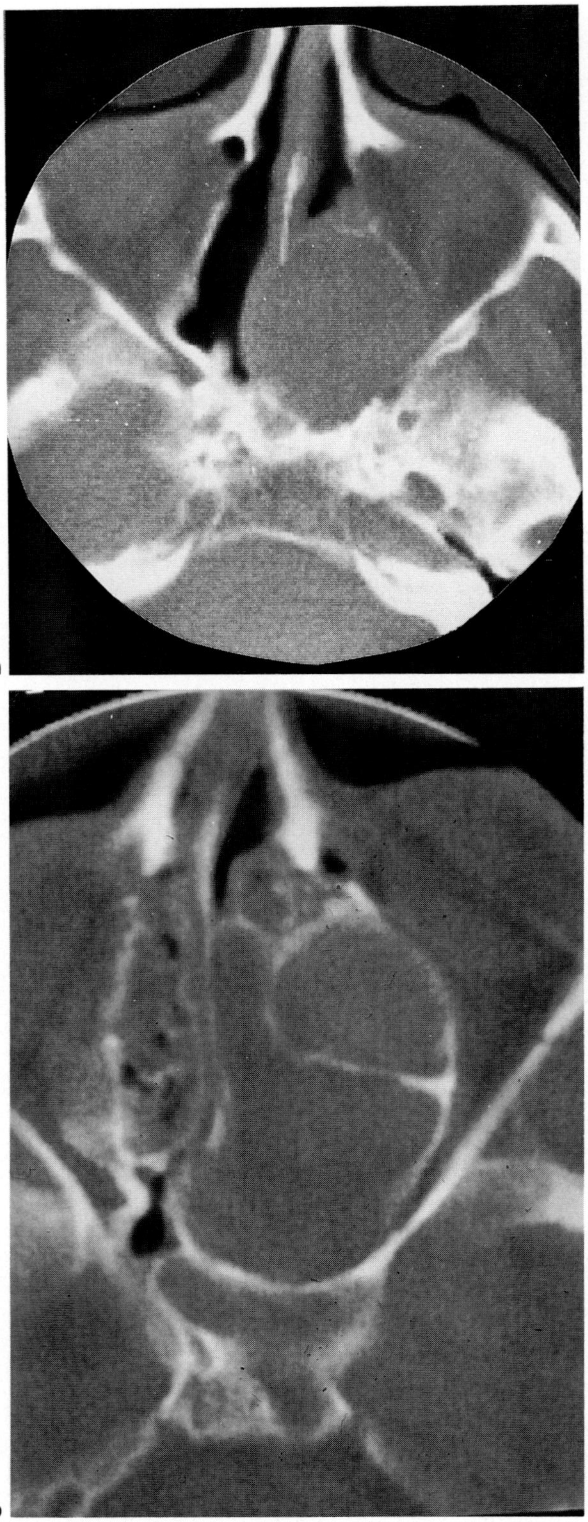

a

b

Fig. 6.26a,b. Examples of sphenoid mucocoeles shown on axial CT scan.

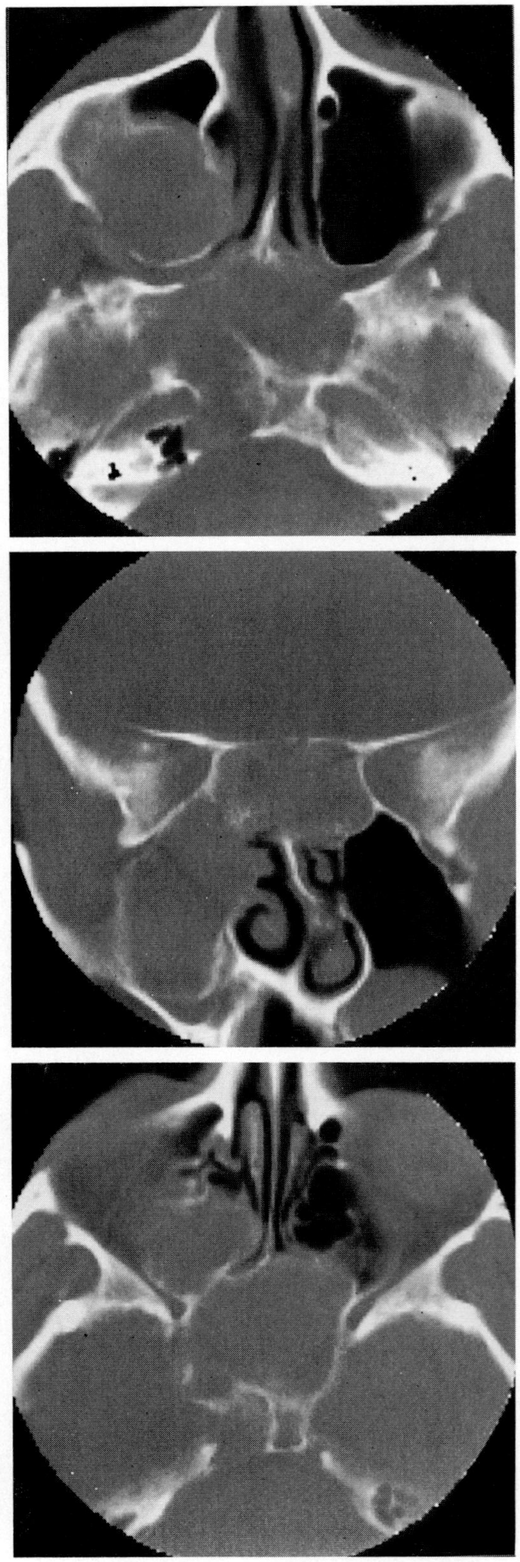

a

b

c

Fig. 6.27a–c. Sphenoid mucocoele showing expansion of a loculus of the sinus within the cavity of the maxillary antrum.

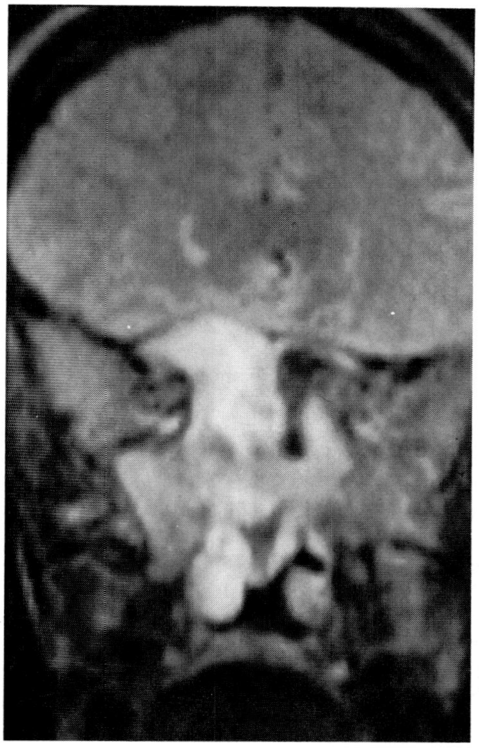

Fig. 6.28. Coronal magnetic resonance scan of a mucocoele of the sphenoid sinus secondary to nasal polyposis. Note the elevation of the planum sphenoidale.

Mucocoeles of the Maxillary Antrum

The maxillary antrum is the least common site in the sinuses for mucocoele formation, and the majority of expansions of the antrum are due to other causes such as odontogenic cysts or cholesterol granuloma. Clinically the symptoms are due to sinus expansion into the nose, mouth and orbit, resulting in upward displacement of the eye, proptosis and swelling of the cheek over the affected side. On plain radiographs the sinus is invariably opaque and expanded (Fig. 6.30), encroaching upon the ethmoid cells. As in the other sinuses enlargement with good preservation of the bulging walls is the clue to the diagnosis, but in some projections the changes may look very like early malignancy. CT or conventional tomography is needed to confirm the bone expansion and magnetic resonance studies will indicate the fluid content of the expanded sinus (Figs. 6.31, 6.32).

Magnetic Resonance and CT

Magnetic resonance is now the imaging method of choice for assessing mucocoele formation, both for primary mucocoeles and secondary mucocoeles associated with sinus tumours, benign or malig-

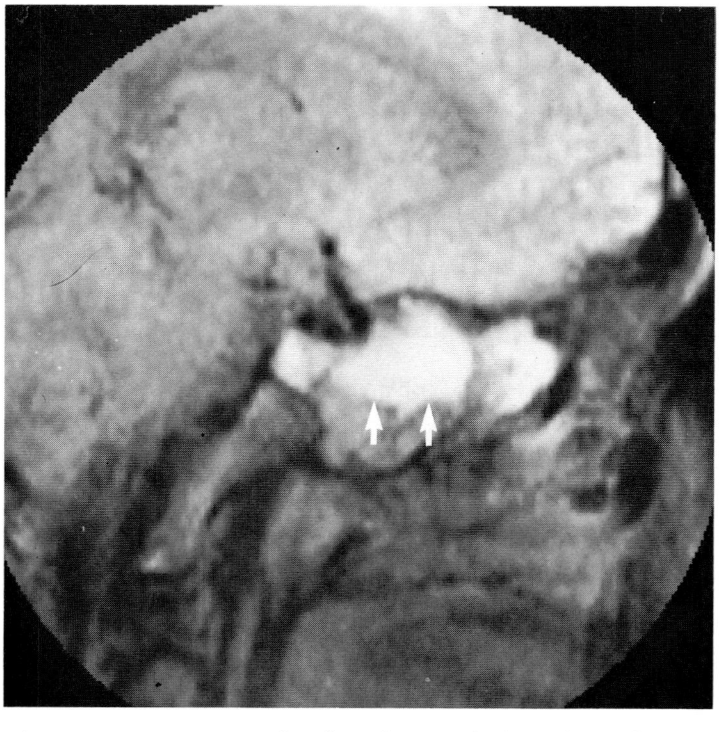

Fig. 6.29. Sagittal magnetic resonance scan of a sphenoid mucocoele (*arrows*) secondary to an angiofibroma.

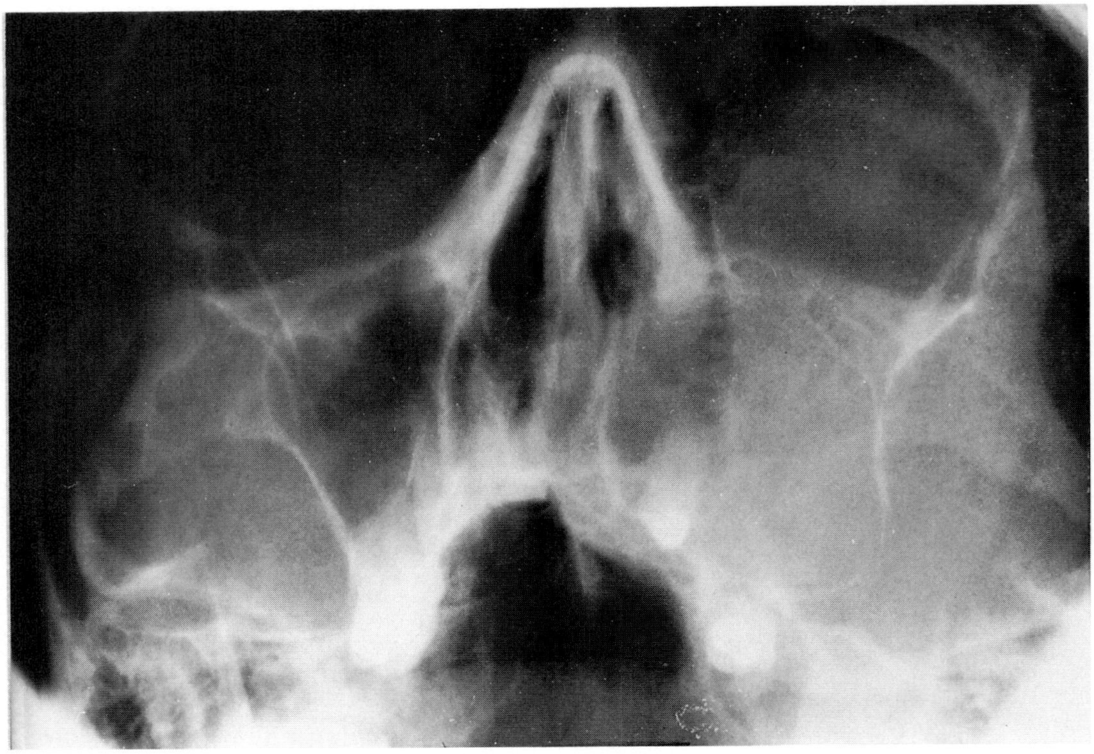

Fig. 6.30. Mucocoele of the maxillary antrum demonstrated on plain radiograph.

nant. Secondary mucocoele has been shown to be far more common at pre-operative assessment since the use of this technique (Figs. 2.28, 6.29). The strong signal shown by retained secretion in the sinuses on T_2-weighted spin echo sequences makes differentiation between tumour and secondary mucocoele much easier than by CT scan.

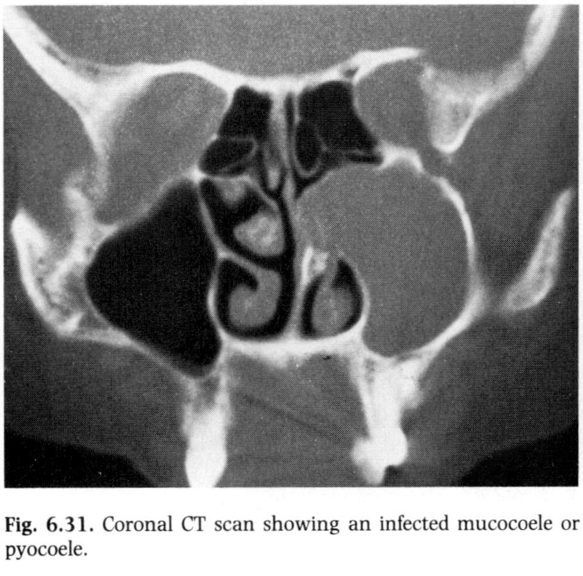

Fig. 6.31. Coronal CT scan showing an infected mucocoele or pyocoele.

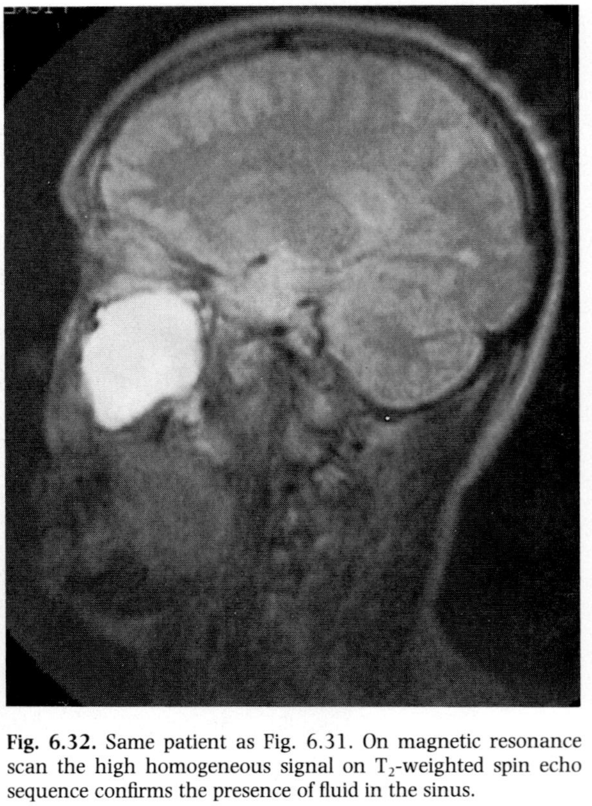

Fig. 6.32. Same patient as Fig. 6.31. On magnetic resonance scan the high homogeneous signal on T_2-weighted spin echo sequence confirms the presence of fluid in the sinus.

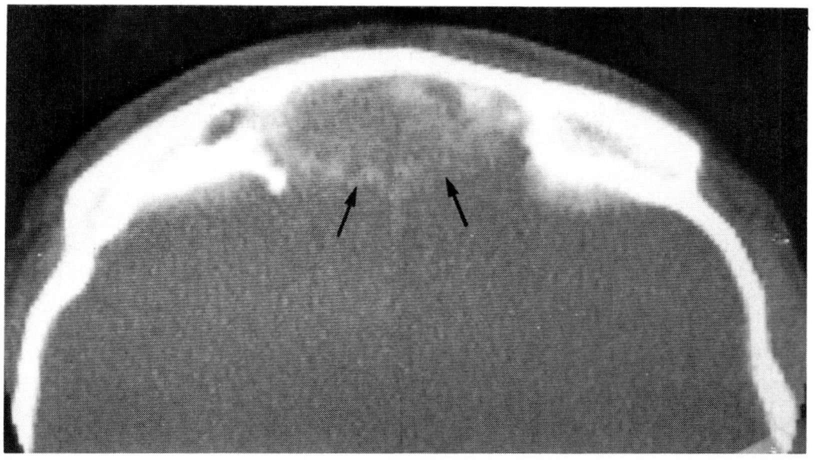

Fig. 6.33. Frontal sinus pyocoele showing ring enhancement on CT scan after intravenous contrast (*arrows*).

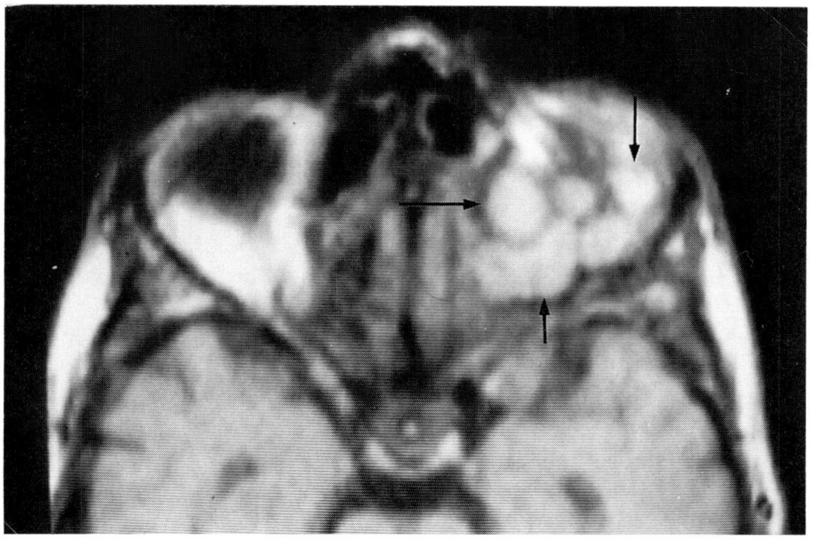

Fig. 6.34. Magnetic resonance scan showing high signal on axial inversion recovery sequence from loculi of a frontal sinus mucocoele (*arrows*) which contained altered blood.

Theoretically with the latter technique mucocoeles should show low attenuation on the Hounsfield scale, without change after contrast administration. However some mucocoeles show initial attenuation values well into the tumour range, so that accurate discrimination between cystic and solid lesions in the sinuses has not always been possible from the attenuation values alone. Very often on CT the diagnosis is better made by recognising the classical features of a mucocoele, namely an overall rounded expansion with some preservation of the sinus boundaries, as opposed to destruction of bone in situ – the hallmark of malignancy in the sinuses. One feature which is diagnostic on CT is the ring enhancement seen after intravenous contrast, which indicates an infected mucocoele or pyocoele (Fig. 6.33).

As outlined in Chap. 2 the signal characteristics of mucocoeles on magnetic resonance differ if there has been previous surgical intervention and blood contamination of the mucus or mucopus. The untouched mucocoele produces very low signal on inversion recovery sequences and high signal on T_2-weighted spin echo sequences (Figs. 6.20, 6.21). In contrast a mucocoele containing blood breakdown products with released methaemoglobin will produce strong signal on both sequences (Figs. 6.34, 6.35).

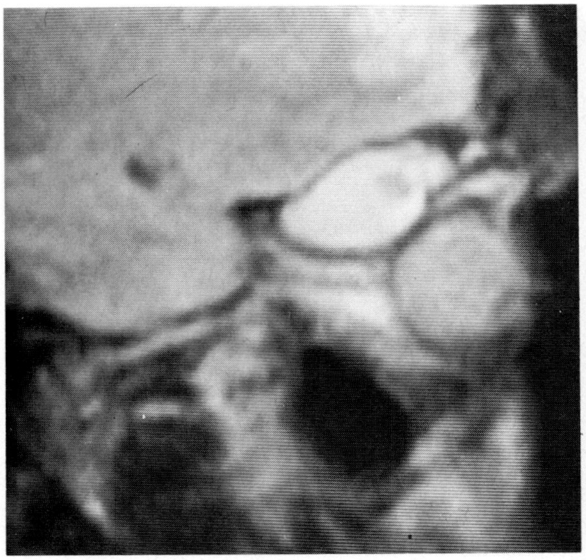

Fig. 6.35. Same patient as Fig. 6.34. Frontal sinus mucocoele giving a high signal on T_2-weighted spin echo sequence.

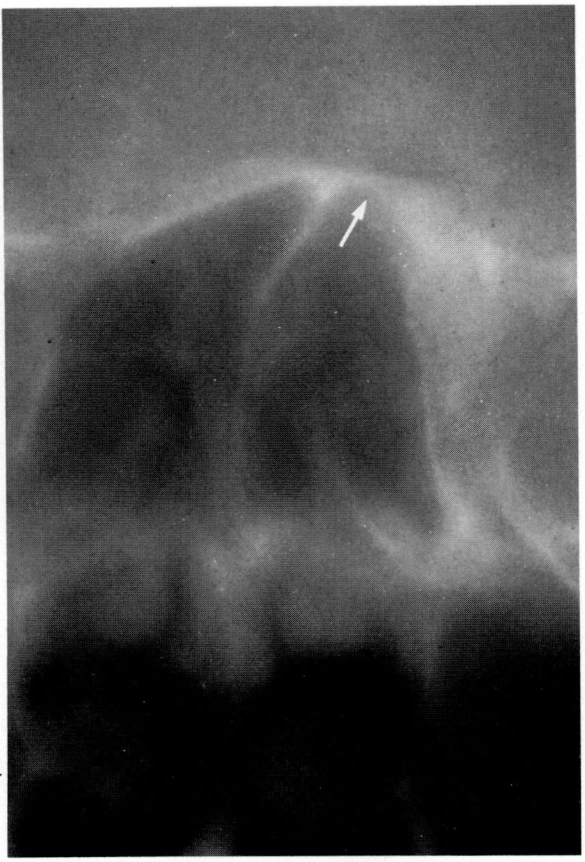

Fig. 6.36. "Blistering", or pneumosinus dilatans, affecting the ▶ sphenoid sinus shown on coronal hypocycloidal tomography. The sinus cavity is elevated and drawn towards the side of a foraminal optic nerve meningioma (*arrow*).

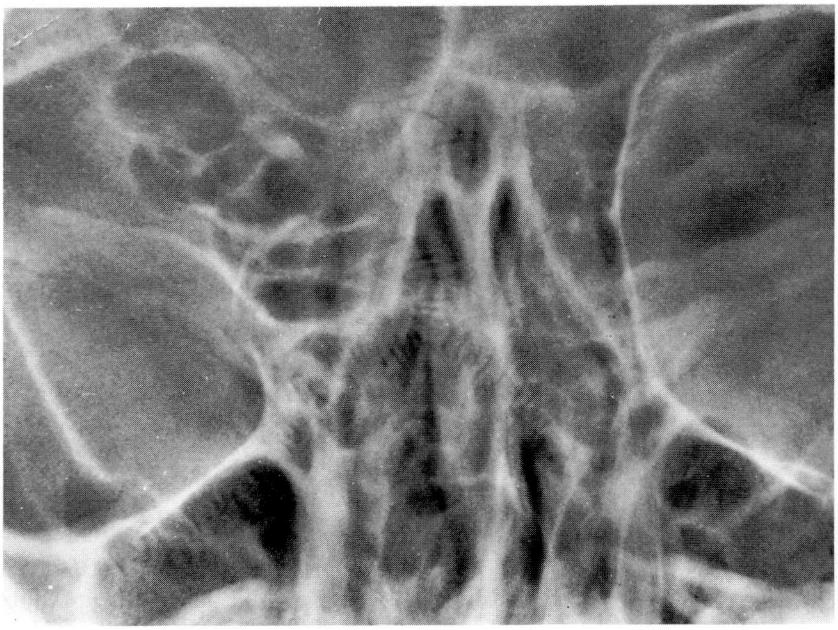

Fig. 6.37. Pneumosinus dilatans due to an extradural meningioma. The macroradiograph shows local dilatation of the right fronto-ethmoidal air cells.

Pneumosinus Dilatans

Pneumosinus dilatans refers to an abnormal dilatation of a local part of the paranasal sinuses which contains air only and is lined by normal epithelium. The change is most commonly observed when it affects the sphenoid sinus as a response to a local intracranial meningioma of the tuberculum sellae or planum sphenoidale (Fig. 6.36). The condition was first described by Benjamins (1918) and was extensively reviewed by Lombardi (1967). More recently the change known as "blistering" of the planum sphenoidale has been described by Wiggli and Oberson (1975) and the same condition (referred to as sphenoid pneumosinus dilatans) described in three patients by Hirst et al. (1982).

Pneumosinus dilatans affecting the sinus walls forming the boundaries of the orbit is less common. Recently six such cases have been described by Lloyd (1985). All six patients presented with a slowly progressive proptosis over a period of months or years. Female patients predominated in a 5 : 1 ratio with an age range of 15–28 years and an average age of 20.7 years. In four patients the ethmoid cells showed unilateral expansion and in three, part of the frontal sinus was also involved (Figs. 6.37, 6.38). In two patients the expansion involved both maxillary antra. Biopsy showed the pneumosinus dilatans to be caused by a meningioma adjacent to the affected sinuses in three patients (Figs. 6.39–6.42); in the remaining three it was associated with fibro-osseous disease of the sinus walls (Figs. 6.43, 6.44, 6.45).

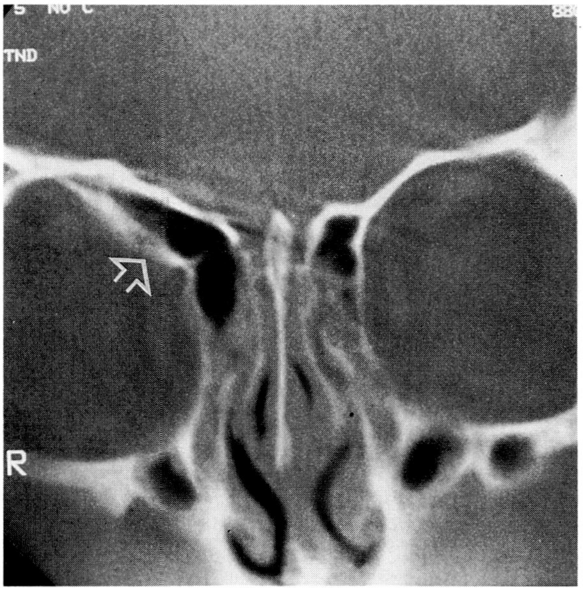

Fig. 6.38. Coronal CT scan showing dilatation of the right fronto-ethmoidal air cells (*arrow*) due to fibro-osseous disease.

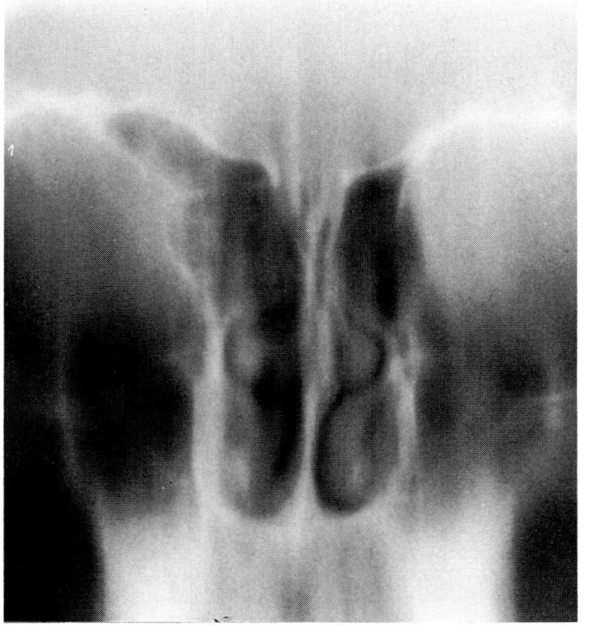

Fig. 6.39. Coronal tomogram showing fronto-ethmoidal dilatation due to an adjacent meningioma.

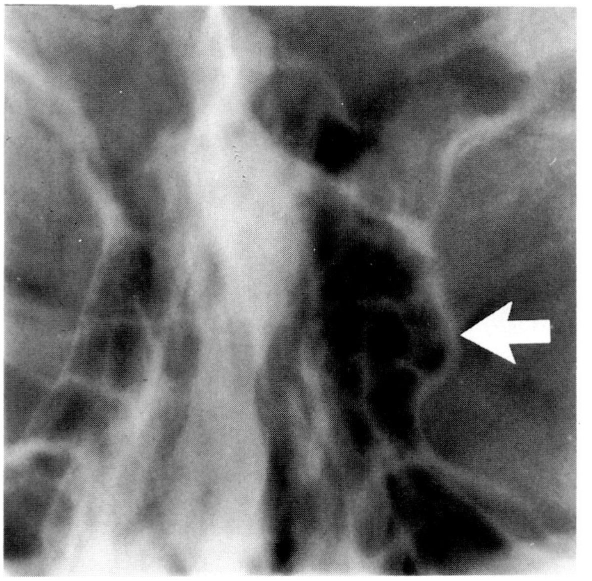

Fig. 6.40. Dilatation of the ethmoid cells (*arrow*) shown on plain radiograph.

Before making the diagnosis of pneumosinus dilatans it is important that the radiologist should keep in mind the variations in the pneumatisation of the sinuses which can occur normally or in some congenital and acquired conditions (see Chap. 1).

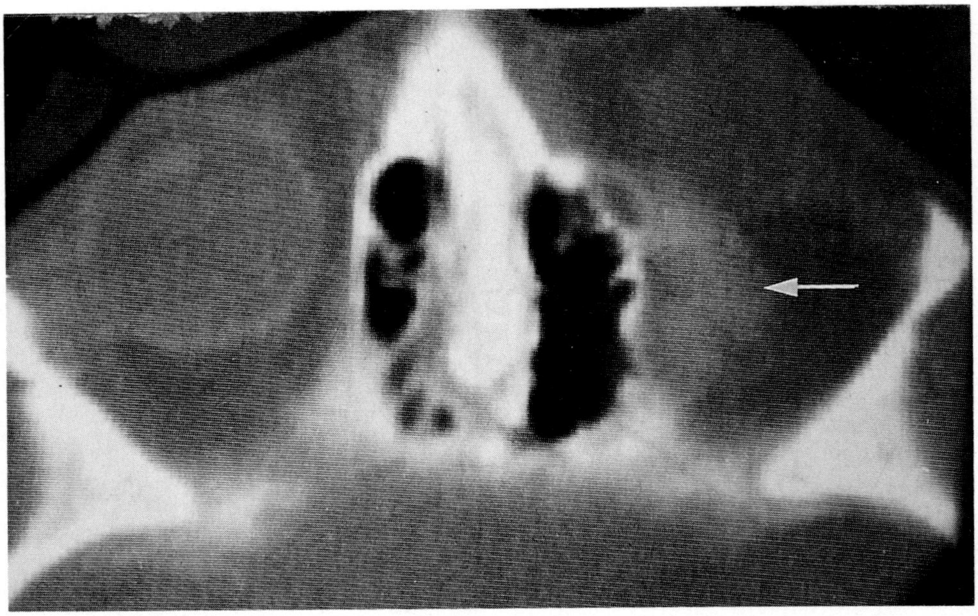

Fig. 6.41. Same patient as Fig. 6.40. Axial CT scan shows the soft tissue mass of a mengioma (*arrow*) adjacent to the dilated ethmoid cells.

a

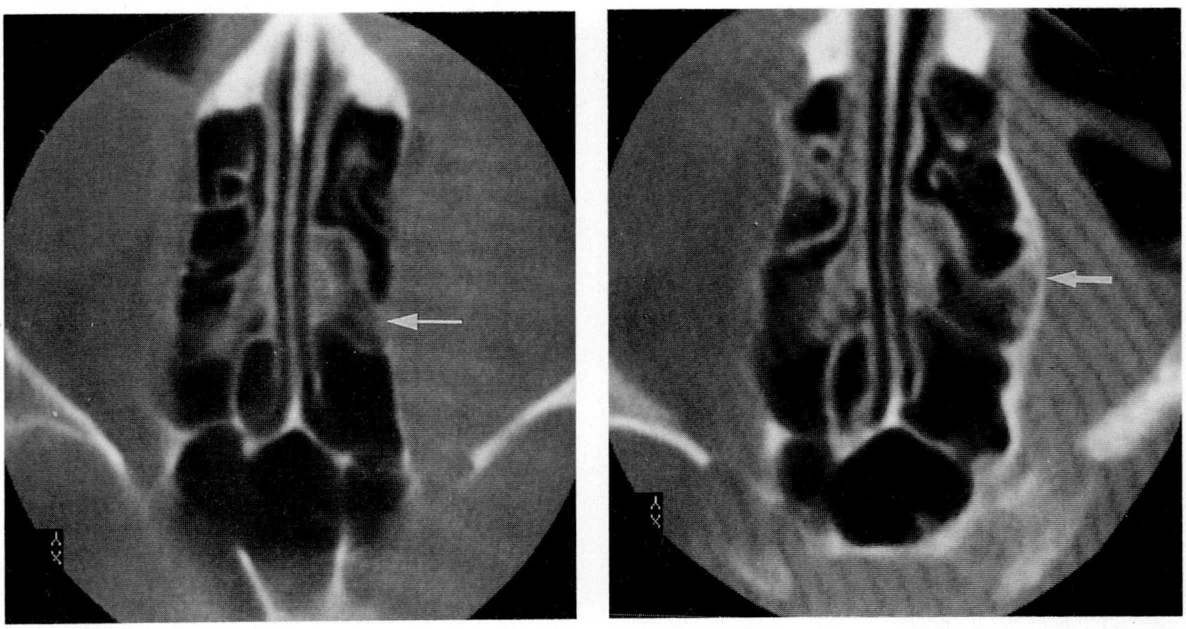

b

Fig. 6.42. a Axial CT scan showing recurrent sheath meningioma in the orbit. The presence of the tumour has expanded the orbit and caused an inward bowing of the medial orbital wall (*arrow*). **b** Further scan 4 years after exenteration of the orbit. Residual meningioma is still present in the orbit and has now provoked a hyperostosis of the medial orbital wall with dilatation of the ethmoid cells locally (*arrow*). Note the stretching of the ethmoid septa accompanying the sinus dilatation.

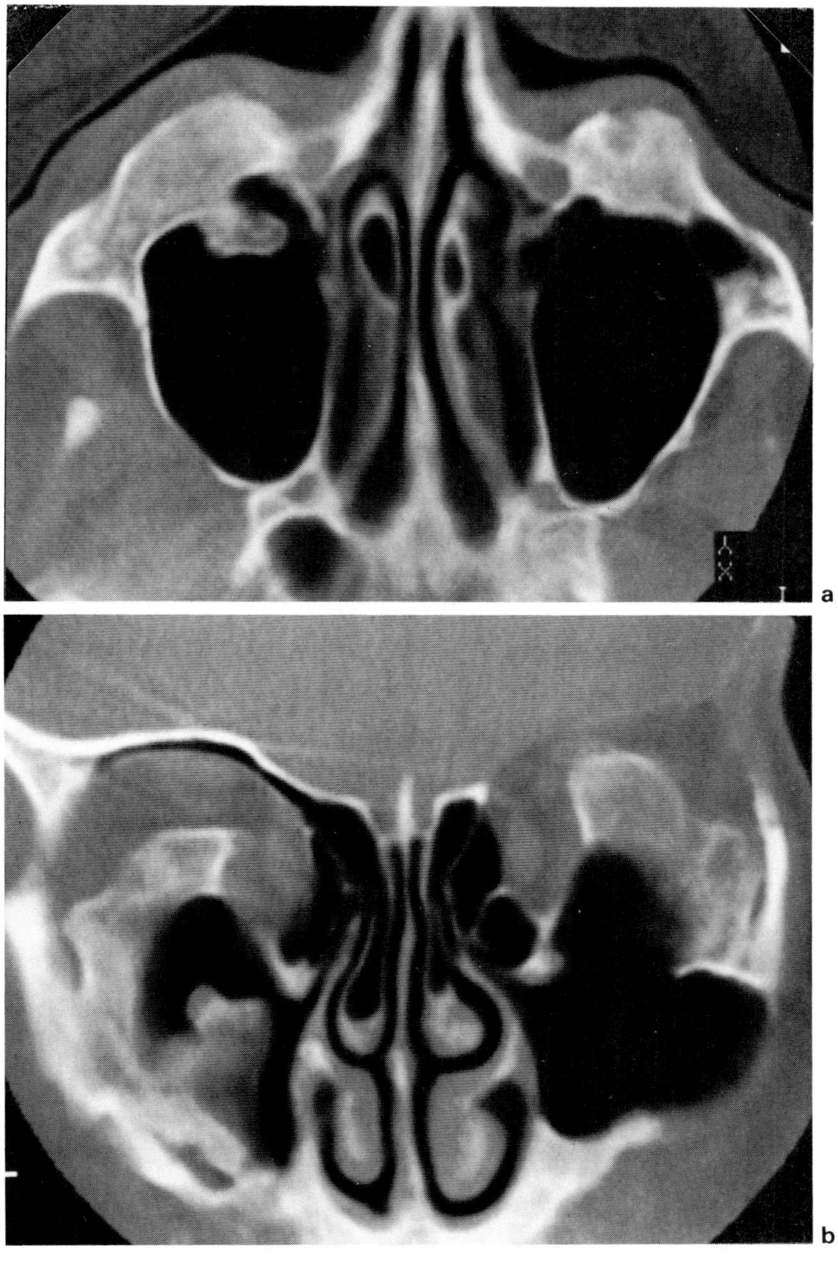

Fig. 6.43. a Axial CT scan showing bony thickening due to fibrous dysplasia of the anterior walls of the maxillary antra. b Coronal CT scan showing bilateral fibrous dysplasia with encroachment into both orbits by thickened bone and dilated maxillary antra.

Confirmation of this change can only be made at surgical decompression, when the criteria for diagnosis laid down by Lombardi (1967) need to be fulfilled: namely a dilated sinus lined with macroscopically normal epithelium and containing air only.

The possible reasons for the sinus dilatation remain speculative and no relevant experimental evidence has been found in the literature. Various possible explanations have been advanced. Sugita et al. (1977) suggest a valve-like obstructive mechanism of the sinus. Hirst et al. (1982) suggest either a congenital abnormality or the stimulating effect of a local meningioma, causing the bone to bulge. The young age of the patients would support a developmental anomaly, probably the result of an

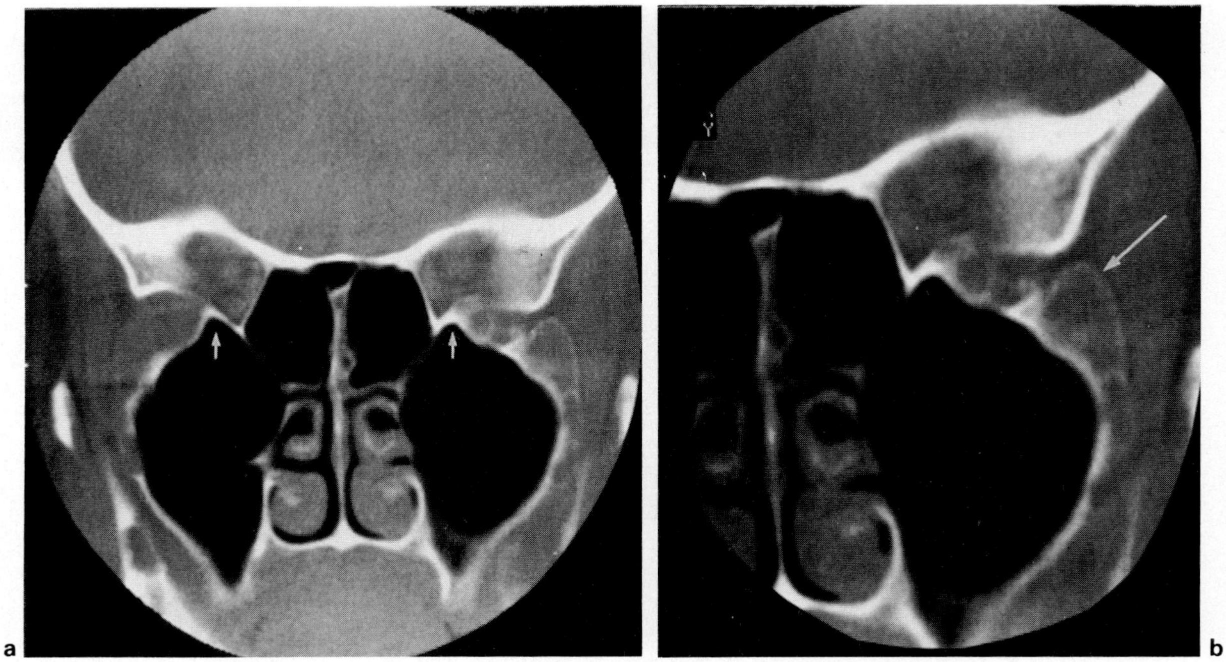

Fig. 6.44. **a** Coronal CT scan showing upward dilatation (*arrows*) of both maxillary antra. **b** Enlarged view of the antrum showing bony thickening due to fibrous dysplasia (*arrow*).

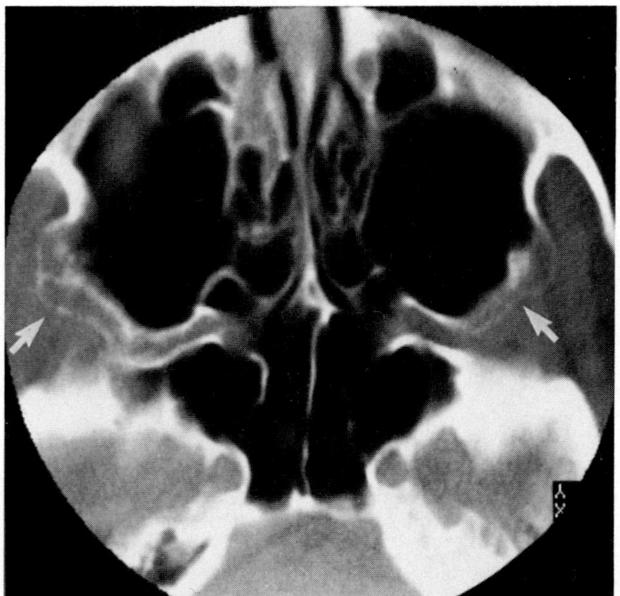

Fig. 6.45. Same patient as Fig. 6.44. Axial CT scan showing irregular dilatation of both maxillary antra posteriorly, associated with bilateral bony thickening (*arrows*).

aberration of bone growth and remodelling provoked by the fibro-osseous disease or meningioma, leading to a weakening of the wall and expansion of the sinus. The practical consideration for the radiologist is to recognise this condition and to appreciate the underlying causes. The changes of fibro-osseous disease in the sinuses are usually obvious, but the hyperostosis produced by a meningioma may be minimal and the presence of pneumosinus dilatans should alert the radiologist to the possibility of an occult meningioma. It is important, therefore, that these patients should be examined by soft tissue imaging techniques – either CT or magnetic resonance.

References

Benjamin CE (1918) Pneumosinus frontalis dilatans. Acta Otolaryngol 1:412–422

Canalis RF, Zajtchuk JT, Jenkins HA (1978) Ethmoidal mucocoeles. Arch Otolaryngol 104:286-291

Evans C (1981) The aetiology and treatment of fronto-ethmoidal mucocoeles. J Laryngol Otol 95:361–375

Hirst LW, Miller MR, Hodges FJ, Corbett J, Thomson S (1982) Sphenoid pneumosinus dilatans. A sign of meningioma originating in the optic canal. Neuroradiology 22:207–210

Lloyd GAS (1985) Orbital pneumosinus dilatans. Clin Radiol 36:381–386

Lloyd GAS, Bartrum CI, Stanley P (1974) Ethmoid mucocoeles. Br J Radiol 47:646–651

Lombardi G (1967) Radiology in neuro-ophthalmology. Williams and Wilkins, Baltimore

Lund VJ (1985) Odontogenic keratocyst of the maxilla: a case report. Br J Oral Maxillofac Surg 23:210–215

Lund VJ (1987) Anatomical considerations in the aetiology of fronto-ethmoidal mucocoeles. Rhinology 23:83–88

Lund VJ, Lloyd GAS (1983) Radiological changes associated with benign nasal polyps. J Laryngol Otol 97:503–510

McIvor J (1972) The radiological features of odontogenic keratocysts. Br J Oral Surg 10:116–125

Minagi H, Margolis MT, Newton TH (1972) Tomography in diagnosis of sphenoid sinus mucocoele. Am J Roentgenol 115:587–595

Osborn DA, Wallace M (1967) Carcinoma of the frontal sinus associated with epidermoid cholesteatoma J Laryngol Otol 81:1021–1032

Philipsen HP (1956) Om Keratocyster (Kolesteatomer) i Koeberne. Tanglaegebladet 60:963–970

Price HI, Danziger A (1980) Computerised tomographic findings in mucocoeles of the frontal and ethmoid sinuses. Clin Radiol 31:169–174

Rollet M (1896) Mucocoele de l'angle superointerne des orbites. Lyon Med 81:573–575

Samuel E, Lloyd GAS (1978) Clinical radiology of the ear, nose and throat. HK Lewis, London

Sugita K, Hirota T, Iguchi I (1977) Transient amaurosis under decreased atmospheric pressure with sphenoid sinus dysplasia. J Neurosurg 46:111–114

Takahashi M, Jingu K, Makayama T (1973) Roentgenologic appearances of spheno-ethmoidal mucocoele. Neuroradiology 6:45–53

Wiggli Q, Oberson R (1975) Pneumosinus dilatans and hyperostosis: early signs of meningiomas on the anterior chiasmatic angle. Neuroradiology 8:217–221

7 Granulomata of the Nose and Paranasal Sinuses

Mid-facial Granuloma Syndrome

Friedmann (1964) has classified the non-healing granulomata of the nose into two basic types: lethal mid-line granuloma (Stewart 1933) and Wegener's granuloma (Wegener 1936, 1939).

Wegener's Granuloma

Wegener's granulomatosis is a multi-system disease characterised by necrotising granulomata of the upper and lower respiratory tract together with a glomerulonephritis and systemic vasculitis (Wolff et al. 1974). It is encountered most commonly in the fourth and fifth decades but may also be found in young subjects. The patient presents with constitutional symptoms: pyrexia, loss of weight and raised sedimentation rate. These symptoms often overshadow a nasal discharge and sinusitis, which is often of long duration and may be indicative of the prodromal stage of the disease. Examination may reveal a crusted granularity of the mucosa of the nasal septum and turbinates, with ulceration and bone destruction. Ulcers may also develop in the mouth and pharynx (McKinnon 1970), and atypical cases may present as rheumatoid arthritis (Pritchard and Gow 1976).

Histologically a necrotising arteritis is an essential component of the microscopy. The vessel walls are infiltrated by acute inflammatory cells and show partial fibrinoid necrosis and giant cell formation. The kidneys show a focal glomerulitis and vasculitis whilst vasculitis and granuloma formation may be found in the lungs and spleen and in the viscera. In general the bone destruction in the nose and sinuses is never as marked in Wegener's granuloma as it is in mid-line (Stewart's) granuloma.

Mid-line Lethal Granuloma

Mid-line lethal granuloma is less common than Wegener's granuloma and has a broadly based age distribution ranging from 15 to 80 years. The condition starts as an indurated swelling of some part of the nose such as the vestibule, septum or more rarely the turbinates. Ulcerations spread inexorably, destroying the soft tissues, cartilage and bone, and involving the hard palate and eventually the pharynx. Death may ensue from cachexia, haemorrhage or intercurrent infection. Histologically the lesions are characterised by non-specific granulation tissue with a dense accumulation of pleomorphic cells in the affected tissues and as a rule no vasculitis. The condition differs from Wegener's granuloma in that the disease is localised and the mid-facial destruction is the most prominent feature, with severe mutilation and bone necrosis. The intense upper airway destruction seen in patients with this condition is rarely if ever found in Wegener's granuloma (Kornblut and Fauci 1982).

Radiological Features

Milford et al. (1986) have reported the radiological findings in 27 patients with mid-facial granuloma: there were 20 patients with Wegener's granuloma

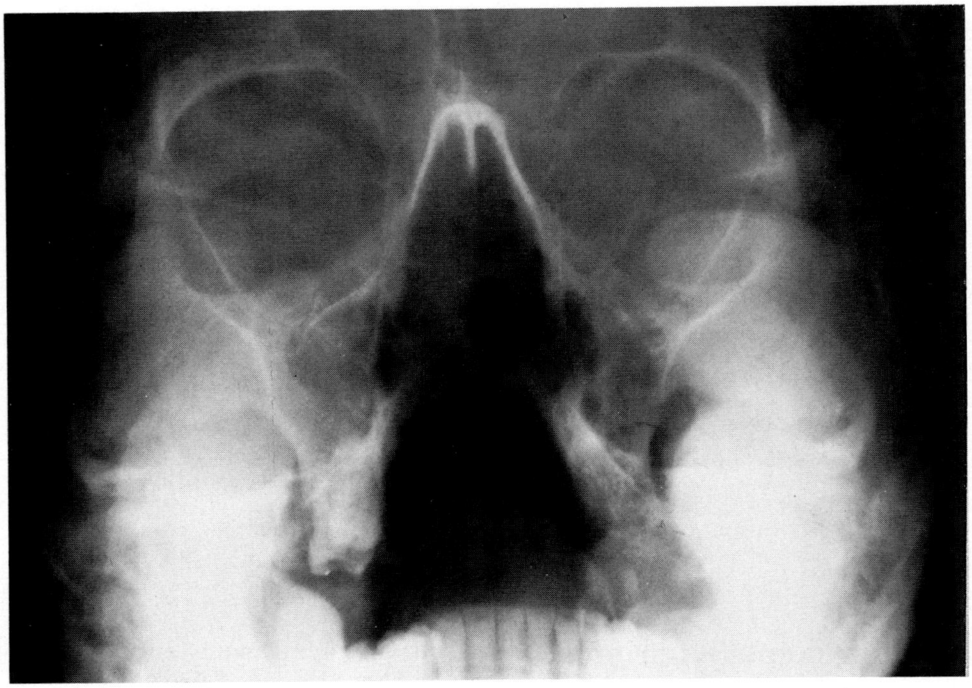

Fig. 7.1. Mid-line lethal granuloma. Massive destruction of the hard palate, nasal bones and ethmoid labyrinth can be seen.

and 7 with mid-line lethal granuloma. The majority of patients with Wegener's granuloma showed either no abnormality on the plain radiograph or non-specific changes; evidence of even minimal bone destruction could only be seen in 25%. In Wegener's granuloma some form of tomography is needed to show the bone changes. In contrast 6 of the 7 patients with lethal mid-line granuloma showed obvious bone erosion on initial plain radiographic examination (Fig. 7.1).

These authors concluded that the difference in the radiological pattern in Wegener's and lethal mid-line granuloma would seem to be one of degree. In Wegener's granuloma the systemic nature of the disease allows little time for bone destruction of the nose and bony walls of the sinuses. The condition is either arrested by treatment or, if this is ineffective, the patient dies of its systemic effects. On the other hand the localised nature of lethal mid-line granuloma allows progressive bone erosion over a period of years.

The essential feature of both diseases is the presence of bone destruction in the nose and sinuses without a large soft tissue mass as in malignant neoplasia: and if bone destruction is easily recognised on plain radiographs the condition is likely to be due to the lethal mid-line granuloma syndrome rather than Wegener's granuloma.

Magnetic Resonance and CT

A combination of magnetic resonance and CT provides the most comprehensive assessment of the mid-facial granuloma syndrome. CT, using wide window widths, provides optimum demonstration of the bone destruction in the nose and sinuses (Figs. 7.2 and 7.3), besides showing the state of the soft tissues and orbital invasion, when present (Figs. 7.2, 7.3 and 7.4).

Three patients with mid-facial granuloma syndrome have been examined by magnetic resonance imaging. In a single example of a patient with lethal mid-line granuloma the technique made little contribution to the diagnosis: the changes were almost entirely confined to bone and were better demonstrated by CT (Fig. 7.2).

Two patients with Wegener's granuloma have also been examined by magnetic resonance. The first was a long-term survivor of the condition, the diagnosis having been established for 16 years. The magnetic resonance scan was dominated by a massive central bone necrosis in the nose and sinuses, interpretation being complicated by previous orbital decompression surgery. The resultant cavity was lined by granulations giving a strong signal on T_2-weighted sequences (Fig. 7.5). This patient also had massive infiltration of both orbits.

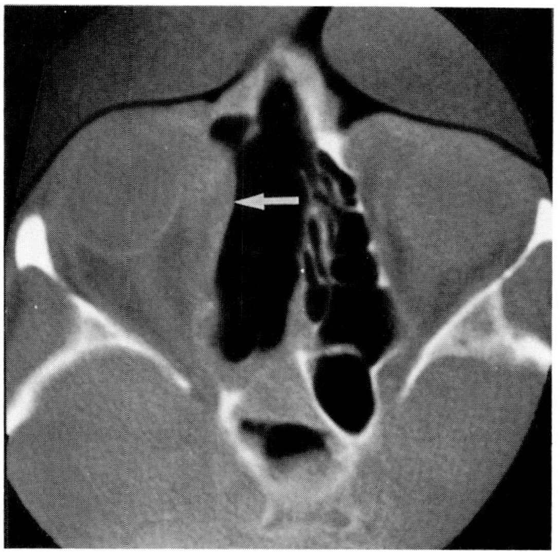

Fig. 7.2. Mid-line lethal granuloma. Axial CT scan showing total destruction of the ethmoid cells on the right side (the patient presented with right-sided proptosis). There was also infiltration of the medial orbit by the granuloma (*arrow*).

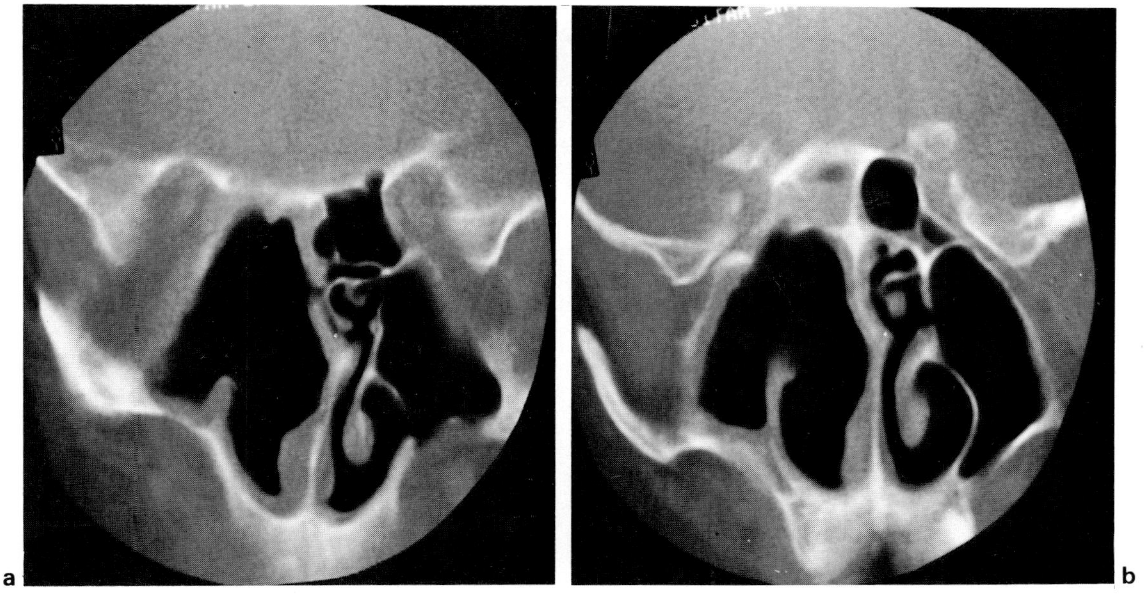

▲
Fig. 7.3a, b. Same patient as Fig. 7.2. Massive bone destruction in the right side of the nose, ethmoids and maxillary antrum can be seen.

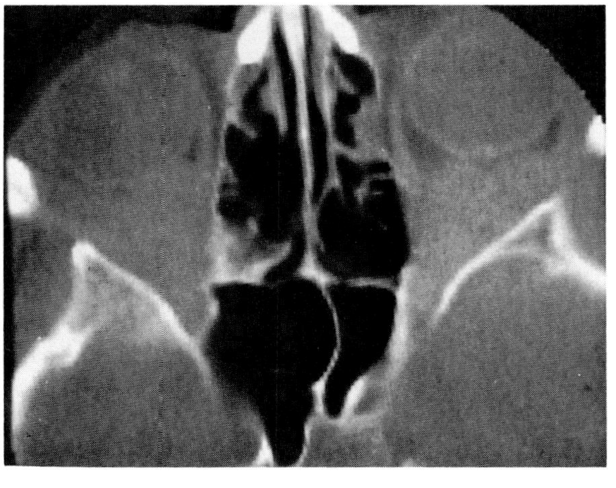

Fig. 7.4. Wegener's granuloma. There is massive bilateral orbital infiltration. At decompression surgery the tissue was shown to be densely fibrotic. Note the destruction of the posterior ethmoid septa.

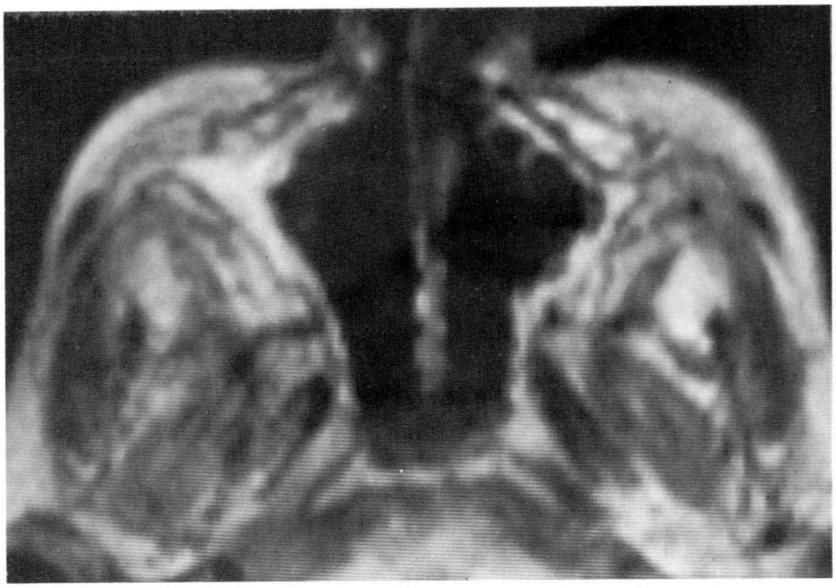

Fig. 7.5. Wegener's granuloma. On magnetic resonance scan T_2-weighted spin echo sequences on axial section showed a massive central cavity in the nose and sinuses lined by irregular inflammatory tissue giving high signal.

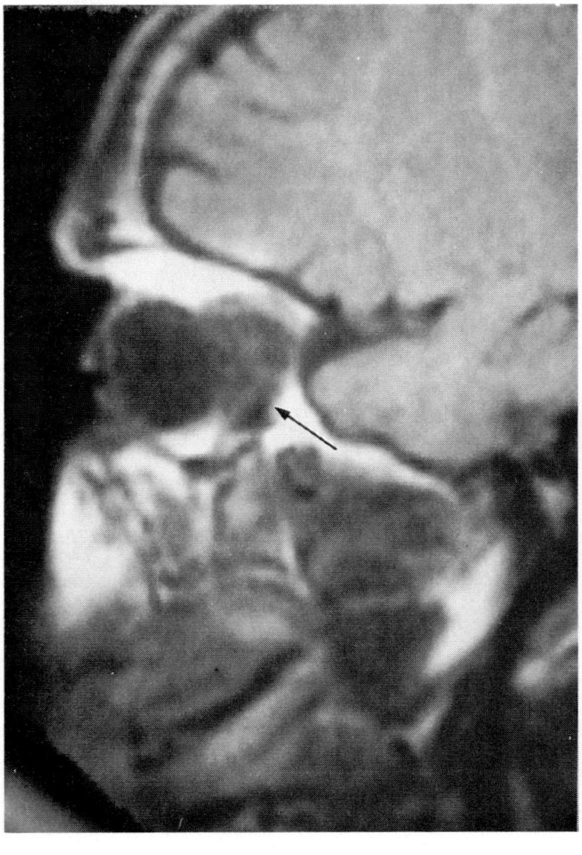

Fig. 7.6. Same patient as Fig. 7.5. On sagittal magnetic resonance scan the tissue infiltrating the orbit (*arrow*) showed low signal on both T_1- and T_2-weighted spin echo sequences. This is characteristic of dense fibrous tissue from whatever cause.

At decompression this was shown to be due to dense fibrosis. The signal characteristics were quite different from those of the nose and sinus granulation tissue, giving a low signal on both T_1- and T_2-weighted spin echo sequences (Fig. 7.6). The second patient with Wegener's granuloma produced exactly similar magnetic resonance changes: there was a high-intensity signal on T_2-weighted spin echo sequences from the granulations in the nose and sinuses (Fig. 7.7) and a hypointense signal from the tissue infiltrating the left orbit on both spin echo

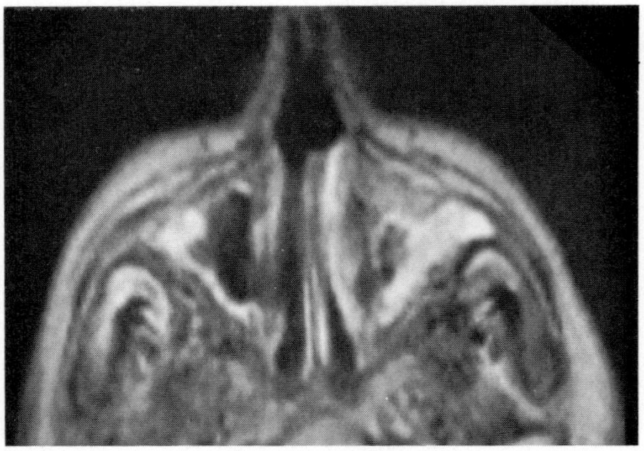

Fig. 7.7. Wegener's granuloma. On axial magnetic resonance scan the T_2-weighted spin echo sequence showed high signal from the granulomatous tissue lining the maxillary antra. The irregularity of the inflammatory tissue is characteristic of the necrotising granulomata.

and inversion recovery sequences. This was again due to the presence of dense fibrous tissue as opposed to the more active granulomatous process in the mucosa of the nose and sinuses.

Sarcoidosis of the Nose and Sinuses

Sarcoidosis is a generalised disease which affects many organs and the generalised condition may be associated with lesions of the nose or paranasal sinuses. Geraint James et al. (1982), in a series of 818 patients with histologically confirmed multi-system sarcoidosis, found that 6% had upper respiratory tract sarcoid involving the nose and/or the paranasal sinuses. In this situation the disease is commonly associated with lupus pernio and other manifestations of chronic fibrotic sarcoidosis. The nasal mucosa of the septum and inferior turbinates is most commonly involved and the microscopical picture is that of non-caseating epithelioid tubercles. In some patients there may be polypoid hypertrophy of the mucosa causing nasal obstruction and in a minority the paranasal sinuses are also involved.

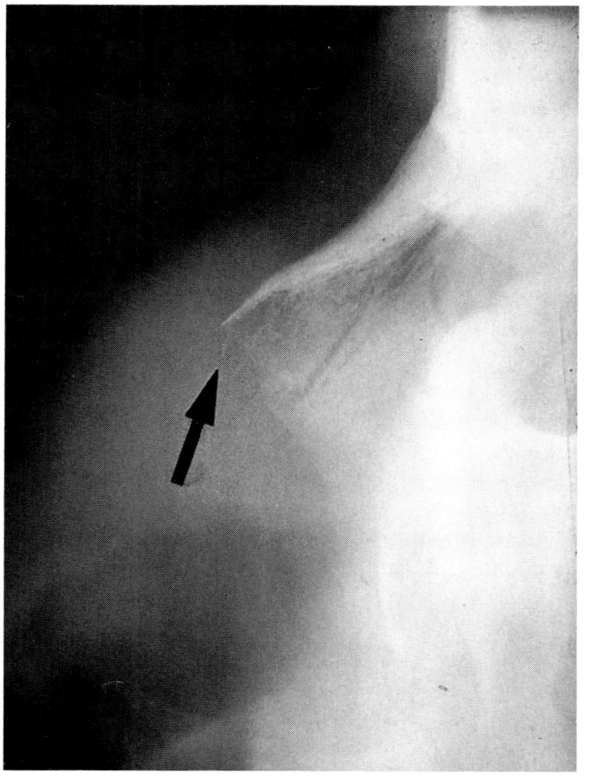

Fig. 7.9. Nasal sarcoid. There is soft tissue swelling with osteolysis of the tip of the nasal bones (*arrow*).

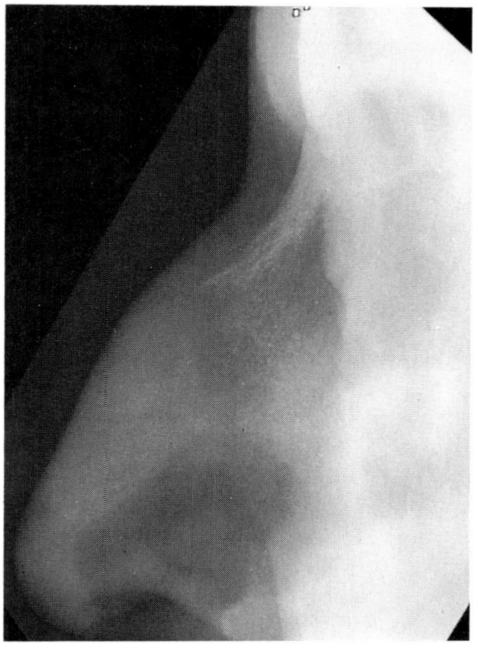

Fig. 7.8. Typical features of nasal sarcoid affecting the nasal bones: a soft tissue mass associated with typical "lacework" rarefaction.

Radiological Features

In the nasal bones sarcoid produces alterations in the bone texture (Fig. 7.8). Scattered areas of punctate osteolysis or zones of frank destruction may be seen (Curtis 1964) (Fig. 7.9). In longstanding involvement a wide-meshed trabecular pattern may form due to the presence of granulomata within the bone, and in such lesions the suture lines may disappear (Geraint James et al. 1982).

The changes recorded in the paranasal sinuses are less definite and are usually non-specific, such as mucosal thickening or clouding of one or more of the sinuses, sometimes with a soft tissue mass in the sinus or nasal cavity. Bone destruction has also been observed, however (Bordley and Proctor 1942; Dodd and Bao-Shan Jing 1977).

Tuberculosis

Tuberculosis of the nose may be cutaneous or mucosal. The former represents lupus vulgaris, a tuberculous infection of the skin which may extend into the nasal vestibule. The mucosal form appears in the nose as a small lobulated mass or granulation on the mucosa of the cartilaginous part of the nasal septum; or it may affect the mucosal lining of the adjacent maxillary antra. The condition is usually associated with pulmonary tuberculosis. The lining membrane is affected first, with polypoid thickening followed by bone involvement. The radiographic features consist of an opacity of the affected sinuses with erosion and destruction of the bony walls, and active tuberculosis in the lung fields.

The sinuses may also be involved in orbital tuberculosis when the bony walls of the orbit are eroded (Evans et al. 1963); and the sphenoid sinuses may also be affected in skull base tuberculosis. Witcombe and Cremin (1978) have described lytic bone changes in the sphenoid occurring in four black children – emphasising the susceptibility of blacks to this form of skull tuberculosis.

Syphilis

Primary syphilis is very rare in the nose and is usually the result of inoculation resulting from scratching or picking the nose with an infected finger after attending an infected patient. Tertiary lesions have been more commonly observed and may represent either acquired or congenital disease. The pathological lesion is a syphilitic gumma which invades mucosa, periosteum or bone and may spread from the nose to the paranasal sinuses when the lateral nasal wall is involved. Radiologically there is an irregular mucosal thickening in the sinus cavity involved, accompanied by bone destruction and some sclerotic change in the sinus walls.

Leprosy

Leprosy is an infective disease of the skin, the mucosa of the upper respiratory tract and the peripheral nervous system. The disease may infiltrate the mucosa of the nose with nodular thickening and may result in perforation of the cartilaginous septum, although the nasal bone is seldom involved. Destruction of the nasal septum results in a saddle nose similar to that seen in congenital syphilis.

Cholesterol Granuloma

In the head and neck cholesterol granuloma is most frequently found in the mastoid antrum and middle ear following haemorrhage, but it has also been described in other parts of the skull, including the maxillary antrum (Graham and Michaels 1978;

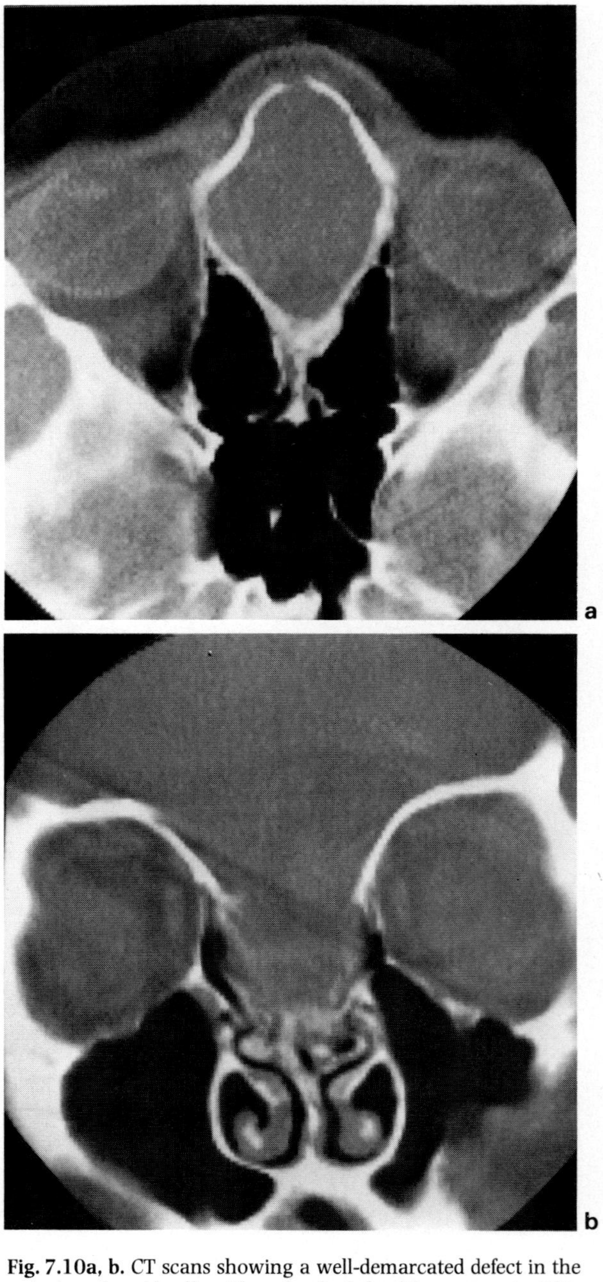

a

b

Fig. 7.10a, b. CT scans showing a well-demarcated defect in the anterior ethmoid cells with a clearly defined bony margin. This was initially diagnosed as a meningo-encephalocoele.

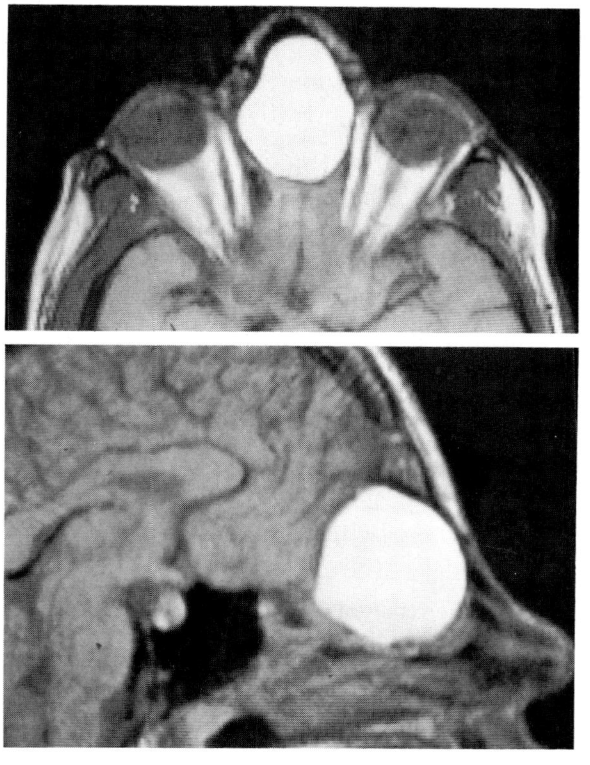

a

b

Fig. 7.11 a, b. Same patient as Fig. 7.10. Axial and sagittal magnetic resonance scans showing a large cystic lesion occupying the anterior ethmoid and cribriform plate area. At surgery this was shown to be due to a cystic cholesterol granuloma. The cystic cavity was packed with cholesterol crystals, and typical cholesterol clefts were shown in the lining membrane of the cyst.

Hellquist et al. 1984). In this situation it may produce cystic expansions within the sinus cavity. Although the majority of sinus cholesterol granulomata occur in the maxillary antrum, other sinuses may also be involved (Figs. 7.10 and 7.11) and a histologically identical lesion may be seen in the frontal bone in the supero-temporal quadrant of the orbit (Lloyd 1986).

The condition is totally unrelated to cholesteatoma; the latter is a post-inflammatory response with the formation of squamous epithelium and keratin, while cholesterol granuloma is essentially a granulomatous change provoked by the formation of cholesterol crystals (Fig. 7.12). Graham and Michaels (1978), in reviewing a series of 5 antral cholesterol granulomata, postulated that when blood stagnates in a closed space, cholesterol comes out of solution as crystals, which then provoke the granulomatous process; the key factor in the aetiology is the presence of a closed cavity containing exudate and blood. This would imply that poor drainage from the site of haemorrhage and stagnation is an important aetiological factor, because cholesterol is relatively insoluble and needs time to become dissociated from the lipoprotein complex and be precipitated as crystals (Hellquist et al. 1984). This hypothesis has been supported by the work of Hiraide et al. (1982), who were able to induce middle ear cholesterol granuloma experimentally in monkeys by prolonged obstruction of the eustachian tube.

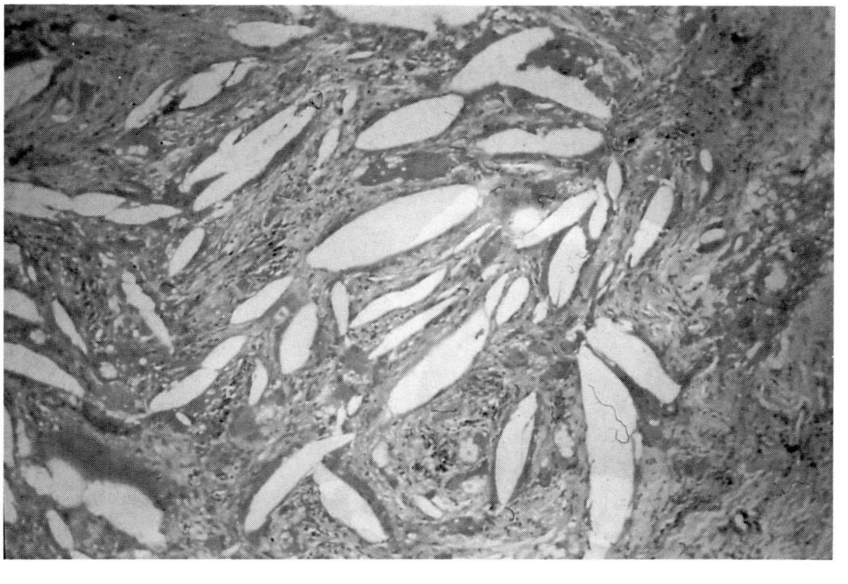

Fig. 7.12. Histological appearance of cholesterol granuloma showing typical cholesterol clefts.

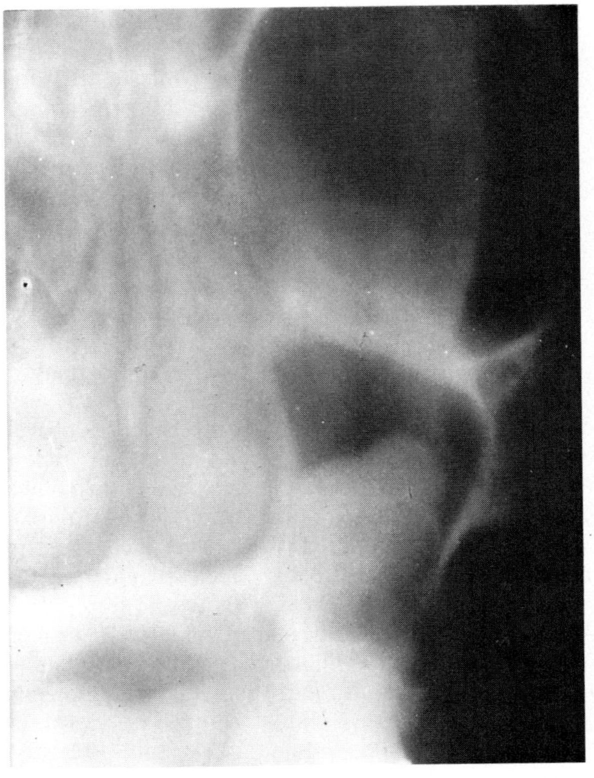

Fig. 7.13. Cholesterol granuloma of the maxillary antrum showing a polypoid soft tissue mass in the antral cavity.

Nine examples of cholesterol granuloma in the maxillary antrum have recently been reported (Lloyd 1986). The commonest presenting symptoms were nasal obstruction, rhinorrhoea and facial pain. The duration of symptoms ranged from 3 months to 10 years with an average duration of 2.75 years. In addition there was a past history of sinus disease in 5 patients, 4 of whom had undergone previous surgery.

The radiological features in those antra with proven cholesterol granuloma were of two types: one variety in which the changes were entirely non-specific and could not be differentiated from inflammatory or allergic sinus disease (Fig. 7.13), and another in which there was a more characteristic appearance of a cyst-like expansion within the sinus cavity accompanied by expansion of the bony walls of the antrum involved (Figs. 7.14, 7.15).

Cholesterol granuloma is now readily diagnosed by magnetic resonance scan; the presence of cholesterol has the effect of shortening the T_1 and T_2 relaxation times, producing a strong signal on both T_1- and T_2-weighted pulse sequences. The diagnostic feature is the very strong signal on inversion recovery sequences (Fig. 7.16).

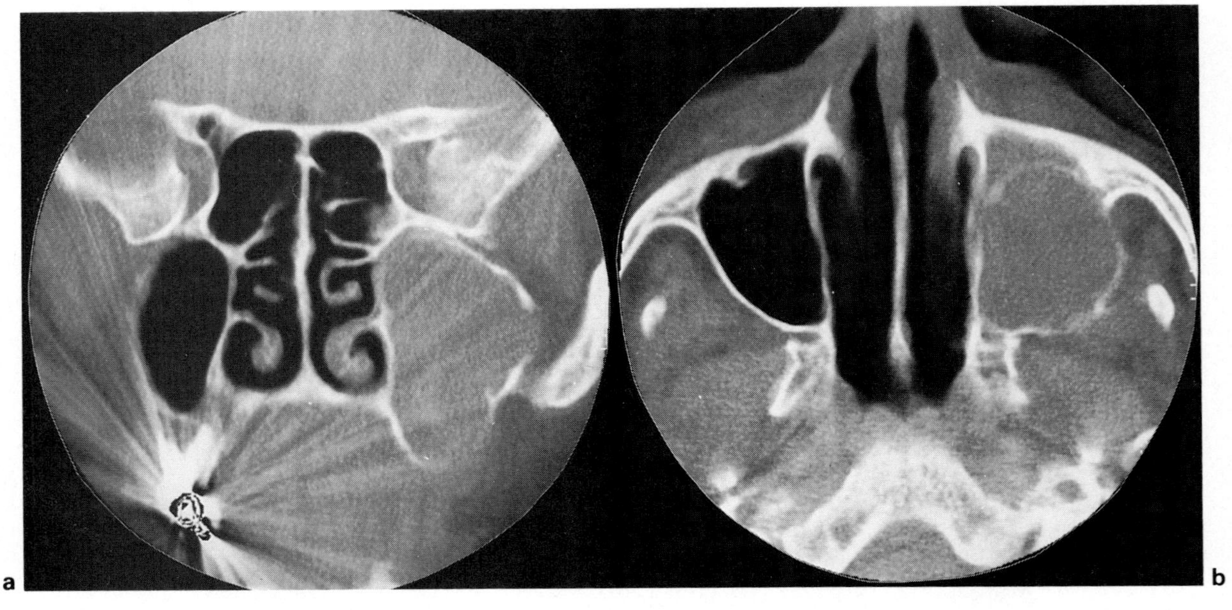

Fig. 7.14a, b. Antral cholesterol granuloma. a Coronal CT section showing gross expansion of the maxillary antrum. b Axial CT section showing expanded maxillary antrum and cyst-like expansion within the antral cavity.

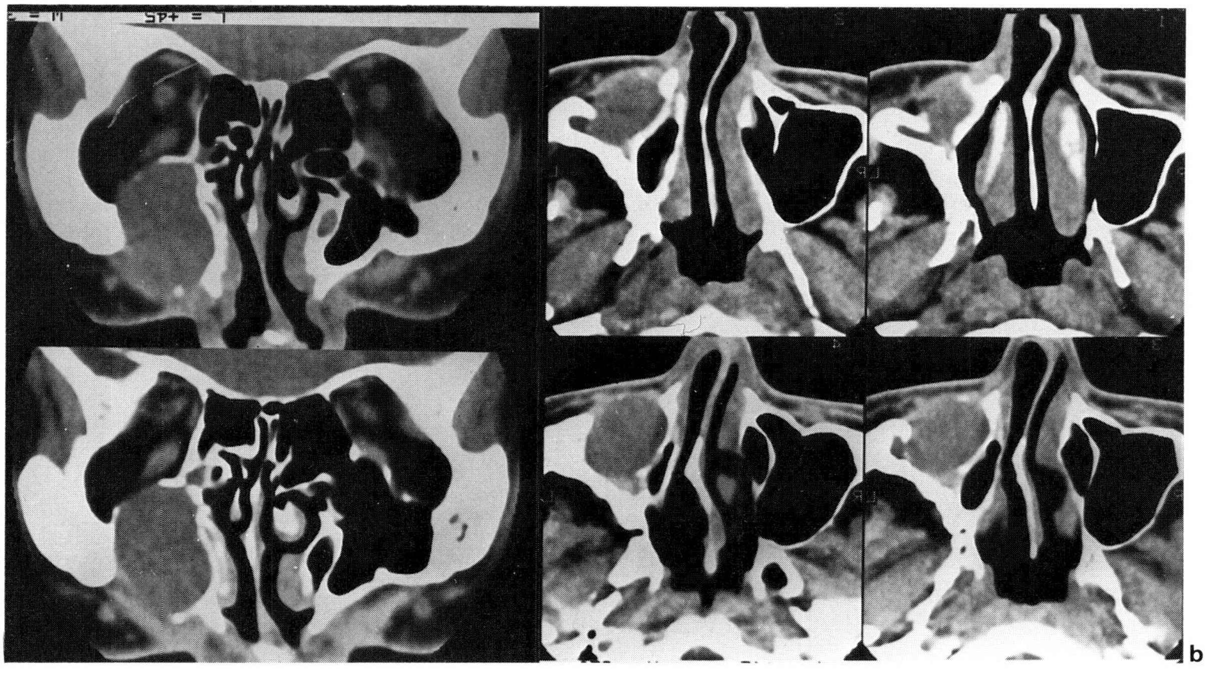

a b

Fig. 7.15a, b. Coronal (**a**) and axial (**b**) CT sections showing an expanded loculus of the right maxillary antrum due to a cholesterol granuloma. The patient had undergone a Caldwell–Luc procedure on the affected antrum many years previously.

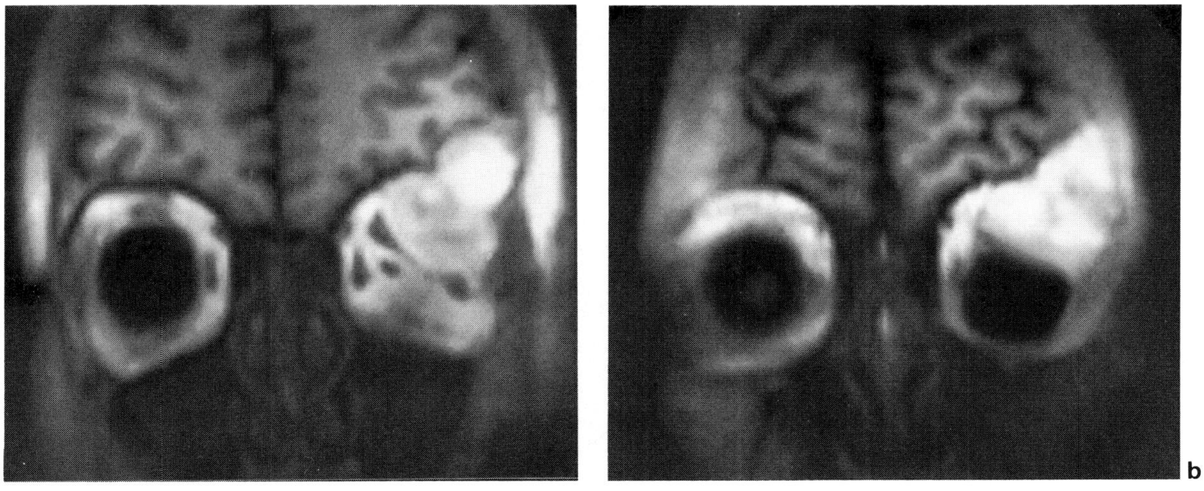

a b

Fig 7.16a, b. Inversion recovery magnetic resonance sequence showing a cholesterol granuloma of the frontal bone. The two sections show high signal due to the presence of cholesterol shortening the T_1 relaxation time – in this situation diagnostic of cholesterol granuloma.

References

Bordley JE, Proctor DF (1942) Destructive lesion in the paranasal sinuses associated with Boeck's sarcoid. Arch Otolaryngol 36:740–742

Dodd GD, Bao-Shan Jing (1977) Radiology of the nose, paranasal sinuses and nasopharynx. Williams and Wilkins, Baltimore

Curtis GT (1964) Sarcoidosis of the nasal bones. Br J Radiol 37:68–70

Evans RA, Schwartz JF, Chutorian AM (1963) Radiologic diagnosis in pediatric ophthalmology. Radiol Clin North Am 1:459–495

Friedmann I (1964) Mid-line granuloma. Proc R Soc Med 57:289–297

Geraint James D, Barter S, Jash D, McKinnen DM, Carstairs LS (1982) Sarcoidosis of the upper respiratory tract (SURT). J

Laryngol Otol 96:711–718

Graham J, Michaels L (1978) Cholesterol granuloma of the maxillary antrum. Clin Otolaryngol 3:155–160

Hellquist H, Lundgren J, Olofsson J (1984) Cholesterol granuloma of the maxillary and frontal sinuses. Otorhinolaryngology 46:153–158

Hiraide F, Inouye T, Miyakogawa H (1982) Experimental cholesterol granuloma. J Laryngol Otol 96:491–501

Kornblut AD, Fauci AS (1982) Idiopathic mid-line granuloma. Otolaryngol Clin North Am 15:685–692

Lloyd GAS (1986) Cholesterol granuloma of the facial skeleton. Br J Radiol 59:481–485

McKinnon DM (1970) Lethal mid-line granuloma of the face and larynx. J Laryngol Otol 84:1195–1203

Milford CA, Drake-Lee AB, Lloyd GAS (1986) Radiology of paranasal sinuses in non-healing granulomas of the nose. Clin Otolaryngol 11:199–204

Pritchard MH, Gow PJ (1976) Wegener's granulomatosis presenting as rheumatoid arthritis. Proc R Soc Med 69: 501–504

Stewart JP (1933) Progressive lethal granulomatous ulceration of the nose. J Laryngol Otol 48:657–701

Wegener F (1936) Über generalisierte, septische Gefaserkrankungen. Verh Dtsch Pathol Ges 29:202–212

Wegener F (1939) Über eine eigenartige rhinogene Granulomatos mit besonderer Beteiligung des Arteriensystems und der Nieren. Beitr pathol Anat allgem Pathol 102:36–38

Witcombe JB, Cremin BJ (1978) Tuberculous erosion of the sphenoid bone. Br J Radiol 51:347–350

Wolff SM, Fauci AS, Horn RG, Dale DC (1974) Wegener's granulomatosis. Ann Intern Med 81:513–525

8 Mycotic Disease

Phycomycosis

By definition phycomycosis is caused by fungi belonging to the class Phycomycetes. There are two important orders in this class: Mucorales, which cause mucormycosis, and entomophthorales, which give rise to entomophthorosis. There are significant clinical and histopathological differences between these two conditions.

Mucormycosis

Mucormycosis appears principally in the immune-suppressed patient and has a particular predilection for those with acidosis secondary to diabetes mellitus (Berthier et al. 1982). The disease presents with an acute onset, the duration of symptoms being numbered in days. The clinical patterns are variable but the naso-orbital-cerebral form is the most common. When the nasal infection has been established it spreads through the sinuses to the vasculature of the conjunctiva and invades orbital structures giving rise to orbital cellulitis and ophthalmoplegia. Subsequent spread from the orbits may cause thrombosis of the cavernous sinus. Extension to the cranial cavity may occur either via the orbital roof (Parmentier et al. 1965) or cavernous sinus, or the fungus may invade the anterior fossa directly through the cribriform plate. This gives rise to a meningo-encephalitis with necrosis of brain tissue predominantly in the inferior part of the frontal lobes. Histologically aseptate fungal hyphae are seen within inflammatory granulation tissue. The fungus has a predilection for invading vessels, resulting in thrombosis, infarction and areas of necrosis (Martin et al. 1954; Carpenter et al. 1968).

The radiological features consist initially of an opacity of the sinuses involved (Fig. 8.1) (Becker et al. 1968), followed by bone destruction mimicking a malignant neoplasm. CT demonstrates, in addition to the bone changes, the exact degree of soft tissue invasion, particularly in the orbit, and abscess formation can be shown in the frontal lobes (Berthier et al. 1982). Vessel invasion can be demonstrated angiographically (Courey et al. 1972), with filling defects in the arteries concerned and total occlusion leading to embolism and brain infarction.

Entomophthorosis

Entomophthorosis differs from mucormycosis in that it is more likely to occur in otherwise healthy subjects and develops more slowly, the clinical presentation being less dramatic than that of mucormycosis. The general condition of the patient is good and the common form of presentation is nasal obstruction together with swelling of the nose and cheek often of many months' duration – which contrasts sharply with the rapid progress of mucormycosis. Spread from the nose to the paranasal sinuses often occurs but the infection is usually confined to these regions. Invasion of the anterior fossa can occur, however. CT may show a soft tissue mass in the sinuses involved and in the nose, with massive invasion of the orbit and erosion of bone and sinus walls (Fig. 8.2).

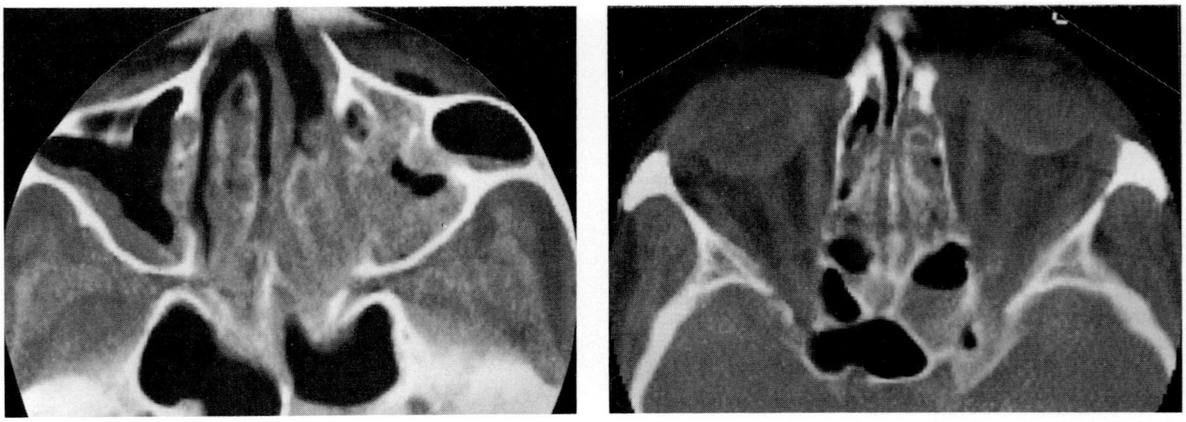

Fig. 8.1a, b. Mucormycosis. Axial CT sections through the antra and ethmoids showing an irregular soft tissue mass with destruction of the medial antral wall, invasion of the pterygo-palatine fossa and early encroachment on the orbital apex.

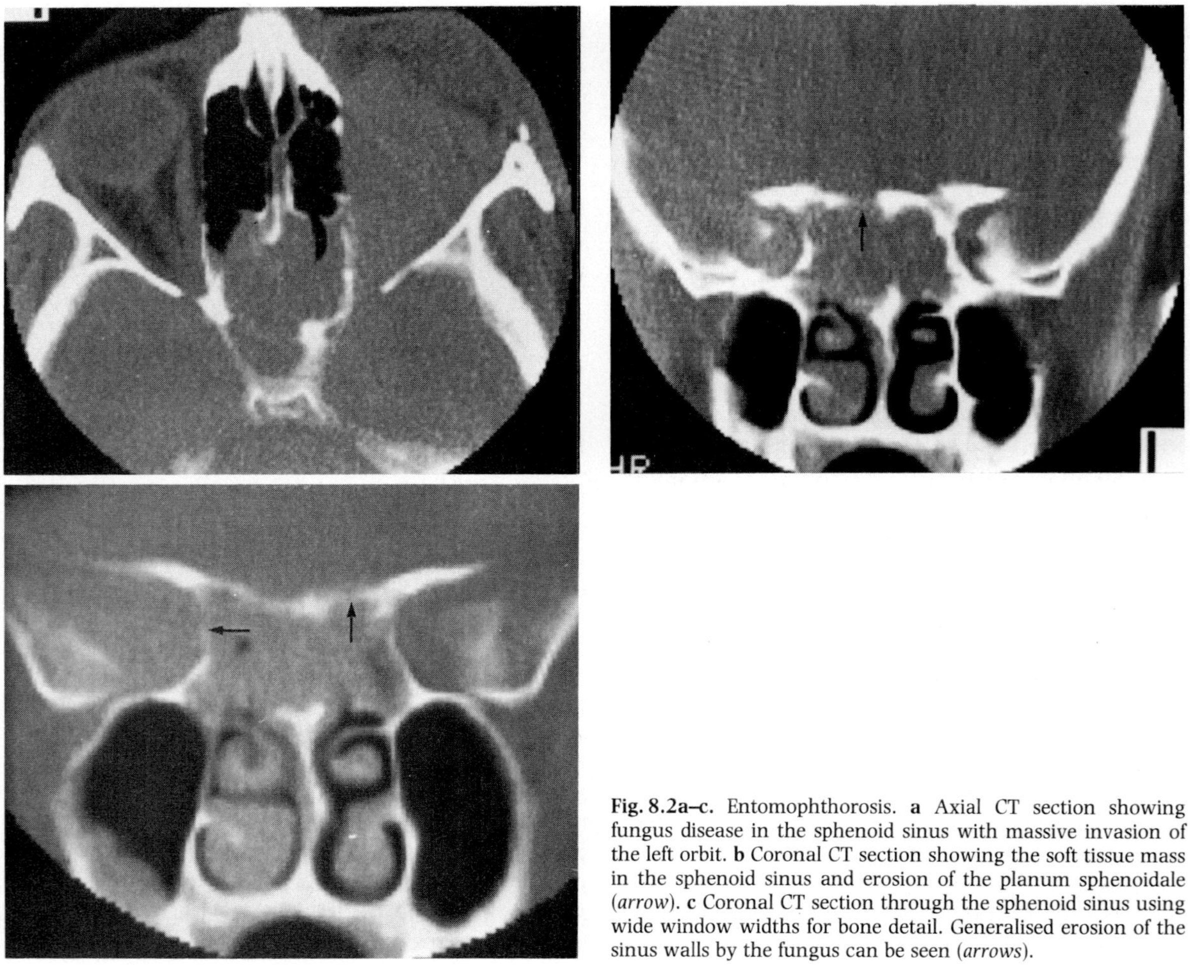

Fig. 8.2a–c. Entomophthorosis. **a** Axial CT section showing fungus disease in the sphenoid sinus with massive invasion of the left orbit. **b** Coronal CT section showing the soft tissue mass in the sphenoid sinus and erosion of the planum sphenoidale (*arrow*). **c** Coronal CT section through the sphenoid sinus using wide window widths for bone detail. Generalised erosion of the sinus walls by the fungus can be seen (*arrows*).

Aspergillosis

The fungus *Aspergillus* is a common saprophyte found in soil, dust, and decaying organic matter such as fruits and grains. It may be pathogenic in man, animals and birds. The usual point of entry is the nose, from where it may spread to the sinuses and bronchi or disseminate into lung tissue. The condition may occur in the immune-compromised host or in those debilitated by other primary conditions; it may also affect otherwise healthy subjects with no underlying disease. *Aspergillus fumigatus* is the species which most often causes paranasal sinus disease; *Aspergillus flavus* is less common, but in the Sudan, where the disease is endemic, it is the predominant organism.

There are two forms of the disease: an invasive form which behaves similarly to a malignant neoplasm, with bone destruction, orbital involvement and occasional intracranial extension; and a non-invasive form which may present much like a bacterial sinusitis, with rhinorrhoea, nasal congestion and an opaque antrum on plain radiography. The distinction between these two varieties of the disease is not entirely clear-cut and it may well be that the invasive form follows the non-invasive type in some patients.

More recently McGill et al. (1980) have described fulminant aspergillosis, a new variety of the disease that occurs in individuals with a severely depressed immune system and is characterised by a rapidly progressive gangrenous mucoperiostitis with destruction of the nasal cavity and paranasal sinuses.

Imaging Features

The maxillary antrum is the commonest site of *Aspergillus* infection (Jahrsdorfer et al. 1979), producing a simple opacity in the sinus on plain radiography. This of course is a non-specific sign, but when there is involvement of two adjacent sinuses in a patient from a particular locality such as the Sudan, or in a patient on immunosuppressive drugs, the possibility of aspergillosis needs to be excluded. According to Rudwan and Sheik (1976) fluid levels in the sinuses are not seen in this condition, so this must be a point of differentiation from bacterial infection. Bone destruction is a common feature in the invasive form of the disease and the antrum is most often involved. This may result in a mass which is continuous from the antrum into the nasal cavity, with a clearly defined erosion of the medial wall (Fig. 8.3). At this stage the lesion may mimic an antro-choanal polyp or a tumour such as an inverted papilloma. With further extension the fungus may destroy the lateral wall of the antrum and invade the orbital floor. In some patients the fungus may reach the orbit via the ethmoid cells or the naso-lacrimal duct (Bahadur et al. 1983). Intracranial extension is not uncommon from the orbit and may result in areas of curvilinear calcification on skull radiographs (Rudwan and Sheik 1976).

Recent publications (Glass et al. 1984; van Haake 1984) have emphasised the use of CT to demonstrate aspergillosis in the paranasal sinuses. In

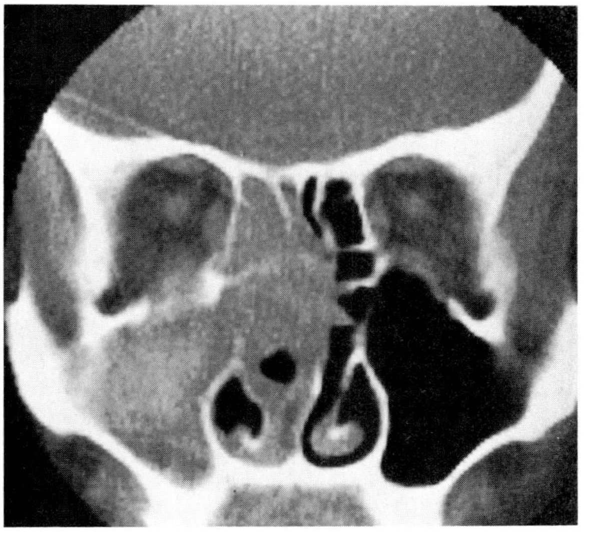

Fig. 8.3. Aspergillosis. Coronal CT section showing a soft tissue mass in the right antrum, right nasal cavity and ethmoids with erosion of the medial antral wall.

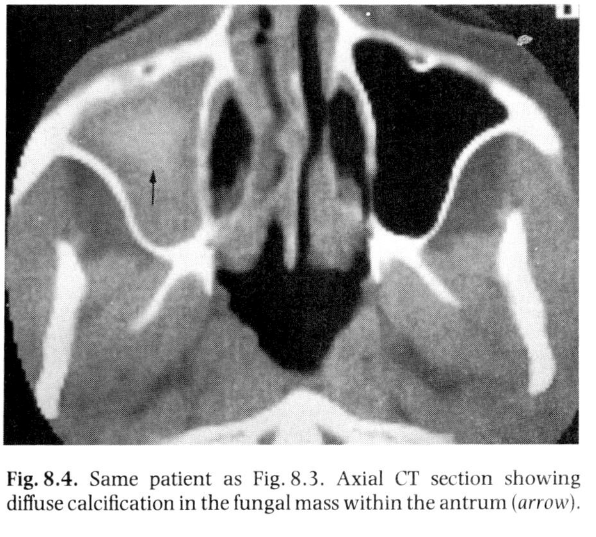

Fig. 8.4. Same patient as Fig. 8.3. Axial CT section showing diffuse calcification in the fungal mass within the antrum (*arrow*).

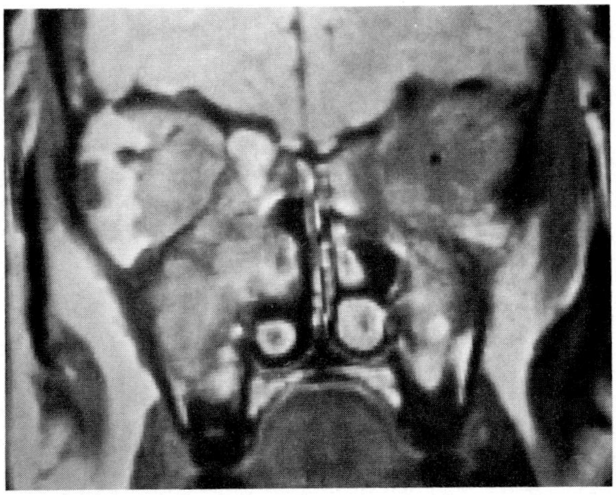

Fig. 8.5. Coronal magnetic resonance scan showing aspergillosis of the sinuses with massive invasion of both orbits.

Fig. 8.6a, b. Invasive aspergillosis. Axial (**a**) and sagittal (**b**) magnetic resonance scans showing both orbits massively infiltrated by dense fibrous tissue containing microscopic calcium deposits and *Aspergillus*. These T_1-weighted spin echo sequences showed low signal from the intraorbital fibrous tissue.

▼

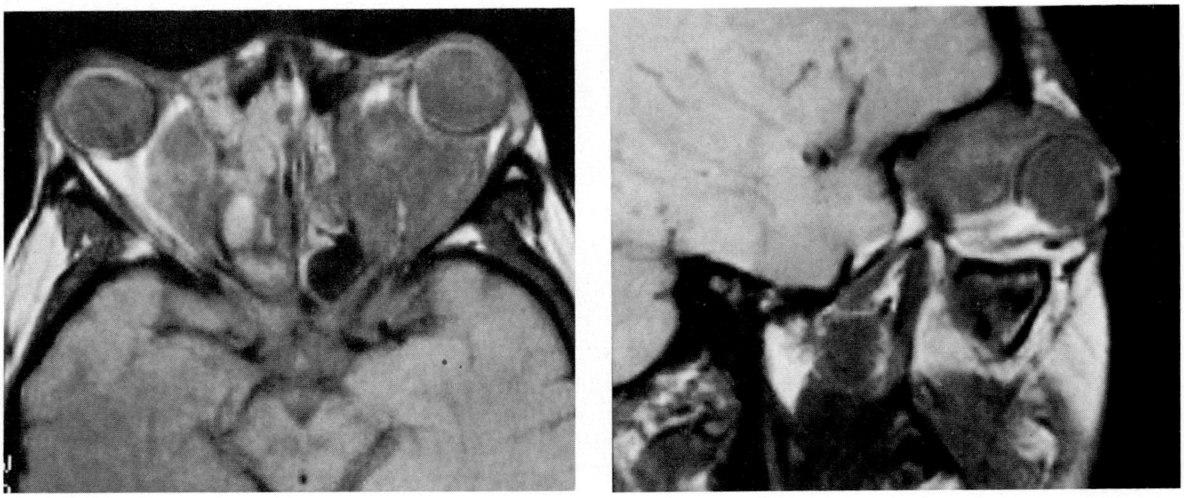

a b

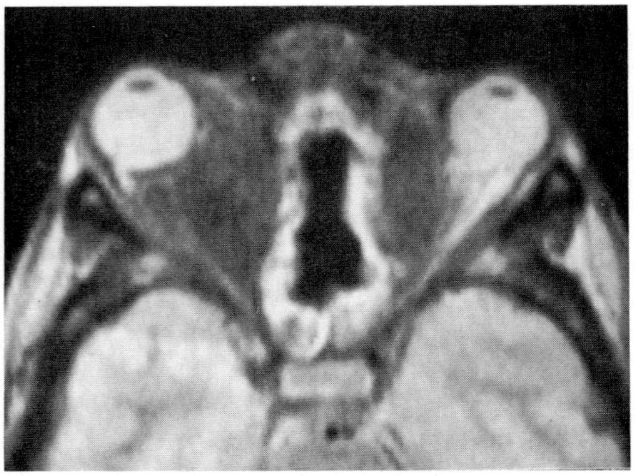

Fig. 8.7. Post-operative magnetic resonance scan of the same patient as Fig. 8.6 after cranio-facial resection. The axial T_2-weighted spin echo sequence shows very low signal from the fibrous tissue within the orbits and high signal from the tissue lining the cranio-facial cavity.

one patient examined by the author using high-resolution CT a mass was present in the right side of the nasal cavity continuous with a mass in the antrum (Fig. 8.3). Within the antral mass a diffuse calcification was shown (Fig. 8.4) which was not appreciated on conventional tomography but was clearly demonstrated on CT because of the inherently high contrast of the latter technique. This type of calcification is unlike that seen in any neoplastic process and is suggestive of aspergillosis when present.

Calcification can also be nodular and very dense on radiography. Stammberger et al. (1984) reported the presence of calcification in 27 patients out of 59 with sinus aspergillosis. The calcification was seen on conventional X-ray studies and was often of metal radiopacity. By means of light and electron microscopy and X-ray fluorescence analysis, these authors showed that the areas of cal-

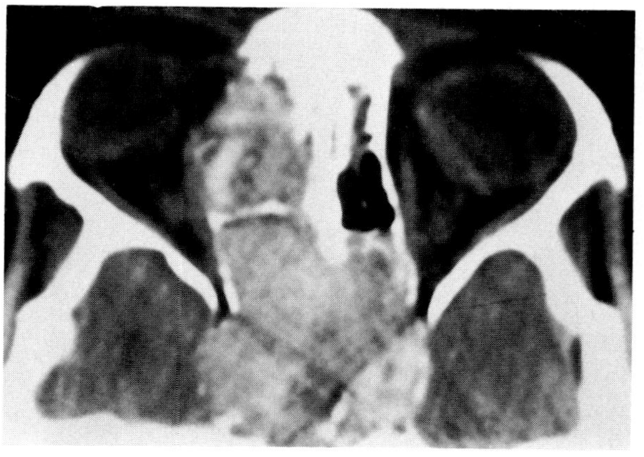

Fig. 8.8. Aspergillosis. High density in the sphenoid and ethmoids is seen on axial CT scan prior to intravenous contrast.

cification were due to the deposition of calcium phosphate in the centre of the fungal mass.

Magnetic Resonance

Both the invasive and the non-invasive forms of aspergillosis have been studied by magnetic resonance scan. The invasive form showed a mass in the nose and paranasal sinuses with massive invasion of both orbits (Fig. 8.5). A distinctive feature of this case was the different spin characteristics of the intranasal tissue and the tissue invading the orbits. The latter consisted of very hard fibrous tissue on biopsy with microscopic evidence of small calcium deposits. It gave a low signal on T_2-weighted spin echo sequences similar to the dense fibrous reaction in the orbits provoked by other conditions such as Wegener's granuloma. In comparison the intranasal tissue gave a fairly strong signal (Figs. 8.6, 8.7).

A single example of the non-invasive form of aspergillosis has also been investigated by magnetic resonance. This presented initially as a sphenoid sinus mucocoele and showed a generalised high density on CT prior to intravenous contrast (Fig. 8.8). The material evacuated from the sphenoid and ethmoid sinuses and from the maxillary antrum at surgery was of a soft consistency and brownish-black in colour and gave no signal either on T_1-weighted or T_2-weighted spin echo sequences (Fig. 8.9, 8.10 and 8.11). The explanation for this absence of signal is speculative. From the evidence of the CT scan the material almost certainly had a high calcium content, which would contribute to the lack of signal. Stammberger et al. (1984) found traces of several heavy metals (silver, lead, copper, cadmium and mercury) in mycotic concrements of aspergillosis, some of which may have been responsible for the signal void. Information derived from nuclear magnetic resonance spectroscopy studies suggests that tight molecular binding of hydrogen might also be a causative factor.

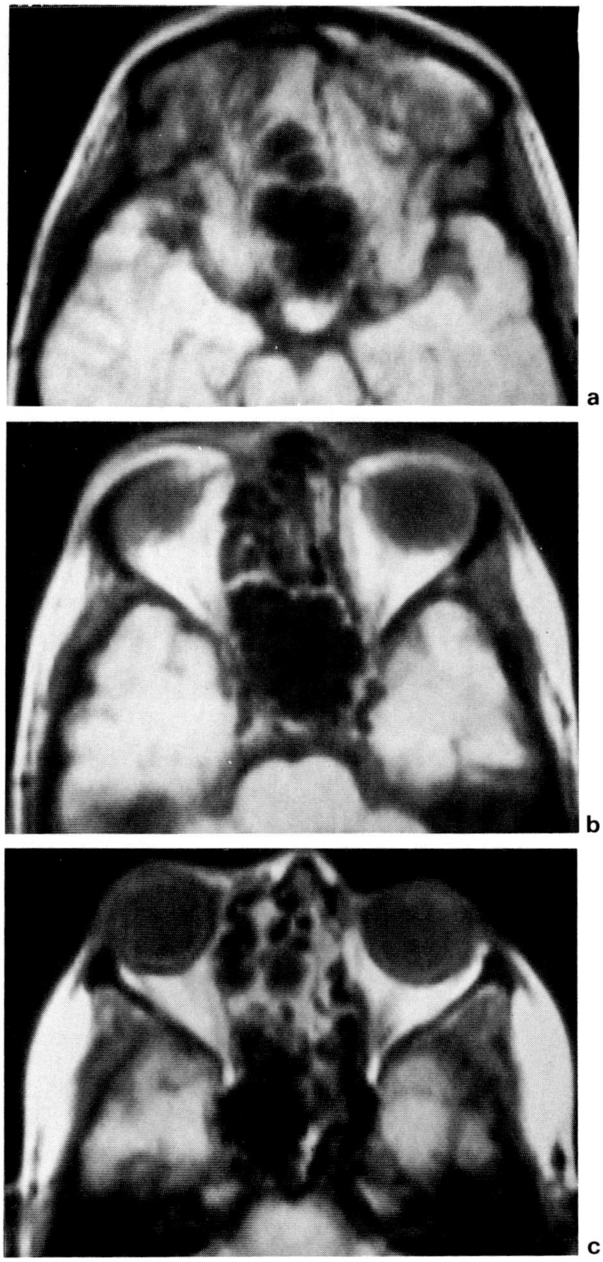

a

b

c

Fig. 8.9a–c. Same patient as Fig. 8.8. Axial T_1-weighted spin echo magnetic resonance scans showing total absence of signal from the substance occupying the expanded sphenoid sinus. This was non-invasive aspergillosis with secondary mucocoele formation.

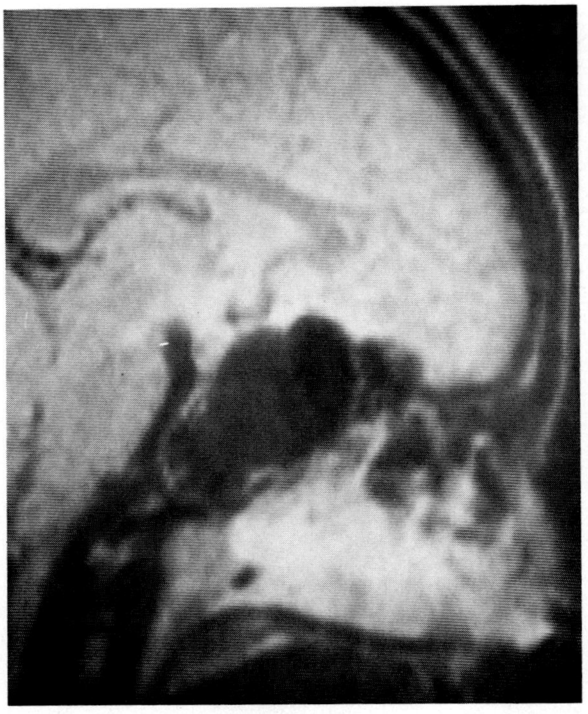

Fig. 8.10. Same patient as Fig. 8.8. The sagittal T_2-weighted spin echo sequence again shows a total absence of signal from the fungus within the expanded sphenoid sinus.

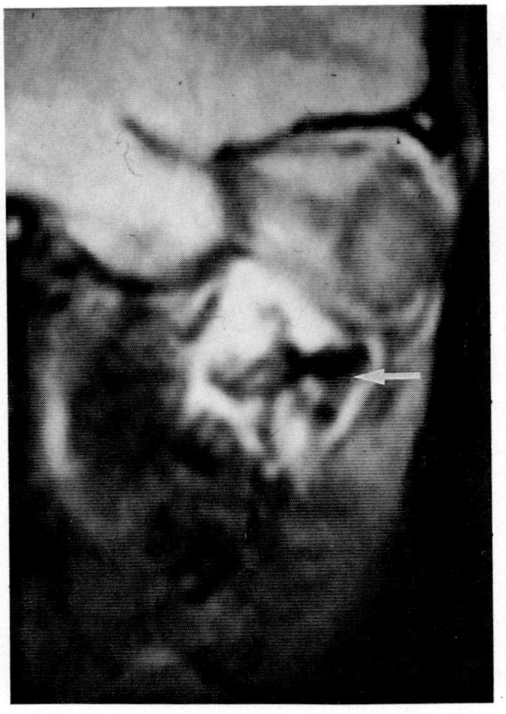

Fig. 8.11. Same patient as Fig. 8.8. Sagittal T_2-weighted spin echo sequence through the maxillary antrum showing fungus (*arrow*) within the antrum giving no signal.

References

Bahadur S, Kacker JK, D'Souza B, Chopra P (1983) Paranasal sinus aspergillosis. J Laryngol Otol 97:863–867

Becker MH, Ngo N, Berambaum SL (1968) Mycotic infection of the paranasal sinuses. Radiology 90:49–51

Berthier M, Palmieri O, Lylyk P, Leiguarda R (1982) Rhino-orbital phycomycosis complicated by cerebral abscess. Neuroradiology 22:221–224

Carpenter DF, Brubaher LH, Powell RA, Valsamis MP (1968) Phycomycotic thrombosis of basilar artery. Neurology 18:807–812

Courey WR, New PFJ, Price DL (1972) Angiographic manifestations of craniofacial phycomycosis. Radiology 103:329–334

Glass RBJ, Hertzanu Y, Mendelsohn DB, Posen J (1984) Paranasal sinus aspergillosis. J Laryngol Otol 98:199–205

Jahrsdorfer RA, Ejercito VS, Johns MME, Cantrell RW, Sydnor JB (1979) Aspergillosis of the nose and paranasal sinuses. Am J Otolaryngol 1:6–14

Martin FR, Lukeman JM, Ranson RF et al. (1954) Mucormycosis of the nervous system associated with thrombosis of the internal carotid artery. J Paediatr 44:437–444

McGill TJ, Simpson G, Healy GB (1980) Fulminant aspergillosis of the nose and paranasal sinuses. A new clinical entity. Laryngoscope 90:748–754

Parmentier M, Belasse E, Pirart J, Van der Haegen JJ (1965) Mucormycose orbitaire. Arch Ophthalmol (Paris) 25:689–704

Rudwan MA, Sheik HA (1976) Aspergilloma of the paranasal sinuses – a common cause of unilateral proptosis in Sudan. Clin Radiol 27:497–502

Stammberger H, Jakse R, Beaufort F (1984) Aspergillosis of the paranasal sinuses. X-ray diagnosis, histopathology and clinical aspects. Ann Otol Rhinol Laryngol 93:251–256

van Haake N (1984) Aspergillosis of the paranasal sinuses. J Laryngol Otol 98:193–197

9 Giant Cell Lesions of the Nose and Paranasal Sinuses

The existence of tumours or tumour-like lesions showing multinucleate giant cells has been known for over 100 years and the giant cell tumour of bone is a well-recognised entity occurring at the ends of long bones and characteristically presenting in the third and fourth decades. The histological similarity of this lesion to the so-called brown tumour of hyperparathyroidism is also well known. Giant cell tumours have also been reported in the jaw and cranium since the mid-nineteenth century, and in this situation they have generally been regarded as behaving in a more benign manner than those in the long bones.

In 1953 Jaffe introduced the term giant cell reparative granuloma to designate a rare condition involving the jaw bones. He described it as non-neoplastic and a separate entity from the giant cell tumour, which he regarded as extremely rare in the skull. Undoubtedly most of the early and even later reports of giant cell tumour of the cranio-facial bones were in fact examples of reparative granuloma, thus accounting for their relatively benign behaviour (Friedmann and Osborn 1982). Some authorities question whether true giant cell tumours occur at all in the skull, but there would appear to be a few authentic cases, particularly those which have developed frankly malignant characteristics (Potter and McLennan 1970; Fu and Perzin 1974).

There are thus three entities in the nose and sinuses presenting histologically as a dispersion of multinucleate giant cells in a stroma of spindle cells:

1. Giant cell granuloma
2. A giant cell lesion identical to the above but associated with hyperparathyroidism and subperiosteal bone resorption and referred to as a brown tumour (Figs. 9.1, 9.2)
3. An active neoplasm with locally aggressive activity similar to the giant cell tumour of long bones

These conditions are very similar histologically. Friedberg et al. (1969) have suggested histological differentiation based on the appearance of the

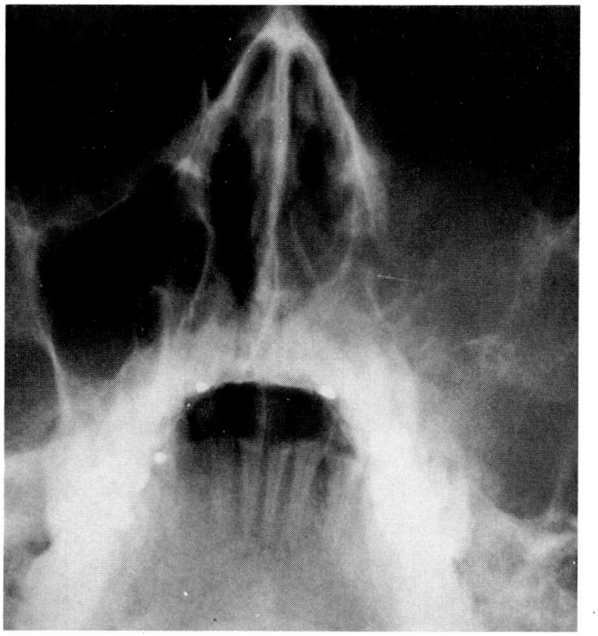

Fig. 9.1. Expansion of the left maxillary antrum and bone destruction due to a brown tumour.

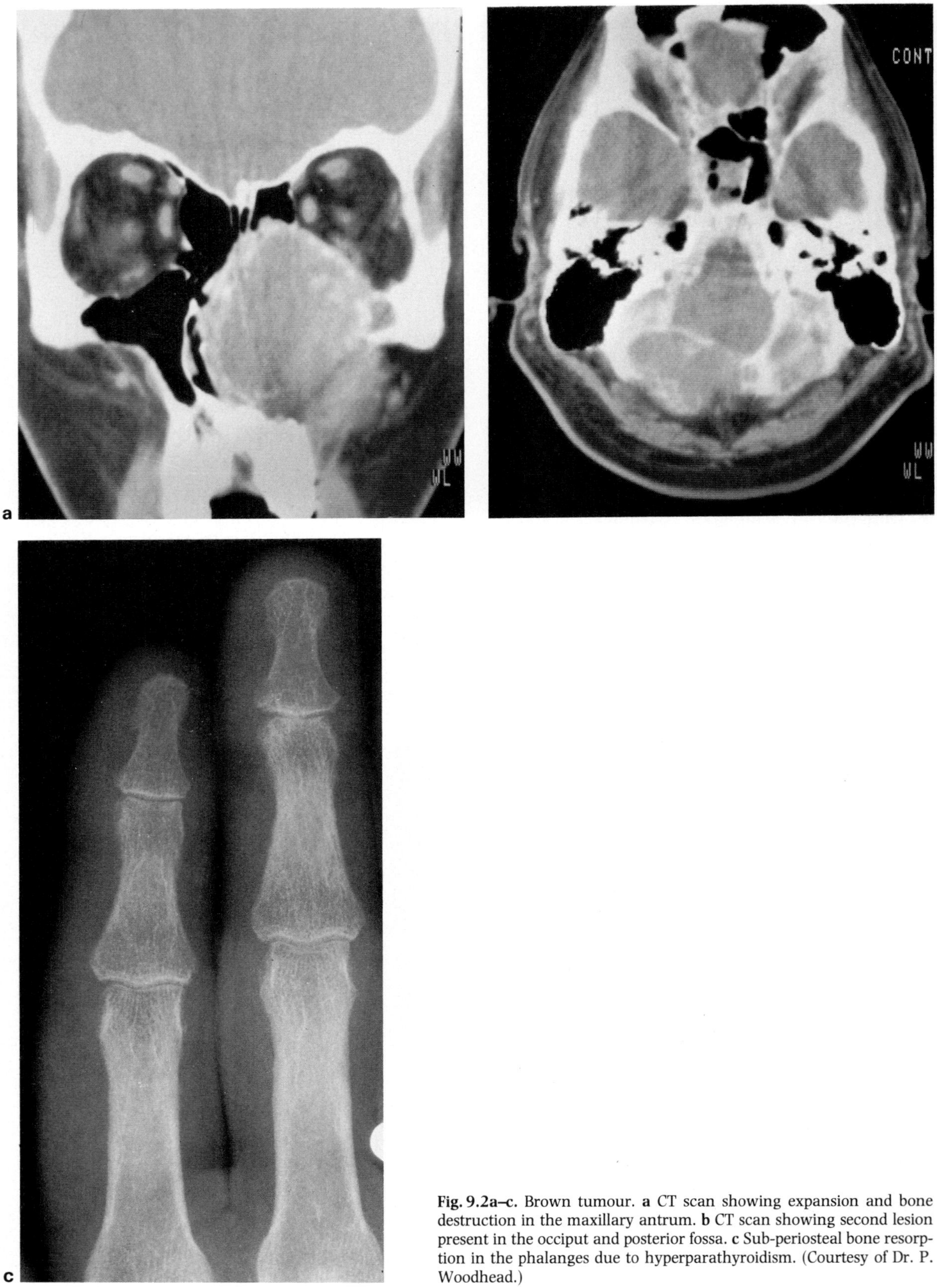

Fig. 9.2a–c. Brown tumour. **a** CT scan showing expansion and bone destruction in the maxillary antrum. **b** CT scan showing second lesion present in the occiput and posterior fossa. **c** Sub-periosteal bone resorption in the phalanges due to hyperparathyroidism. (Courtesy of Dr. P. Woodhead.)

spindle cells and the distribution of giant cells: giant cell granuloma shows cytoplasmic preponderance in the spindle cells and the giant cells are variable in number and show irregular dispersion; in giant cell tumour the stroma shows nuclear preponderance and the giant cells are numerous and evenly distributed.

In spite of their histological similarity these two lesions behave in a different manner. Giant cell granuloma is a benign, slow-growing lesion most common in the second decade of life that is usually controlled by simple curettage. Giant cell tumour presents in the third and fourth decades and is characterised by extremely aggressive behaviour with invasion of neighbouring structures and a high recurrence rate.

Radiological Features

The lesions may present, on plain radiographs, as a simple opacity of the affected sinus resembling a localised thickening of the lining membrane. Hlavacek and Jolma (1974) described a giant cell tumour arising in the upper part of the frontal sinus and presenting as a mid-line circumscribed opacity. More often bone expansion or destruction is present. In the maxillary sinus erosion and destruction of the alveolar ridge may be seen (Dodd and Bao-Shan Jing 1977) and nine out of ten patients reviewed by Potter and McLennan (1970) in whom the sphenoid was involved showed destruction of the sellar floor.

Seven patients with an initial histological diagnosis of giant cell lesion in the nose or paranasal sinuses have been investigated radiologically. Two of these cases were brown tumours of hyperparathyroidism and were diagnosed on the evidence of the blood chemistry and bone survey. Both occurred in the maxillary antrum (Figs. 9.1, 9.2). Of the remaining five lesions three were localised in the sphenoid sinus and posterior ethmoid cells, one was situated in the nose and anterior ethmoid cells (Fig. 9.3) and one in the maxillary antrum. All five presented on plain radiographs as a soft tissue mass in the sinuses – some with simple expansion of the bony walls, as in a benign process, others with frank bone destruction as in a malignant tumour.

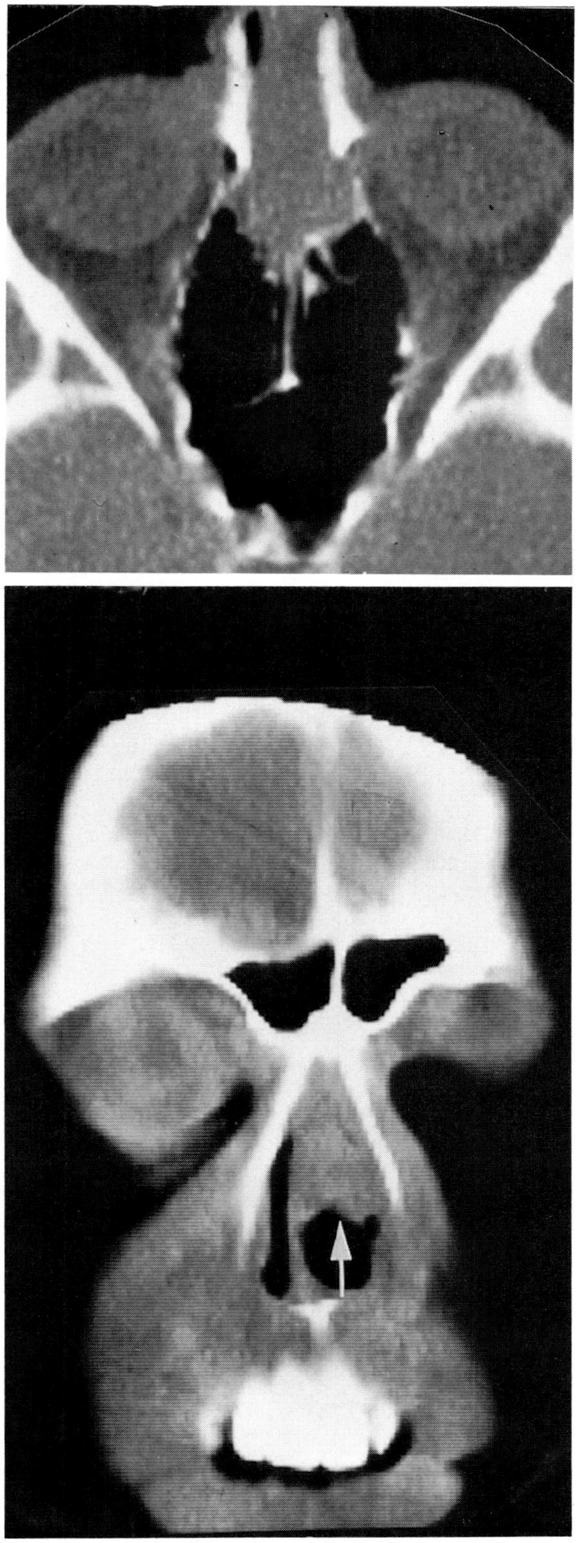

a

b

Fig. 9.3a, b. Giant cell reparative granuloma in the anterior nasal cavity (*arrow*) shown on axial (**a**) and coronal (**b**) CT scans.

Magnetic Resonance and CT

Five patients were investigated by CT and one by

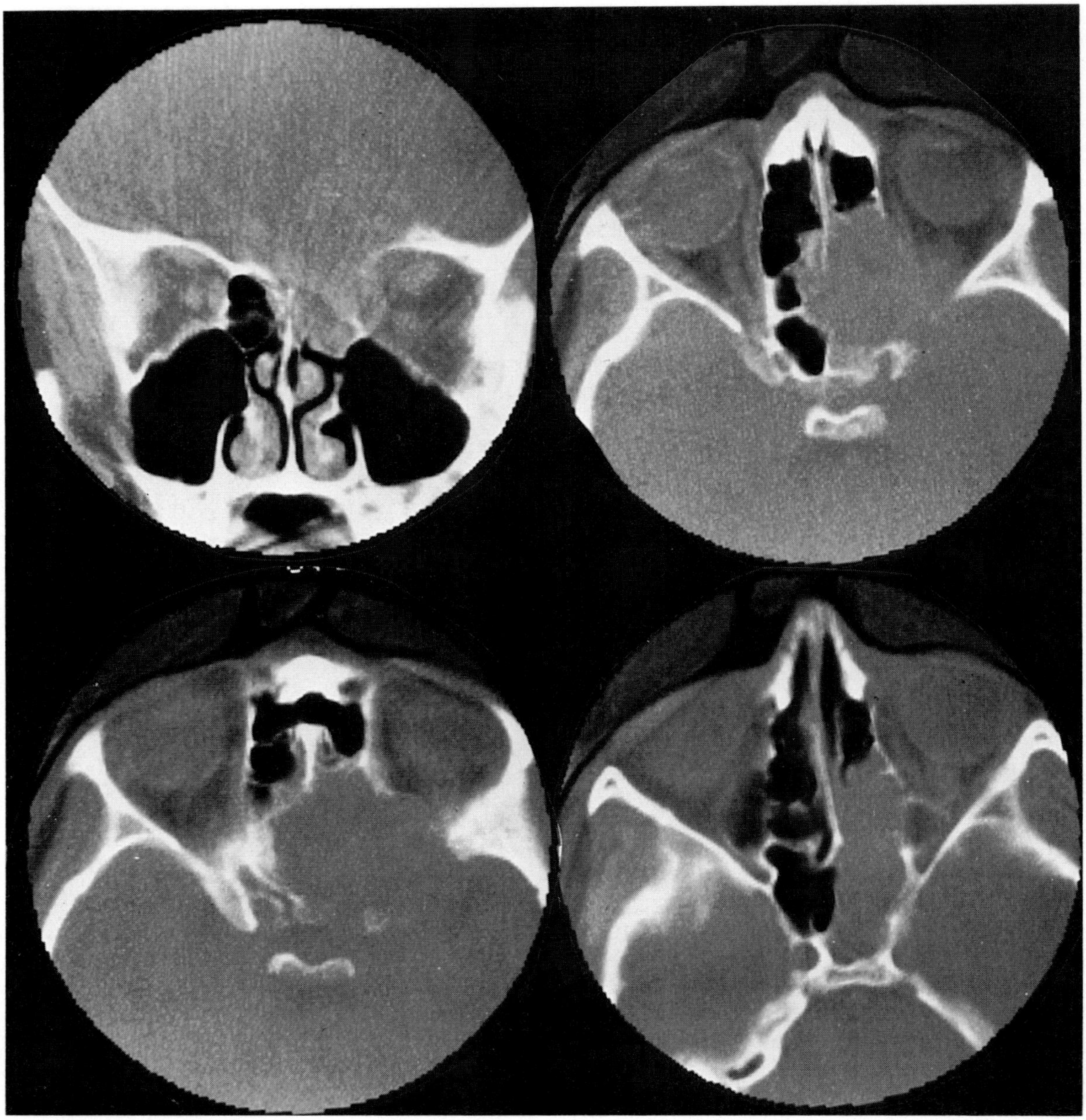

Fig. 9.4. Giant cell reparative granuloma in a 14-year-old male. Three axial CT sections and one coronal section show a soft tissue mass in the sphenoid sinus with destruction of the posterior orbit and optic canal and massive upward extension into the anterior and middle fossa.

magnetic resonance. As on plain radiography, CT demonstrated either expansion or destruction of the sinus walls. In one instance an enhancing mass was shown on CT arising in the sphenoid sinus with massive invasion of the anterior fossa and total destruction of the adjacent optic canal (Fig. 9.4). The single case examined by magnetic resonance showed a generalised expansion of the sphenoid sinus by a soft tissue mass which gave high signal on T_2-weighted spin echo sequences and which expanded into the middle fossa (Fig. 9.5). The type of bony change seen on initial examination gave no clue to the eventual clinical outcome; nor did it allow any differentiation between giant cell tumours and granulomata. The lesion most destructive of bone (Fig. 9.4) has not recurred 5 years after cranio-facial resection, while the lesion shown in Fig. 9.5 has recurred twice in 2 years following initial surgery and its histological appearance has become unequivocally malignant.

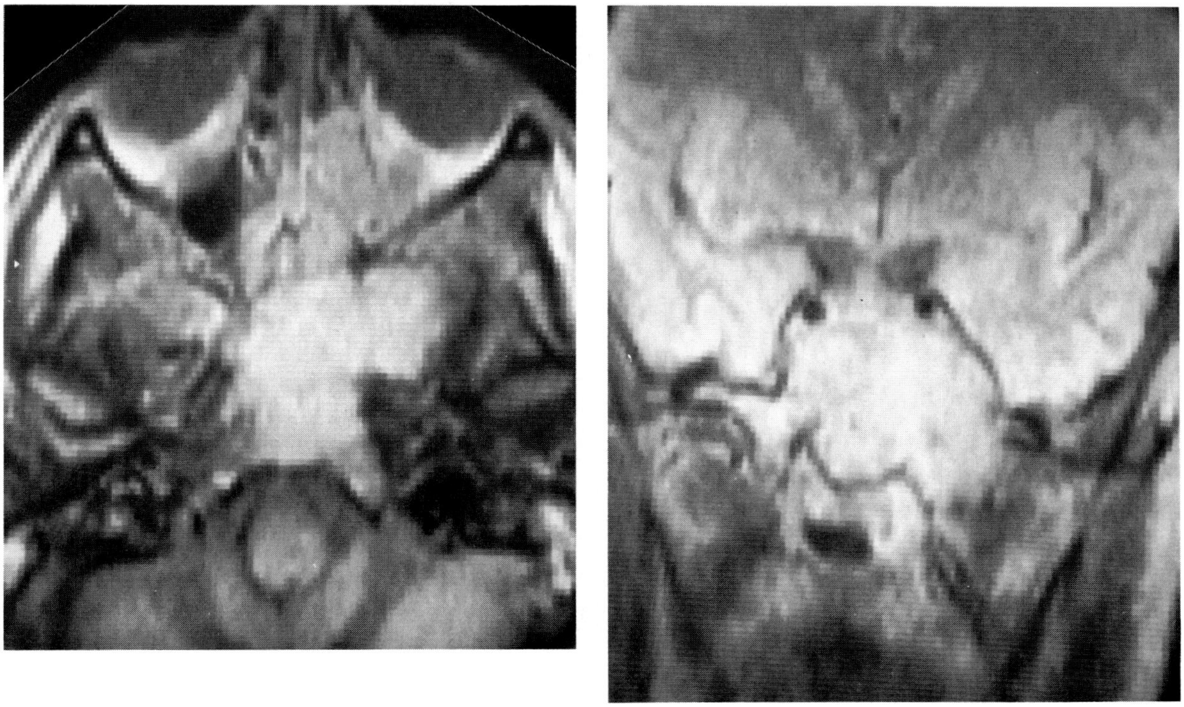

a b

Fig. 9.5a, b. Axial (**a**) and coronal (**b**) magnetic resonance scans showing a giant cell reparative granuloma of the sphenoid sinus. There is a generalised expansion of the sphenoid with encroachment on the middle fossa and high signal on T_2-weighted spin echo sequences.

References

Dodd GD, Bao-Shan Jing (1977) Radiology of the nose, paranasal sinuses and nasopharynx. Williams and Wilkins, Baltimore

Friedberg SA, Eisenstein R, Wallner LJ (1969) Giant cell lesions involving the nasal accessory sinuses. Laryngoscope 79:763–776

Friedmann I, Osborn DA (1982) Pathology of granulomas and neoplasms of the nose and paranasal sinuses. Churchill Livingstone, Edinburgh

Fu Y, Perzin KH (1974) Non-epithelial tumours of the nasal cavity, paranasal sinuses and nasopharynx. Cancer 33:1289–1305

Hlavacek V, Jolma VH (1974) Giant cell tumours of bone in ENT organs. Acta Otolaryngol (Stockh) 77:374–380

Jaffe HL (1953) Giant cell reparative granuloma, traumatic bone cyst and fibrous (fibro-osseous) dysplasia of the jaw bones. Oral Surg Oral Med Oral Pathol 6:159–175

Potter GD, McLennan BL (1970) Malignant giant cell tumour of the sphenoid bone and its differential diagnosis. Cancer 25: 167–170

10 Epithelial Tumours

Inverted Papilloma

Inverted papilloma is an uncommon epithelial tumour of the nose and paranasal sinuses which has been the subject of much interest and considerable debate since it was first described by Ringertz (1938). Although the earliest references to nasal papilloma were by Ward (1854) and Billroth (1855), Ringertz first drew attention to the "inverted" nature of the tumour and since then considerable controversy has surrounded the nomenclature, natural history and treatment of these lesions.

Inverted papillomata arise from the lateral wall of the nose in the region of the middle turbinate. They are almost always unilateral and produce a nasal mass with involvement of the maxillary antrum in the majority of patients at first presentation. The ethmoids are also frequently affected but involvement of the frontal and sphenoid sinuses is uncommon. On histological examination the lesions present a characteristic infolded appearance and are covered by alternating layers of columnar and squamous epithelium which has given rise to the alternative name of transitional papilloma. In 200 cases described by Friedmann and Osborn (1982) the patients' ages ranged from the second to the ninth decades, peaking in the sixth decade, with a 5 : 1 male predominance. Patients present clinically with nasal obstruction and discharge; the lesions are polypoid in appearance and may be mistaken both clinically and radiologically for an antrochoanal polyp.

Radiology and Imaging

The radiological features of inverted papilloma derive from its usual point of origin in the nasal cavity – the mucosa of the middle meatus or turbinate. Initially, when small and confined to the nose, the inverted papilloma is difficult to detect on plain radiographs, often producing non-specific features due to concomitant infection or allergic change in the sinuses. Alternatively the sinuses may show no abnormality. The nasal mass needs to be identified by pluri-directional tomography, CT scan or magnetic resonance. All three techniques will demonstrate the characteristic appearance of a mass continuous from the middle meatus into the antrum through a well-defined bone defect at the level of the antral ostium (Lund and Lloyd 1984) (Figs. 10.1, 10.2 and 10.3).

CT has advanced the possibilities of diagnosis in sinus pathology by its ability to detect minor degrees of increased density not readily demonstrable by conventional radiographic techniques. In addition to showing the extent of the soft tissue mass, areas of calcification may be demonstrated within the tumour (Fig. 10.4), accompanied in some patients by sclerotic bone shown either in the turbinates or in the sinus walls adjacent to the papilloma (Figs. 10.5, 10.6).

Inverted papillomata present two aspects of clinical importance: firstly, a strong tendency to recur after local removal, and secondly, occasional malignant change. Recurrence is common and may occur in more than 30% of cases (Friedmann and Osborn 1982), with some patients having multiple recur-

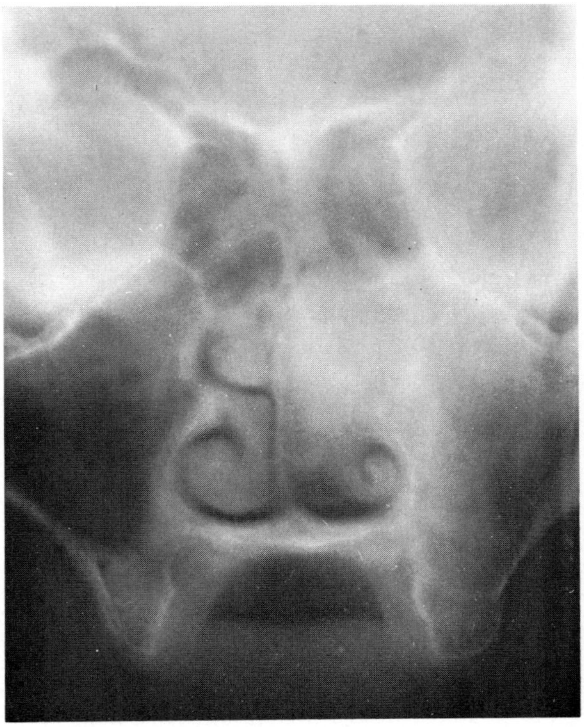

Fig. 10.1. Typical appearance of inverted papilloma shown on hypocycloidal tomography. There is a nasal mass occupying the middle meatus and extending into the adjacent maxillary antrum through a well-demarcated bone defect.

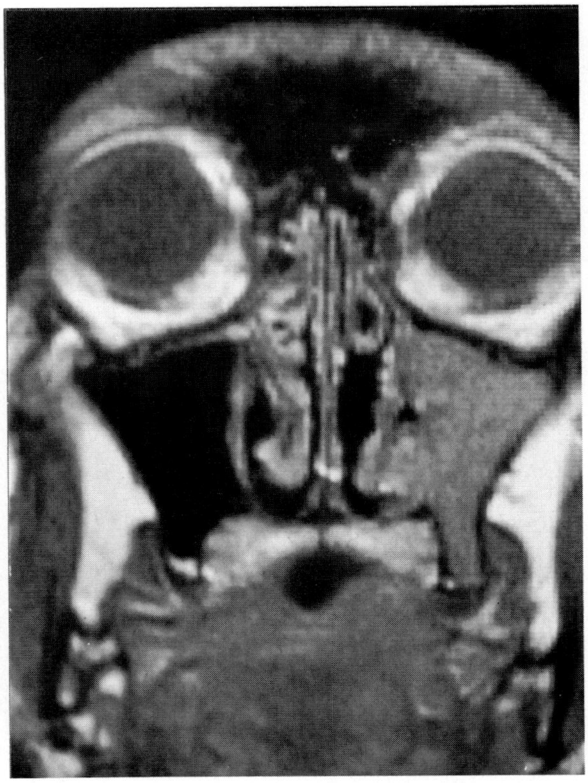

Fig. 10.3. Recurrent inverted papilloma demonstrated by coronal magnetic resonance scan.

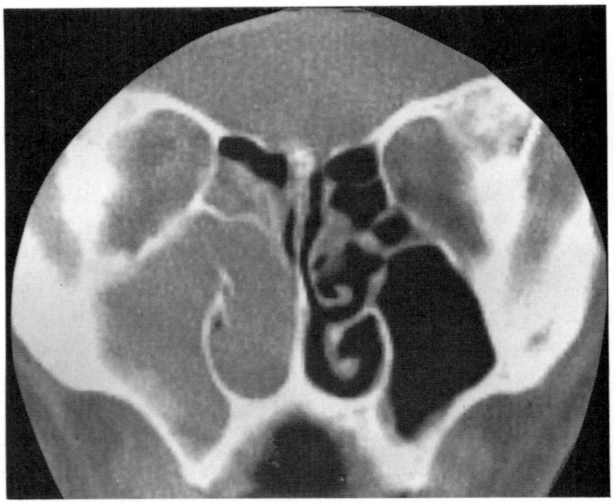

Fig. 10.2. Coronal CT scan showing an inverted papilloma occupying the right side of the nasal cavity and filling the maxillary antrum.

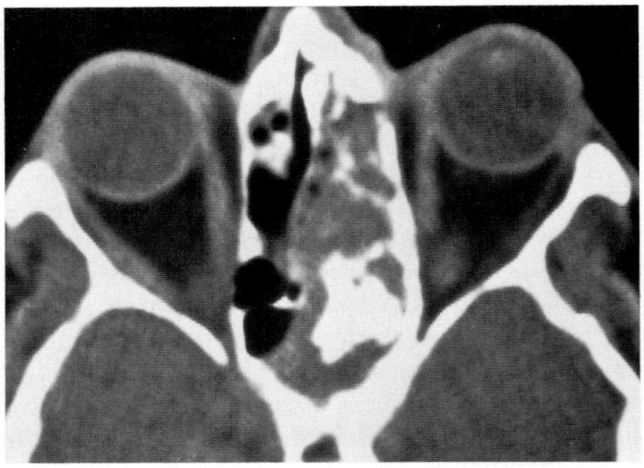

Fig. 10.4. Axial CT scan showing plaques of calcification in an inverted papilloma of the ethmoids.

rences over a period of years. Radiological assessment in cases of recurrent disease is especially difficult when there has been repeated resection of tissue, and a combination of CT and magnetic resonance studies is needed in these patients.

The problem of malignant change is somewhat confused by the simultaneous finding of inverted papilloma and carcinoma at initial and subsequent histological examination. Such an observation is strong circumstantial evidence of malignant change

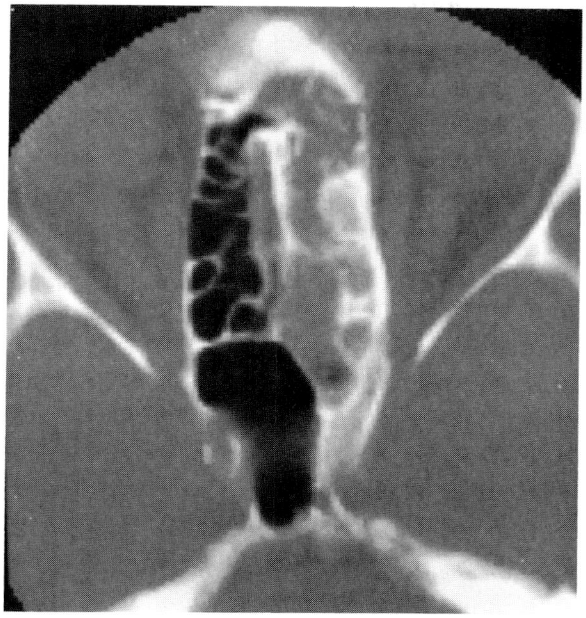

Fig. 10.5. Axial CT scan of an inverted papilloma involving the ethmoids and showing sclerosis of the sinus walls and septa.

but it is not proof. However, unequivocal evidence of malignant change provided by long-term observation and histological control has been recorded in the literature (Kramer and Som 1935; Ringertz 1938; Mabery et al. 1965; Worgan and Hooper 1970) and there is no doubt that malignancy does occur in a minority of patients (from 2% to 13% depending on the series). Proof of malignant change depends upon biopsy and at an early stage radiological differentiation is not usually possible. At a later stage, however, when the tumour becomes frankly invasive, it will show features which are indistinguishable from those of a squamous cell carcinoma or adenocarcinoma in which there is widespread bone erosion and invasion of adjacent areas such as the infratemporal fossa, orbit or anterior cranial fossa.

Hyams (1971), in a clinico-pathological study of 315 cases of papillomata, described thinning or erosion of the bony sinus wall by the papillomatous mass, but no definite invasion of bone unless there was associated malignancy. Other authors (Brown 1964) have stressed the distinction between pressure erosion by the benign papillomata and the development of bone destruction with the presence of malignant change.

a

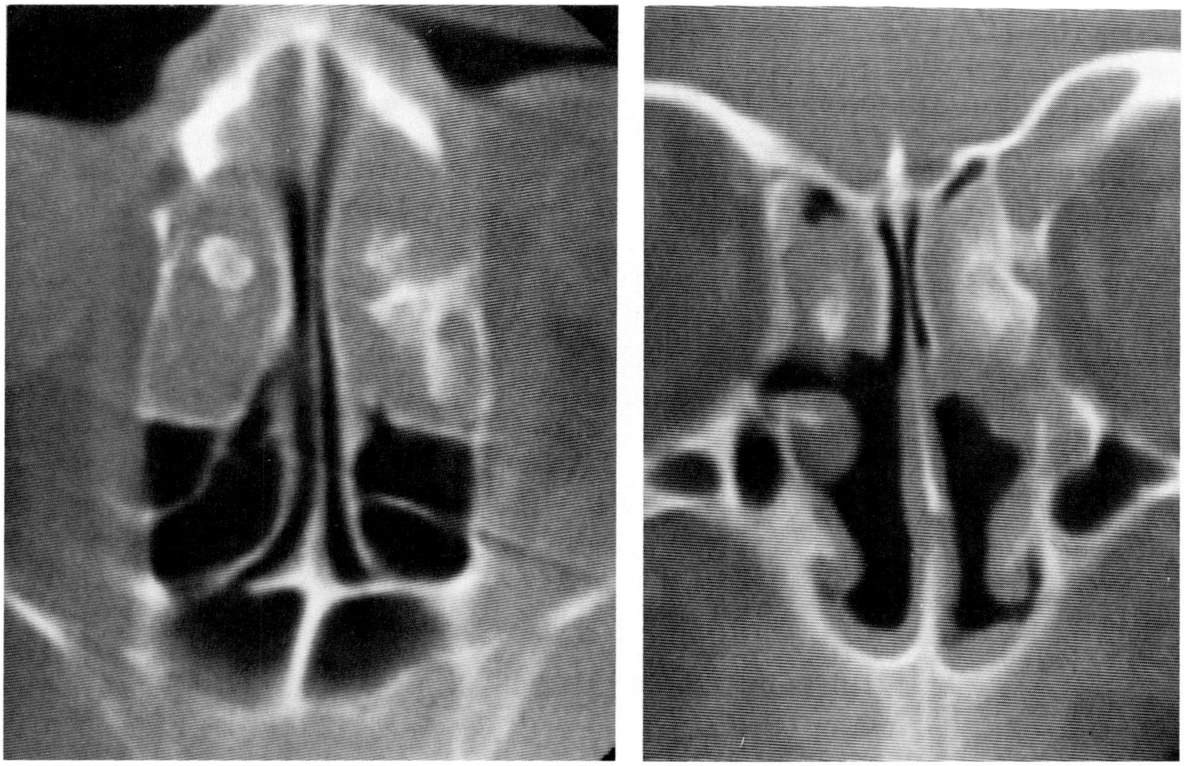

b

Fig. 10.6a, b. Axial (**a**) and coronal (**b**) CT scans of a recurrent inverted papilloma. There are changes in the anterior ethmoid cells similar to those seen in Fig. 10.5.

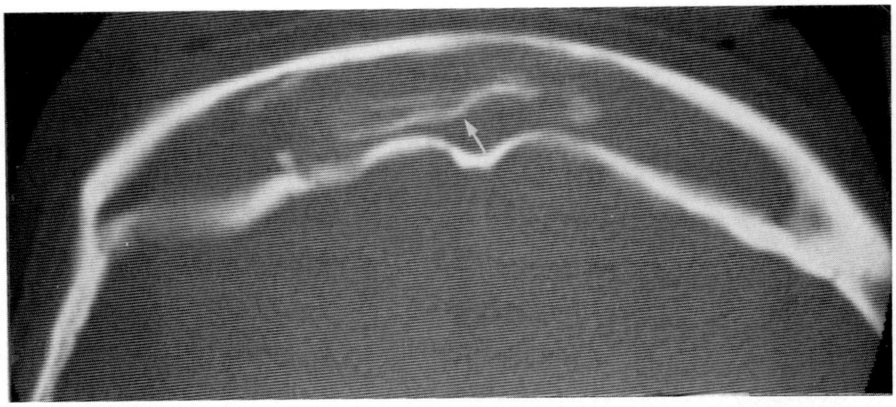

Fig. 10.7. Axial high-resolution CT scan of the frontal sinuses showing linear calcification (*arrow*) within the soft tissue mass of an inverted papilloma occupying the sinus cavity.

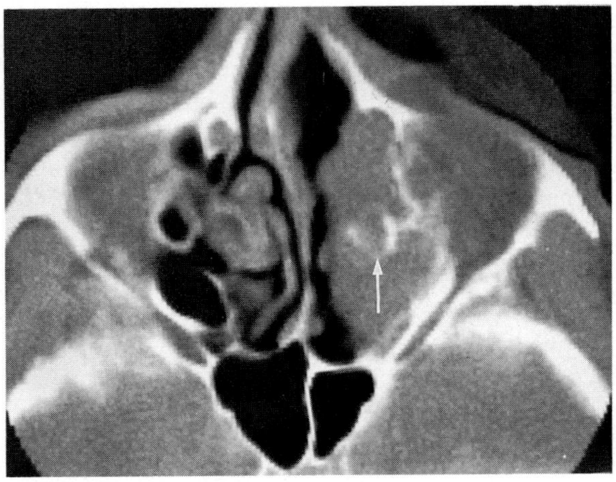

Fig. 10.8. Axial CT scan demonstrating ring-like calcification (*arrow*) in a recurrent inverted papilloma.

Differential Diagnosis

When a nasal mass is shown with complete opacity of the ipsilateral maxillary antrum and erosion of bone around the sinus opening, an inverted papilloma or antro-choanal polyp are the two most likely diagnoses. At a stage in the development of squamous cell carcinoma, adenocarcinoma or lymphoma, similar appearances may arise, although there is usually more widespread bone destruction on initial radiological examination of patients with primary sinus malignancy. Fungus infection must also be considered in the differential diagnosis and may present as an antro-nasal mass (see Chap. 8). Calcification is a common feature of aspergillosis but is most often diffuse and unlike the linear (Fig. 10.7), curvilinear or circular (Fig. 10.8)

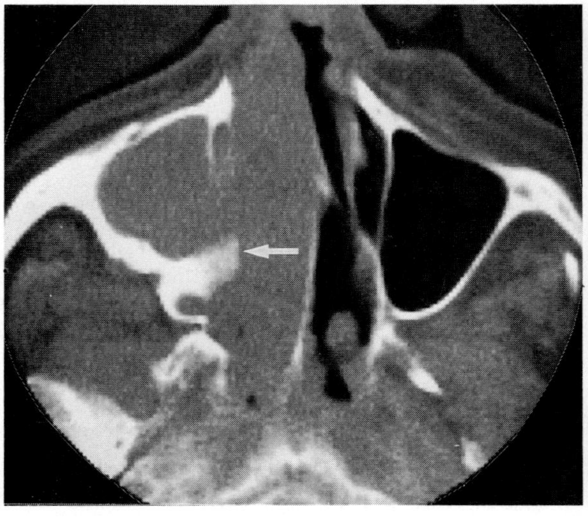

a

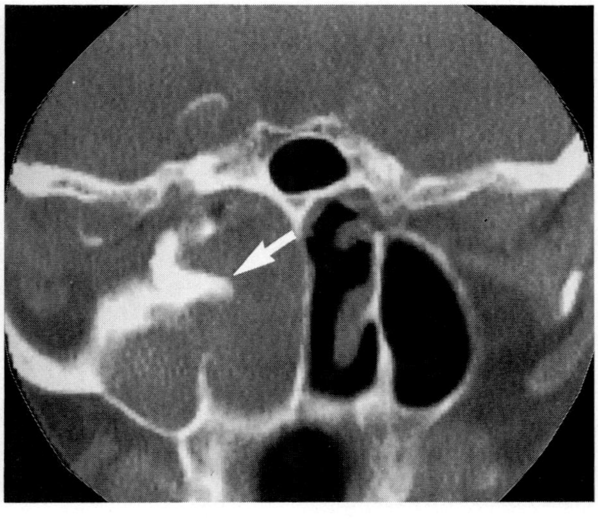

b

Fig. 10.9a,b. Axial (**a**) and coronal (**b**) CT scans showing an extensive inverted papilloma associated with sclerosis and deformation of the posterior maxillary antrum (*arrows*).

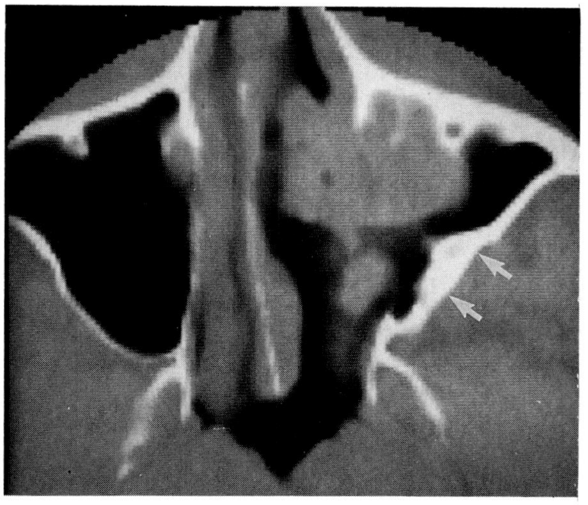

Fig. 10.10. Axial CT scan showing the soft tissue mass of a recurrent inverted papilloma with a deformed and sclerotic posterior wall of the maxillary antrum (*arrows*).

calcification seen in inverted papilloma. Sclerosis of the sinus walls is a non-specific change most frequently seen in chronic sinus infection. However, in these patients the new bone formation is distributed evenly along the sinus walls without deformation. The combination of bone deformity and sclerosis suggests a slow-growing tumour such as an inverted papilloma (Figs. 10.9, 10.10).

Malignant Epithelial Tumours

Squamous Cell Carcinoma

Squamous cell carcinoma is the most common malignant epithelial tumour of the nose and paranasal sinuses, comprising 63% of nasal and 44% of paranasal carcinomata. Most arise in the nose or maxillary antrum, but when first seen the tumours are mostly naso-antro-ethmoidal in distribution (see Fig. 10.17). The majority of patients are male, the male to female ratio being approximately 2 : 1; and most patients are over 65 years of age at the time of onset of the disease. An occupational basis for the origin of this disease has been established in the nickel industry, but this would appear to be the only known aetiological agent. Histologically, as in squamous carcinoma in other locations, there is a wide range of cellular differentiation with varying degrees of keratin production, but in 80%–85% of cases the tumour is well to moderately well differentiated (Michaels 1987).

When the patient presents the tumour is usually widespread in the nose and paranasal sinuses, but occasionally a more localised neoplasm may be seen, and if largely confined to the nasal cavity nasal obstruction may draw early attention to it. However, in carcinoma confined to the maxillary antrum early symptoms of pain and swelling of the cheek may be first diagnosed as sinusitis, and it is not until dental symptoms or those related to orbital spread become apparent that the true nature of the disease is realised. Proptosis is in fact the commonest sign of sinus malignancy after nasal obstruction and epistaxis (Shaw 1964).

Diagnostic imaging has a threefold part to play in the management of these patients: at an early stage in the development of sinus malignancy the tumour may be inaccessible to clinical methods of examination and imaging studies may demonstrate the lesion comparatively early when surgery or radiotherapy have a reasonable chance of success; it may help to form a prognosis by establishing the site of the tumour; and above all it is the optimum means of determining the extent of disease prior to treatment. In this respect the role of CT and magnetic resonance is especially important: a combination of these techniques will show the best area for biopsy of the tumour, will demonstrate the presence of bone destruction and will delineate the precise outlines of the mass and its relationship to vital structures.

Plain Radiographic Changes

In most patients both the antrum and ethmoids are already involved at the time of first examination and will show complete loss of translucence on plain radiographs. Bone destruction in the sinus walls may be demonstrable either by direct radiography or by tomography, and these changes are frequently associated with a soft tissue mass in the nasal cavity with bowing of the nasal septum in a convexity away from the side of the tumour (Fig. 10.11). Destruction of bone in situ is characteristic of malignant tumours as opposed to the simple expansion of the sinus cavity seen with benign tumours and mucocoeles. In the antrum the medial wall is most frequently involved by squamous cell carcinoma, but is difficult to show on plain radiographs due to overlap of other structures. Destruction of bone in the roof of the antrum, in the lateral recess and in the alveolar recess is usually identifiable on plain radiographs.

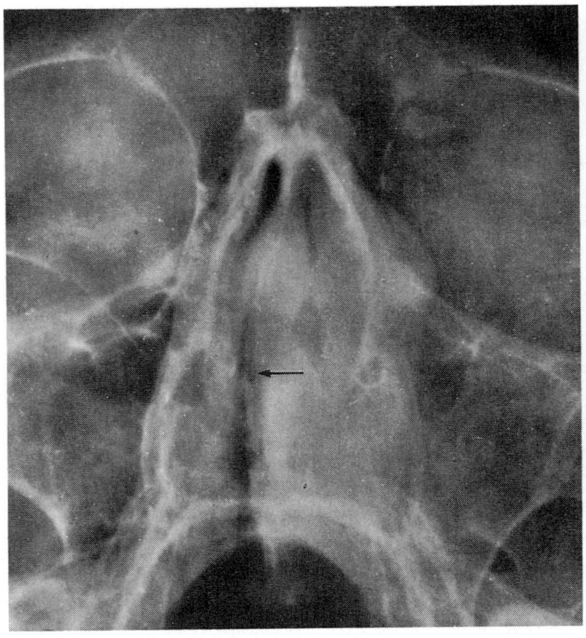

Fig. 10.11. Typical appearance on plain radiography of an advanced squamous cell carcinoma. There is a mass in the left nasal cavity, ethmoids and antrum and displacement of the nasal septum by the tumour (*arrow*).

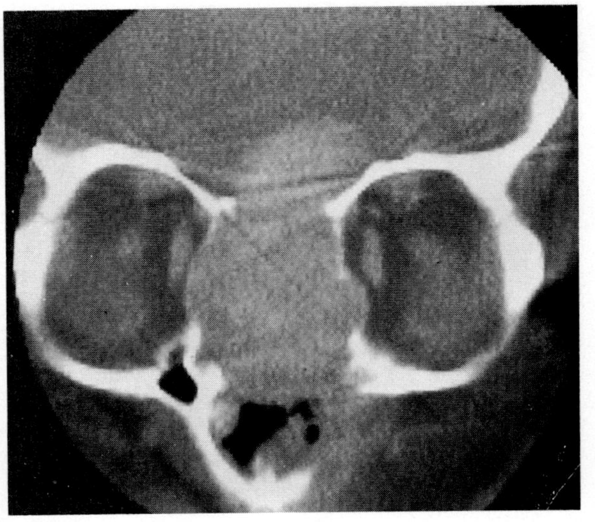

Fig. 10.12. Coronal CT scan following intravenous contrast medium, showing a carcinoma invading the anterior fossa.

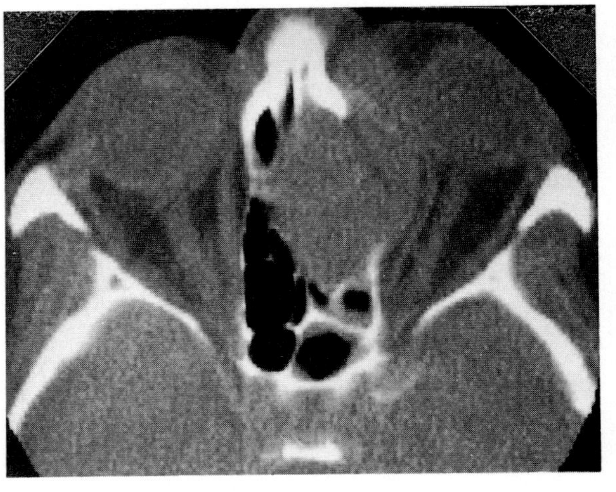

Fig. 10.13. Axial CT scan showing invasion of the orbit from an ethmoid carcinoma. The tumour is simply displacing the structures rather than infiltrating them. This is by far the most common type of orbital invasion seen in both benign and malignant sinus tumours.

Magnetic Resonance and CT

Magnetic resonance and CT are of particular value in assessing several anatomical sites:

1. Extension of disease through the cribriform plate area into the anterior cranial fossa is now best demonstrated using sagittal magnetic resonance scans following intravenous gadolinium DTPA. It is also well demonstrated by direct coronal CT sections following intravenous contrast medium. When a tumour of the paranasal sinuses has invaded the anterior or middle cranial fossa it will normally enhance against the non-enhancing brain substance (Fig. 10.12).

2. Extension of disease into the soft tissues of the orbit is also optimally demonstrated by these techniques. There are two varieties of orbital involvement which may be demonstrated. The first is more common and is seen in both benign and malignant sinus tumours: in this the orbital contents are displaced laterally without direct invasion, the periorbita acting as a barrier and resisting tumour infiltration for a long time (Fig. 10.13). The second less common type of orbital invasion is a direct infiltration of the orbital structures by tumour so that the normal anatomical landmarks are obliterated (Fig. 10.14). In both varieties coronal scanning is mandatory to show the presence and degree of orbital extension.

3. Evidence of invasion of the pterygo-palatine fossa and infratemporal fossa can also be demonstrated by these techniques. Invasion of the posterior wall of the maxillary antrum is particularly well shown on high-resolution CT. It is possible to show simple expansion of the posterior wall; expansion with early erosion; or direct invasion of the soft tissues and musculature (Figs. 10.15, 10.16). Occasionally a squamous cell carcinoma may show areas of punctate calcification on CT.

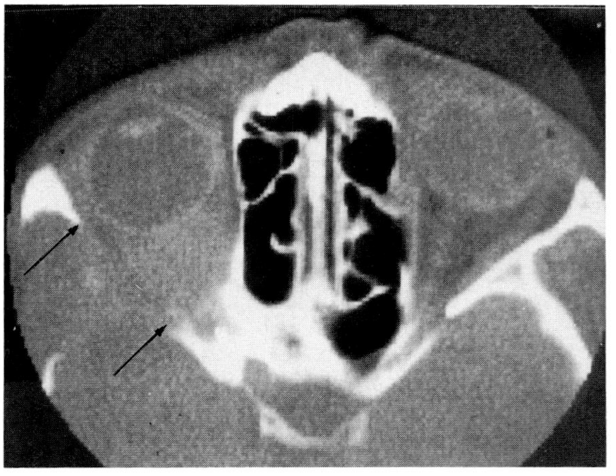

Fig. 10.14. Axial CT scan of a recurrent squamous cell carcinoma after maxillectomy. The tumour has infiltrated the whole orbit, totally obliterated the normal retrobulbar structures, and destroyed the bone of the lateral orbital wall (*arrows*).

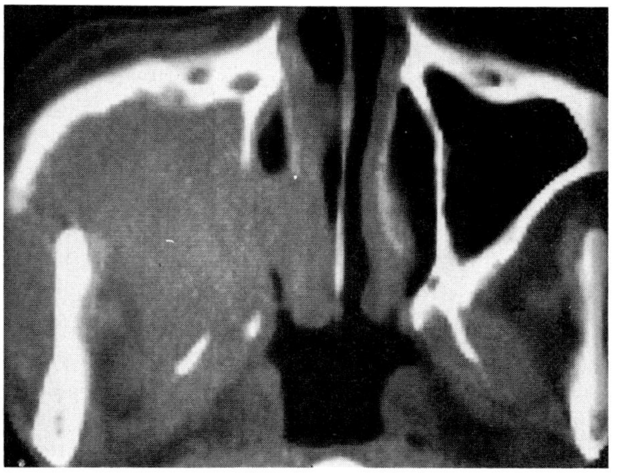

Fig. 10.16. Axial CT scan showing massive invasion of the infratemporal fossa by squamous cell carcinoma.

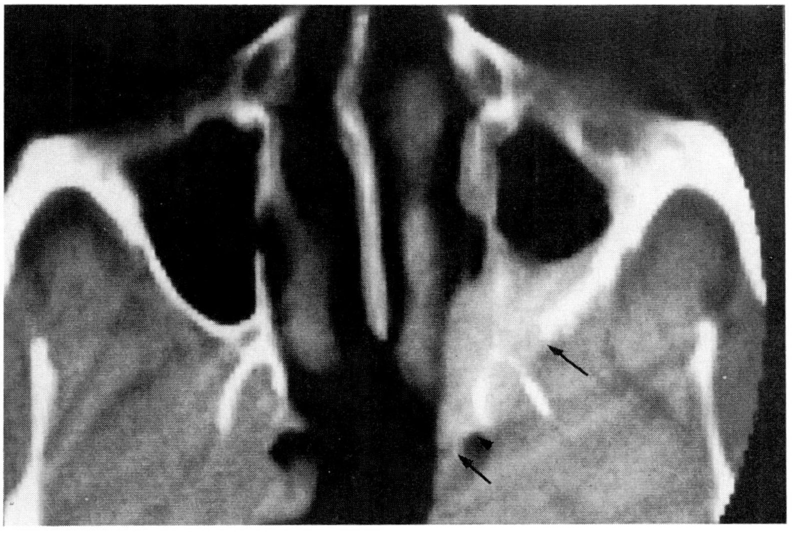

Fig. 10.15. Post-contrast axial CT scan showing an undifferentiated carcinoma invading the pterygo-palatine fossa and nasopharynx (*arrows*).

Adenocarcinoma

Adenocarcinoma is predominantly adenomatous in structure and arises from the glands in the nasal and paranasal mucous membrane. It is not common in this situation. Ringertz (1938) found 10 adenocarcinomata out of 31 glandular tumours involving the nose and paranasal sinuses, representing less than 4% of all carcinomata in this region. This author described two histological types: an alveolar variety, which he regarded as the malignant counterpart of adenoma, showing large cystic spaces with abundant mucus secretion; and a pseudopapillary form resembling adenocarcinoma of the colon in structure and cell type. Other authors have noted the resemblance of some high-grade adenocarcinomata of the nose and sinuses to adenocarcinoma of the colon (Michaels 1987). In some cases there is abundant mucus production and the tumour may then resemble colloid carcinoma of the colon. The age distribution shows a peak occurrence in the sixth and seventh decades with a marked predominance of males.

Patients suffering from this condition may present with nasal obstruction and epistaxis, or pain and local swelling when the tumour becomes large.

Ultimate spread to the anterior fossa is well recorded in the literature and reflects the site of origin of these tumours – which, the majority of authors agree, is high up in the nasal cavity and adjacent ethmoid cells (Ringertz 1938; Batsakis 1970).

Much of the interest in this type of carcinoma has centred on its relation to occupation. Hadfield and her colleagues (Macbeth 1965; Acheson et al. 1967; Hadfield 1970; Hadfield and Macbeth 1971) have demonstrated the prevalence of this disease in woodworkers employed in the furniture industry and it has been established that hardwood dusts are the causative agent. Hadfield has expressed the belief that woodworker's carcinoma probably arises primarily in the middle turbinate, on the anterior end of which the wood dust is found to be deposited. This site of origin would correlate well with the known distribution of this tumour in the nose and sinuses (see Fig. 10.17).

Radiology and Imaging

Thirty-two patients with sinus adenocarcinomata have been investigated: 29 by CT scan, 3 of whom also had magnetic resonance, and the remainder by conventional radiographic studies.

Approximately 37% of these tumours were antro-ethmoidal in location and did not show changes in any way different from those described for squamous cell carcinoma. However a greater number (63%) showed a different location, being principally naso-ethmoidal, the tumour occupying the upper part of the nasal cavity and the adjacent ethmoid cells. This is the common location reported by other authors (see above). The distribution of this tumour is illustrated graphically in Fig. 10.17. From this point of origin the adenocarcinoma will cause erosion of the anterior ethmoid cells with extension to the orbit and cribriform plate and early invasion of the anterior fossa. For this reason these patients need some form of cranio-facial surgery for the disease to be extirpated.

On CT scan the location of the tumour is the most suggestive feature. The majority of this type of carcinoma enhance well with intravenous contrast. Four showed calcification, a feature well recognised in colloid adenocarcinoma of the large bowel with similar histology (Figs. 10.18, 10.19). In some patients a sclerotic reaction was observed in the walls of the affected sinus cavity, but this was usually of low intensity and not distinctive. In three patients the tumour clearly took origin in the frontal sinus – a rare location for any malignant tumour in the sinuses.

The changes shown on magnetic resonance scans were also non-specific. Generally the tumours gave moderately high signal on T_2-weighted spin echo sequences, with inhomogeneous features within the tumour but clearly demarcated contours (Fig. 10.20), the tumour being easy to distinguish from the higher-intensity signal of retained secretion. Failure to demonstrate calcification was a drawback of magnetic resonance: to show this and any associ-

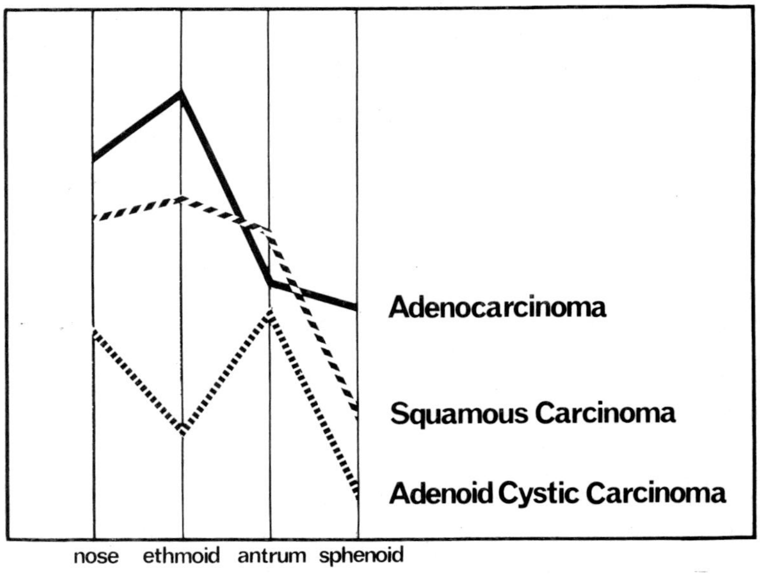

Fig. 10.17. The distribution of three common types of carcinoma seen in the paranasal sinuses, derived from 73 patients examined by CT and magnetic resonance. Note that the adenocarcinomata are predominantly naso-ethmoidal in location, while the distribution graph of adenoid cystic carcinoma is virtually the converse of that for adenocarcinoma.

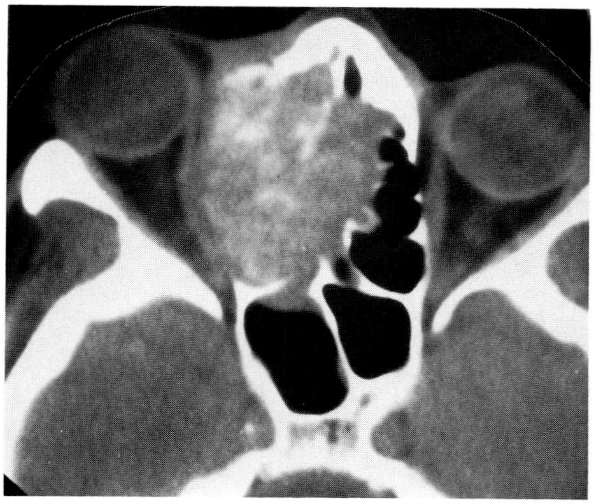

Fig. 10.18. Woodworker's adenocarcinoma. Axial CT scan showing tumour mass with calcification.

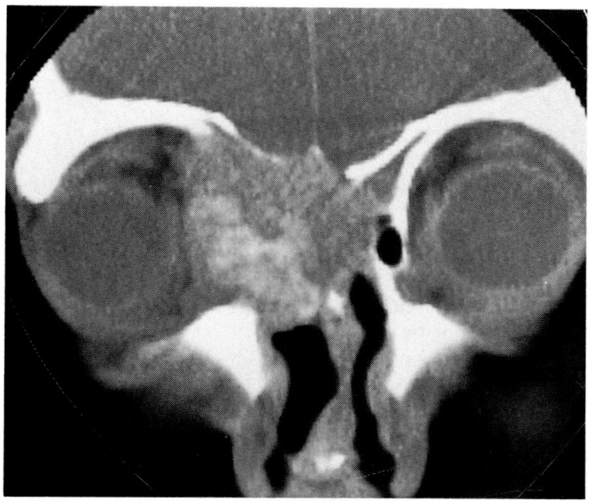

Fig. 10.19. Same patient as Fig. 10.18. Coronal CT scan showing calcification in the adenocarcinoma. Note the early invasion of the anterior fossa.

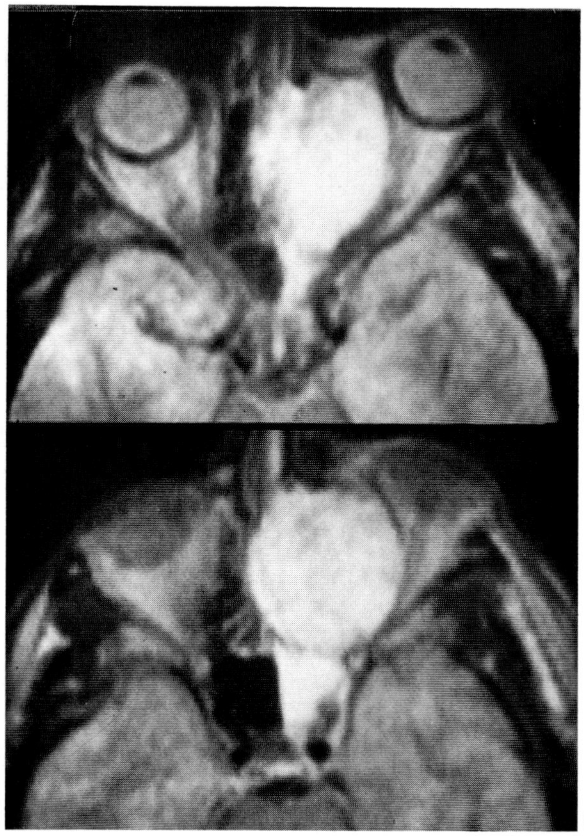

a

b

Fig. 10.20a,b. Magnetic resonance scans (T_2-weighted spin echo sequence) showing a well-demarcated adenocarcinoma of the ethmoids invading the orbit.

ated bone destruction, the magnetic resonance studies need to be augmented by CT (Figs. 10.21, 10.22). However, it should be emphasised that the essential pre-operative requirement is to show the extent of intracranial invasion above the cribriform plate, and this is best achieved by sagittal magnetic resonance with or without an intravenous paramagnetic contrast agent (Fig. 10.23).

Adenoid Cystic Carcinoma

Adenoid cystic carcinoma is a malignant tumour derived from salivary type sero-mucinous glands in the nose and paranasal sinuses. It was first described by Billroth (1859), who gave it the name of "cylindroma"; since that time it has become apparent that this neoplasm has a relentless tendency to recurrence over many years, leading eventually to the death of the patient. The term adenoid cystic carcinoma has been generally adopted and better fits the histological appearance and behaviour of the tumour.

The age incidence is broadly based with a peak distribution in the fourth decade; there is a slight preponderance of female patients. As in other malignant tumours nasal obstruction, epistaxis with facial pain and swelling are the common forms of presentation. To these may be added paraesthesia and anaesthesia in the distribution of branches of the fifth nerve, particularly the infraorbital branch of the second division, due to perineural infiltration by tumour: this is the most constant form of spread of this neoplasm. The tumour may also spread to lymph nodes, and blood stream metastases occur in nearly 40% of cases (Spiro et al. 1973). The tumour has a variable histology but the essential feature is

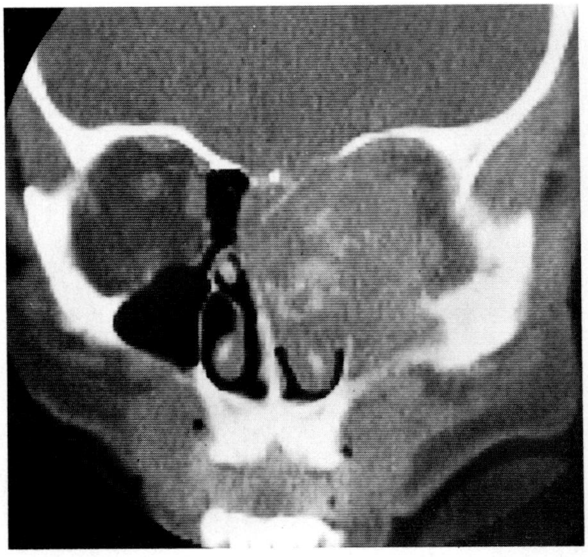

Fig. 10.21. Coronal CT scan showing calcification in an adenocarcinoma.

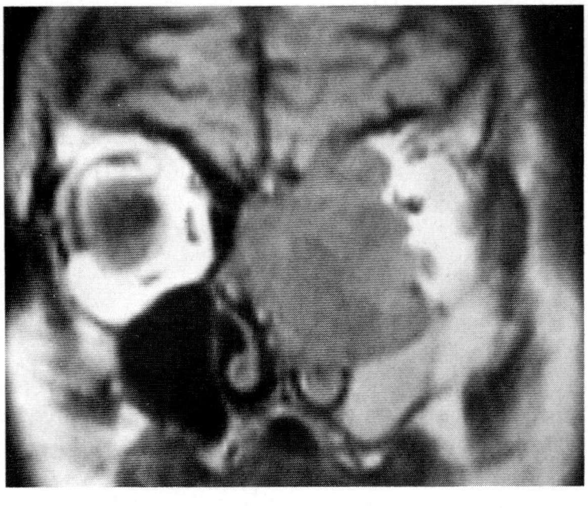

Fig. 10.22. Coronal magnetic resonance scan (T_1-weighted spin echo sequence) of the same patient as Fig. 10.21, showing orbital invasion but no evidence of the calcification seen on CT.

the cribriform or "sieve-like" pattern of the cells without which the diagnosis cannot be established (Friedmann and Osborn 1982). The cribriform pattern is produced by lumina or holes within the cell structure which contain connective tissue mucin or epithelial mucin.

Radiology and Imaging

Twenty-two patients have been investigated with this condition: 17 by CT, 1 by magnetic resonance,

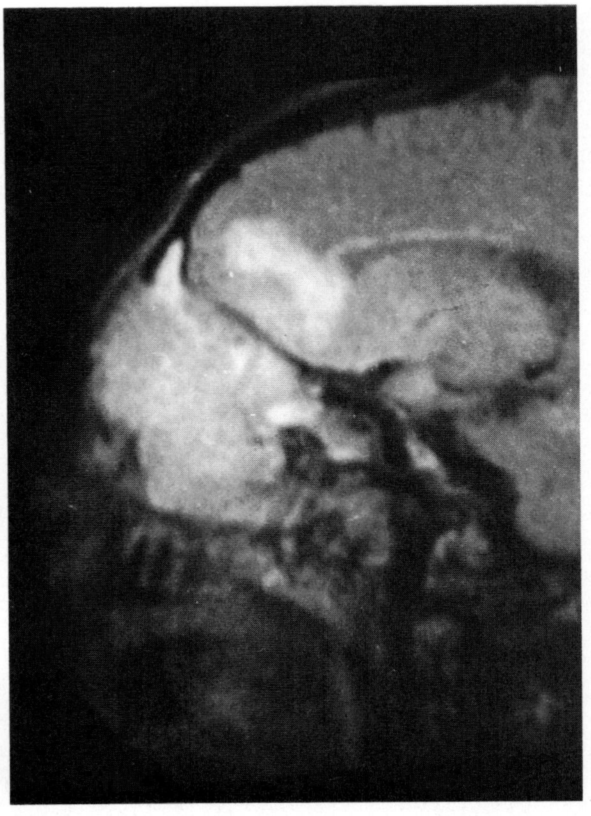

Fig. 10.23. Adenocarcinoma shown on sagittal magnetic resonance scan. Extension of the tumour into the anterior fossa is clearly demonstrated. The high signal seen in the area immediately above the tumour was shown at surgery to be due to necrotic brain tissue.

and the remainder by conventional radiography. Adenoid cystic carcinoma in the sinuses may present as a bulk tumour (Fig. 10.24) occupying an antro-ethmoidal location and may show the usual features of malignancy in the sinuses with bone destruction on plain radiography and CT. However a high proportion do not present a large mass but exhibit a more infiltrative process: they tend to infiltrate tissue planes (Fig. 10.25) and advance along nerve sheaths. Perineural infiltration may show on plain radiographs as an enlargement of the infraorbital canal and local destruction of the floor of the orbit, from an adenoid cystic tumour in the maxillary antrum. In more advanced lesions sequential enlargement of the foramen rotundum and foramen ovale may occur. With the new soft tissue imaging techniques it may be possible to see the nerve enlargement as such rather than its secondary effect on the adjacent bone. The perineural infiltration may be remarkably well demonstrated (Figs. 10.26, 10.27).

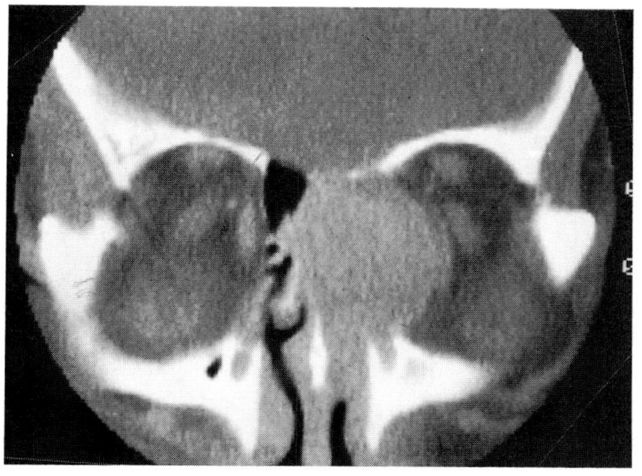

Fig. 10.24. Coronal CT scan of an adenoid cystic carcinoma shown as a rounded tumour mass in the ethmoids.

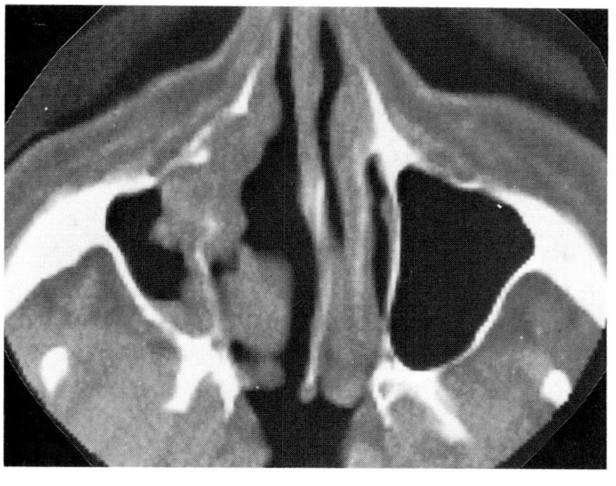

Fig. 10.25. Axial CT scan of an adenoid cystic carcinoma of the right nasal cavity and maxillary antrum. There is infiltration of the mucosa and bone destruction without a large soft tissue mass.

The distribution of these tumours is shown in Fig. 10.17. It can be seen that the shape of the distribution graph for adenoid cystic carcinoma is almost the converse of that for adenocarcinoma and indicates that the majority of these tumours take origin in the lower part of the nasal cavity and maxillary antrum. In the patients investigated by CT the hard palate was shown to be involved in over 40% at first examination, providing supporting evidence for this site of origin.

Ameloblastoma

Ameloblastoma is an uncommon epithelial tumour which may occur in the maxilla and invade the sinuses by direct extension. Malassez (1885) recog-

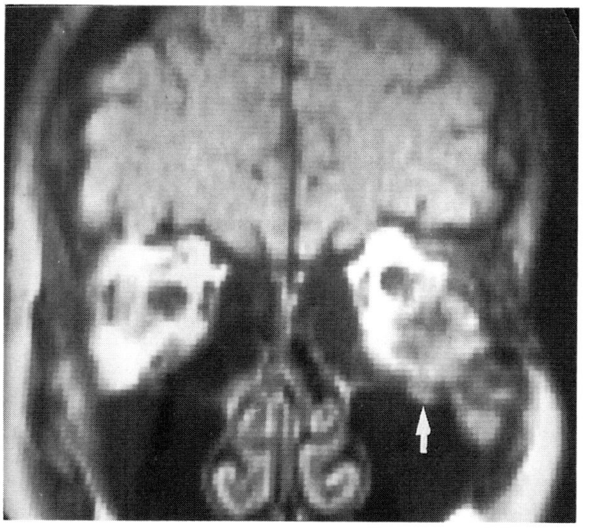

Fig. 10.26. Magnetic resonance scan (T_1-weighted spin echo sequence) of a recurrent adenoid cystic carcinoma of the lacrimal gland. The tumour has invaded the floor of the orbit and the upper part of the maxillary antrum, and there is perineural infiltration and enlargement of the infraorbital nerve (*arrow*).

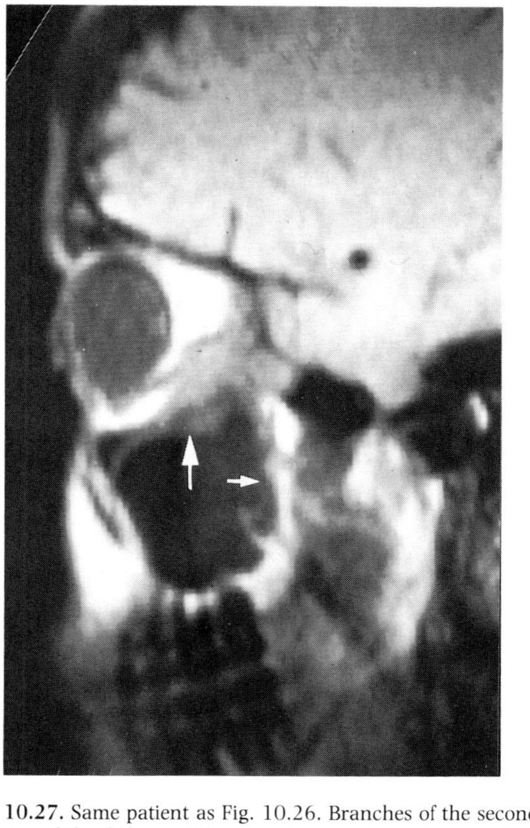

Fig. 10.27. Same patient as Fig. 10.26. Branches of the second division of the fifth cranial nerve show perineural infiltration. The *upper arrow* indicates the infraorbital nerve, the *lower arrow* the posterior superior dental nerve.

nised the origin of this tumour from dental epithelium and was responsible for the introduction of the name adamantinoma. This was, however, a misnomer since enamel is not produced by the tumour and Churchill (1934) gave it the more appropriate title of ameloblastoma. The neoplasm may arise from the remnants of the dental lamina and the enamel organ, the basal layer of the oral mucous membrane, or the epithelial lining of a dentigerous cyst. The tumours are slow-growing and are generally considered to be benign, although they are locally invasive and metastasise on rare occasions.

The radiological features of 16 cases of ameloblastoma have been described by McIvor (1974). In five of his patients a non-specific unilocular radiolucent area was present, in the bone of the alveolar margin or mandible. The remainder of the lesions also showed as a radiolucent area but in combination with at least three of the following additional features: expansion of the overlying cortical plate; a corticated scalloped margin; a multilocular appearance; and resorption of adjacent tooth roots.

When it invades the maxillary antrum the tumour presents either as a dentigerous cyst or more commonly as a solid tumour filling the sinus cavity, expanding and eroding bone (Fig. 10.28). All or part of a tooth may be embodied in the tumour.

In three patients with sinus involvement seen by the author, two had extensive recurrence of the tumour, which spread to other paranasal sinuses from the maxillary antrum (Fig. 10.29).

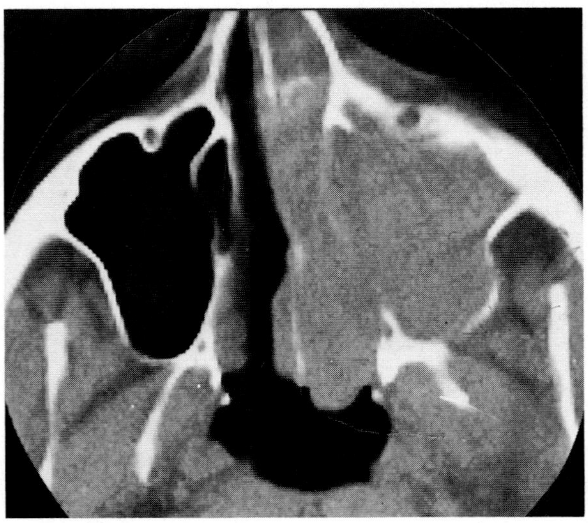

Fig. 10.28. Axial CT scan of an ameloblastoma of the nose and maxillary antrum.

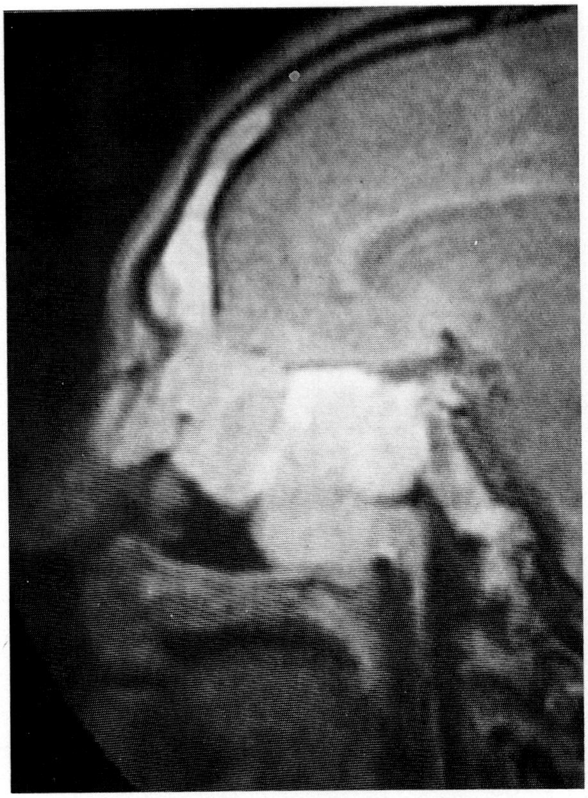

Fig. 10.29. T_2-weighted sagittal magnetic resonance scan showing extensive recurrence of an ameloblastoma in the nose and sinuses.

Malignant Melanoma

Malignant melanoma is relatively rare compared with other malignant neoplasms of the nose and sinuses. Holdcraft and Gallagher (1969) in a review of 1029 neoplasms of the mucosa of the nose and paranasal sinuses found 39 melanomata, representing 3.8% of all malignant tumours in this situation. Melanomata can arise de novo in mucosal tissues which normally are non-pigmented. The patient, after a period in which there may be nasal obstruction and epistaxis, eventually presents with a darkly pigmented or fleshy nasal mass. The majority of tumours arise in the nasal cavity; a minority arise in the sinuses, predominantly the maxillary antrum. Within the nasal cavity the common sites of origin are the septum, the lateral wall and the inferior turbinate. The tumours are probably derived from melanocytes in the nasal epithelium and microscopically the common pattern is the presence of sheets of spheroidal cells and spindle cells with variable degrees of pigmentation.

There is evidence of a constant risk of death from melanoma, no matter how long after treatment the patient lives (Lund 1982). This is unlike the situation with most other malignant tumours where survival for 1 or 2 years reduces the risk of death substantially. This record of survival would imply that in almost all patients the disease has disseminated by the time of initial diagnosis (Harrison 1976), and that the secondaries are controlled immunologically throughout the body. Thus the host–tumour balance is crucial to the final outcome (Lund 1982).

Radiology

In the author's series 25% of cases showed no evidence to suggest the presence of a neoplasm in the sinuses on the initial plain radiographic examination, while another 25% showed a mass in the nasal cavity without other significant abnormality. In 50% of patients bone destruction was recognised at initial examination, usually at the antro-ethmoidal junction and associated with a large antro-nasal mass. These tumours may spread quickly, with a rapid transformation of the radiographic appearance over a short period of observation. Recurrence after surgery is also a common feature.

CT Findings

Eight patients have been examined by CT, seven of whom had a primary melanoma in the nose and sinuses, and one who had a naso-ethmoidal metastasis from a skin melanoma excised previously. The average age of these patients was 56 years and the age range 34–70 years. The tumours were predominantly naso-antro-ethmoidal in location and showed no characteristic attenuation features on CT. The nasal mass was the most prominent element supporting the view that these tumours take origin in the nasal cavity. Associated bone destruction was present in all, involving the maxillary antrum (Fig. 10.30), the ethmoids (Fig. 10.31) or the cribriform plate area (Fig. 10.32). The orbit was invaded in 50% of patients at initial examination (Fig. 10.33), and the overall appearance that of a highly malignant neoplasm showing relentless spread and recurrence.

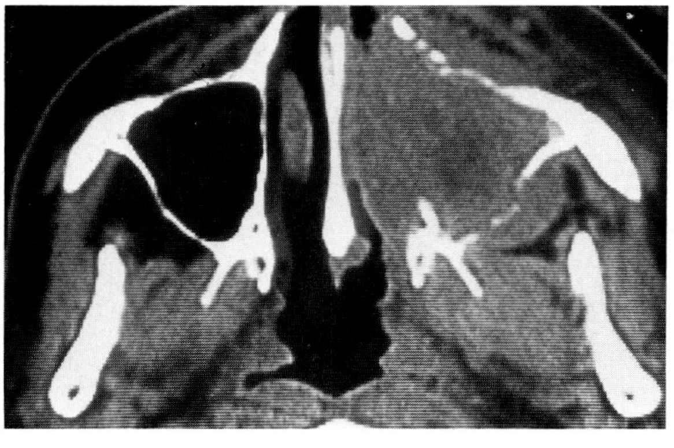

Fig. 10.30. Axial CT scan of a malignant melanoma of the nose and antrum with extension into the infratemporal fossa.

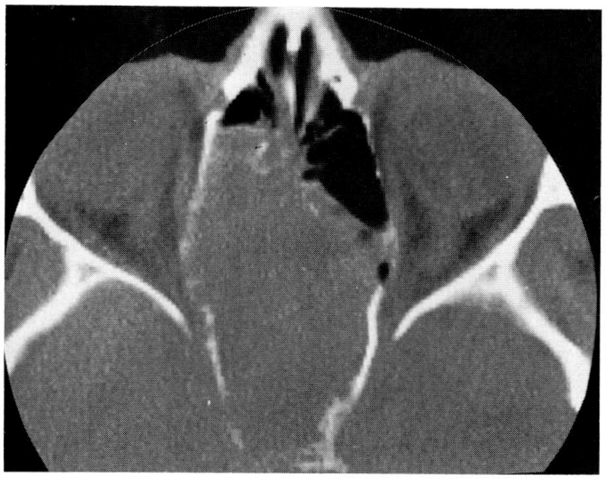

Fig. 10.31. Axial CT scan of a malignant melanoma showing extensive bone destruction in the posterior ethmoid cells. This was a metastasis from a primary skin melanoma which had been removed previously from the scapular region.

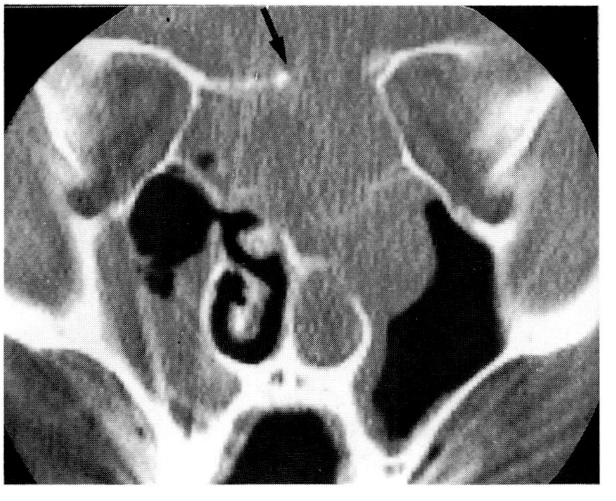

Fig. 10.32. Coronal CT scan of a malignant melanoma invading the floor of the anterior cranial fossa (*arrow*).

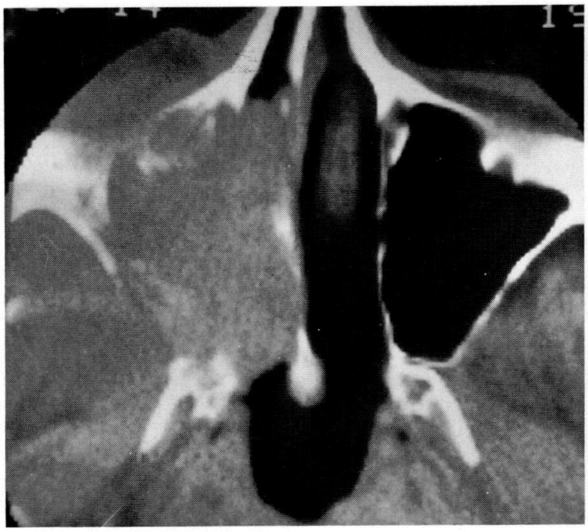

a

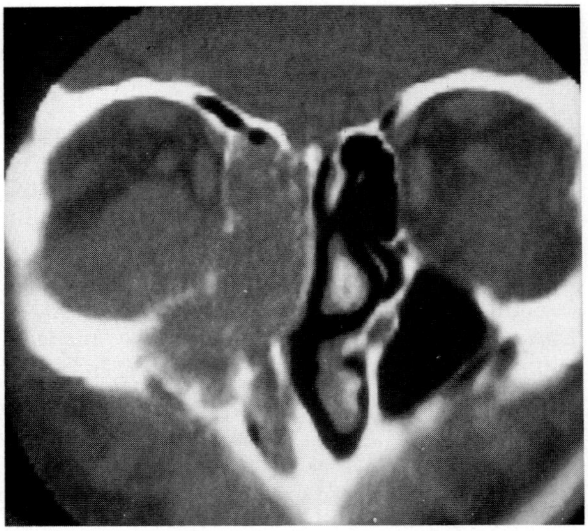

b

Fig. 10.33a,b. Axial (**a**) and coronal (**b**) CT scans showing an extensive antro-nasal melanoma invading the orbit.

Metastatic Carcinoma

Metastases in the nose and paranasal sinuses are very rare and for this reason may be overlooked by the clinician and also by the histopathologist, since microscopically they may closely resemble some forms of primary carcinoma. Almost all are derived from primary tumours of the genito-urinary system, with a strong predominance of renal carcinoma. In a review by Friedmann and Osborn (1965) it was noted that nearly 50% of secondary tumours in the ENT region involved the nose and sinuses and that of these 80% were derived from renal adenocarcinoma. Other primary tumours include seminoma of the testis (Garrett 1959) and uterine chorio-carcinoma (Salimi 1977; Mukherjee 1978). Examples of metastases from carcinoma of the breast, stomach and lung have also been recorded.

An explanation for the predominance of metastases from renal carcinoma in the nose and sinuses has been given by Nahum and Bailey (1963). It is well known that this tumour is prone to grow into renal veins, and these authors quote the work of Batson (1940), who demonstrated that, during periods when the intrathoracic pressure is greatly increased, a retrograde venous flow can occur through the prevertebral and vertebral venous plexus. This retrograde flow may at times progress to the base of the skull and may also be observed in the jugular venous system. With the rich venous anastomoses near the paranasal sinuses – notably the pterygoid plexus – metastases might find their way from the kidney by this route.

The marked vascularity of metastases from renal carcinoma may result in a clinical presentation with severe epistaxis (Eneroth et al. 1961), and in some patients the primary tumour in the kidney may remain occult until the patient presents with a metastasis (Harrison et al. 1964). Like primary tumours in the sinuses, secondaries may also present to the ophthalmologist with proptosis (Fig. 10.34). Other examples of sinus metastases encountered include examples of secondary carcinoma of the pancreas (Figs. 10.35, 10.36) and secondary malignant melanoma (Fig. 10.31). Sinus metastases should be suspected when the histological appearance of a biopsy of a neoplasm does not resemble the usual pattern of primary malignant tumours of the sino-nasal tract (Michaels 1987). In these circumstances an intravenous urogram becomes essential to exclude a primary renal neoplasm.

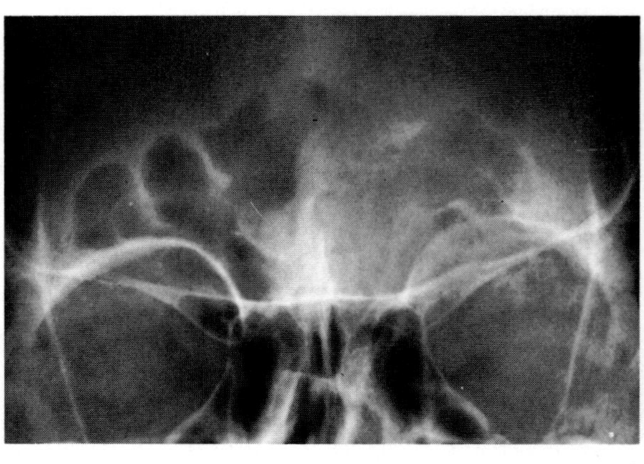

Fig. 10.34. Metastasis from renal carcinoma presenting as a simple opacity of the left frontal sinus on plain radiography.

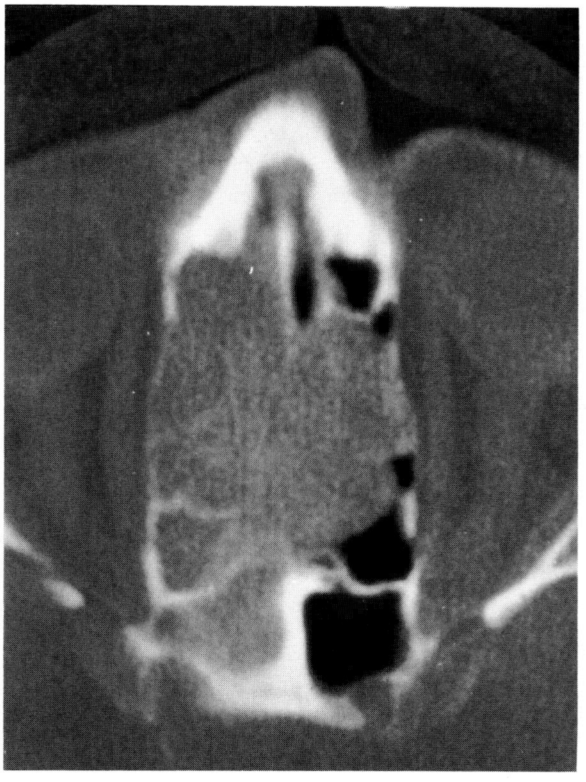

Fig. 10.35. Axial CT scan showing a metastasis in the ethmoid cells from carcinoma of the pancreas.

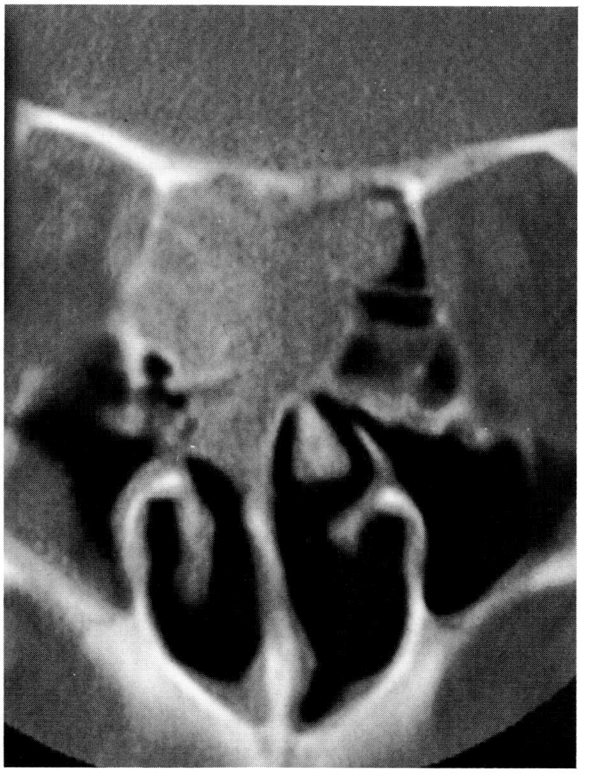

Fig. 10.36. Same patient as Fig. 10.35. Coronal CT scan of the metastasis in the ethmoids.

References

Acheson ED, Hadfield E, Macbeth RG (1967) Carcinoma of the nasal cavity and accessory sinuses in woodworkers. Lancet I:311–312

Batsakis JG (1970) Mucous gland tumors of the nose and paranasal sinuses. Ann Otolaryngol 79:557–562

Batson OV (1940) Formation of the vertebral veins and their role in the spread of metastases. Ann Surg 112:138–147

Billroth T (1855) Ueber dem Bau des Schleimpolyp. Reimer, Berlin, p11

Billroth T (1859) Beobachtungen über Geschwulste der Speicheldrüsen. Virchows Arch Pathol Anat Histol 17:357–375

Brown B (1964) The papillomatous tumours of the nose. J Larngyol Otol 78:889–905

Churchill HR (1934) Histological differentiation between certain dentigerous cysts and ameloblastoma. Dental Cosmos 76:1173–1178

Eneroth CM, Martensson G, Thulin A (1961) Profuse epistaxis in hypernephroma metastasis. Acta Otolaryngol 53:546–550

Friedmann I, Osborn DA (1965) Metastatic tumours in the ear, nose and throat region. J Laryngol Otol 79:576–591

Friedmann I, Osborn DA (1982) Pathology of granulomas and neoplasms of the nose and paranasal sinuses. Churchill Livingstone, Edinburgh

Garrett MJ (1959) Metastatic tumours of the paranasal sinuses simulating primary growths. J Fac Radiol (Lond) 10:151–155

Hadfield EH (1970) A study of adenocarcinoma of the paranasal sinuses in woodworkers in the furniture industry. Ann R Coll Surg Engl 46:301–319

Hadfield EH, Macbeth RG (1971) Adenocarcinoma of ethmoids in furniture workers. Ann Otol Rhinol Laryngol 80:699–703

Harrison DFN (1976) Malignant melanoma arising in the nasal mucous membrane. J Laryngol Otol 90:993–1005

Harrison MS, Doey WD, Osborn DA (1964) Intranasal metastasis from renal carcinoma. J Laryngol Otol 78:103–107

Holdcraft J, Gallagher JC (1969) Malignant melanomas of the nasal and paranasal sinus mucosa. Ann Otol Rhinol Laryngol 78:1–20

Hyams VJ (1971) Papillomas of the nasal cavity and paranasal sinuses. Ann Otol Rhinol Laryngol 80:192–206

Kramer R, Som ML (1935) True papilloma of the nasal cavity. Arch Otolaryngol 22:22–43

Lund VJ (1982) Malignant melanoma of the nasal cavity and paranasal sinuses. J Laryngol Otol 96:347–355

Lund VJ, Lloyd GAS (1984) Radiological changes associated with inverted papilloma of the nose and paranasal sinuses. Br J Radiol 57:455–461

Mabery TE, Devine KD, Harrison EG (1965) The problem of malignant transformation in a nasal papilloma. Arch Otolaryngol 82:296–300

Macbeth RG (1965) Malignant disease of the paranasal sinuses. J Laryngol Otol 79:592–612

McIvor J (1974) The radiological features of ameloblastoma. Clin Radiol 25:237–242

Malassez L (1885) Sur le rôle des débris epitheliaux paradentaires. Arch Physiol Normal Pathol (3rd series) 6:379–449

Michaels L (1987) Ear, nose and throat histopathology. Springer, Berlin Heidelberg New York

Mukherjee DK (1978) Choriocarcinoma of the nose. Ann Otol Rhinol Laryngol 87:257–259

Nahum AM, Bailey BJ (1963) Malignant tumours metastatic to

the paranasal sinuses: case report and review of literature. Laryngoscope 73:942–953
Ringertz N (1938) Pathology of malignant tumours arising in the nasal and paranasal cavities and maxilla. Acta Otolaryngol [Suppl] 27:1–390
Salimi R (1977) Metastatic choriocarcinoma of the nasal mucosa. J Surg Oncol 9:301–305
Shaw H (1964) Clinical importance of orbital signs in cancer of the paranasal sinuses. Proc R Soc Med 57:742–747
Spiro RH, Koss LG, Hajdu SI, Strong EW (1973) Tumors of minor salivary origin. Cancer 31:117–129
Ward N (1854) Follicular tumour involving nasal bones, nasal processes of superior maxillary bone and septum of nose. Removal, death from pneumonia, autopsy. Lancet II:480
Worgan D, Hooper R (1970) Malignancy in nasal papillomata. J Laryngol Otol 84:309–316

11 Tumours of Vascular Origin

Hereditary haemorrhagic telangiectasia (Osler–Weber–Rendu syndrome) is a disease in which groups of dilated vessels are present in the skin and mucosae. The nose is commonly affected and the lesions are important because they frequently give rise to epistaxis. This condition does not require any sophisticated imaging method, and the same is true for another condition which may give rise to epistaxis: capillary haemangioma. Here the lesion is usually localised to the mucosa of the nasal septum or inferior turbinate.

There are six lesions of vascular origin that concern the imaging diagnostician:

1. Juvenile angiofibroma
2. Haemangiopericytoma
3. Angiosarcoma (haemangio-endothelioma)
4. Cavernous haemangioma
5. Venous malformations
6. Angiolymphoid hyperplasia with eosinophilia (Kimura's disease)

Juvenile Angiofibroma

Juvenile angiofibroma is a rare benign but highly vascular tumour occurring in adolescent males. The juvenile nasopharyngeal angiofibroma was so called because it was formerly thought to arise in the nasopharynx. It is not known why these tumours occur almost exclusively in a restricted age group and restricted site at the base of the skull, or indeed whether they should be considered as neoplasms or as hamartomata. Most radiologists are aware of this tumour as a lesion that can be well demonstrated by angiography, because of its highly vascular nature, and which may require preoperative embolisation, but the radiologist may also be called upon to make the initial diagnosis. With such a vascular tumour, severe bleeding may accompany biopsy and for this reason most surgeons are reluctant to undertake biopsy of a nasopharyngeal mass in an adolescent male patient and prefer to rely upon clinical and radiological features to decide whether the mass is likely to be an angiofibroma or a non-vascular lesion such as an antrochoanal polyp. Once the diagnosis is established then the role of the radiologist is to define the limits of the tumour prior to surgery, since this may influence the surgical approach.

According to Tapia Acuna (1956) this tumour was known to Hippocrates, but the first recorded description of juvenile angiofibroma involving the post-nasal region was given by Chelius in 1847. He noted the fibrous nature of the lesion and its occurrence at about the time of puberty. Gosselin (1873) emphasised the occurrence of nasopharyngeal fibrous polyps almost exclusively in young males and noted that while some lesions tend to regress as the patient becomes adult others require surgical removal. The term juvenile nasopharyngeal fibroma was introduced by Chaveau (1906) and Friedberg (1940) suggested the name angiofibroma. Since these descriptions a large volume of literature has accumulated about this relatively rare condition. There is general agreement that in most cases the age of onset is in the second decade of life and that there is an overwhelming predominance of male patients. In the author's series all patients were male and no example of this condition was seen in a female.

The diagnosis is almost always made clinically by the history of a young or adolescent male with nasal obstruction or epistaxis or both; expansion of the lesion may lead to facial deformity with swelling of the cheek and proptosis. Physical examination shows a nasopharyngeal mass. The angiofibroma has a reddish-purple nodular appearance and is composed of vascular and fibrous tissue, the latter varying in texture from a compact hyaline mass to oedematous granulation tissue. The vessels also vary considerably, particularly in the degree of development of the muscular component of the walls.

The site of origin is important. It was first thought that the lesions arise in the nasopharynx, but examples can be found in the literature in which there was no nasopharyngeal mass. Hora and Weller (1961) described an extranasopharyngeal angiofibroma with apparent origin from the pterygo-maxillary space and attachment to the medial pterygoid plate. The tumour extended into the infratemporal fossa and presented as a swelling of the cheek, but there was no associated nasopharyngeal mass. Other authors (Friedmann and Osborn 1983) have described attachments around the margins of the choanae (that is the medial pterygoid plate, the posterior border of the hard palate and the posterior

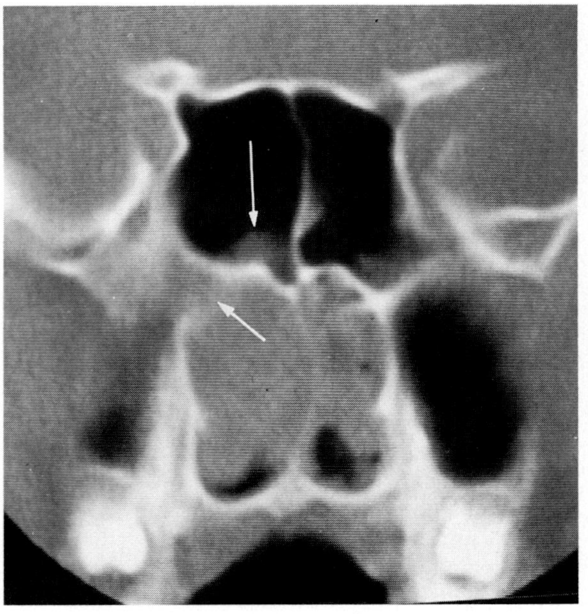

Fig. 11.1. Coronal CT scan showing enlargement of the spheno-palatine foramen (*short arrow*) by early angiofibroma. Early invasion of the sphenoid sinus is also present (*long arrow*).

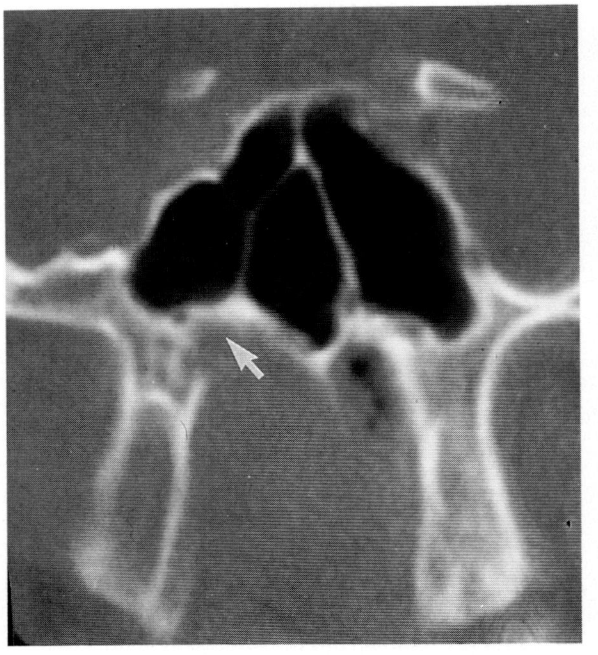

Fig. 11.2. Five-millimetre coronal CT scan taken immediately posterior to Fig. 11.1, showing early erosion of the base of the medial pterygoid plate (*arrow*).

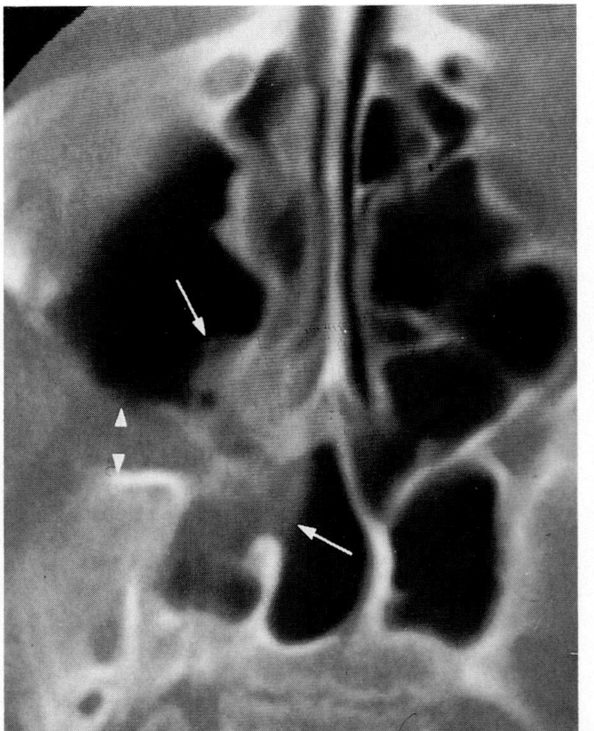

Fig. 11.3. Same patient as Fig. 11.1 and 11.2. Axial CT scan showing early extension into the posterior maxillary antrum and sphenoid sinus (*arrows*) and widening of the pterygo-maxillary fissure (*arrowheads*).

margin of the vomer), but the evidence is that the tumour originates from the region of the spheno-palatine foramen within the pterygo-palatine fossa. This is important not only to the surgeon but also to the radiologist, both for making the initial diagnosis and also for understanding the expected spread of the tumour.

The earliest bone changes seen are illustrated in Figs. 11.1–11.4. These appearances suggest that the site of origin is likely to be within the spheno-palatine foramen (Figs. 11.5, 11.6) at the superomedial extremity of the pterygo-palatine fossa. This point of origin fits with the known extension of the tumour and the early bone changes visible. An expanding tumour in this situation would enlarge the spheno-palatine foramen, grow medially into the area of least resistance – the nose and nasopharynx (Figs. 11.4, 11.7) – erode the root of the medial pterygoid plate (Figs. 11.2, 11.7), invade the sphenoid sinus (Fig. 11.7) and indent the posterosuperior border of the maxillary antrum giving rise to the so-called antral sign described by Holman and Miller (1965). This consists of anterior bowing of the posterior wall of the maxillary antrum, best seen on a lateral projection or on lateral tomography (Fig. 11.8).

The tumour can also gain access to the infratemporal fossa by lateral extension via the pterygomaxillary fissure (Fig. 11.9) and invade the orbit through the inferior orbital fissure. Further extension from these areas may result in middle fossa invasion, either by way of the orbit and superior orbital fissure or directly through the lateral wall of the sphenoid sinus.

Extension into the infratemporal fossa may eventually result in the mass emerging between the upper molar teeth and ascending ramus of the mandible, the tumour coming to lie beneath the skin of the cheek (Fig. 11.10). Thus the tumour comes to present a bilobed shape with a medial component in the nose and nasopharynx and a lateral component in the infratemporal fossa, the two lobes being joined by an isthmus of tumour lying between the maxillary antrum and the pterygoid plates. All these features have been clearly demonstrated by the imaging techniques used.

Essentially, then, the tumour causes an expansion of the spheno-palatine fossa, and the early bone changes to be looked for by the radiologist will consist of pressure erosion and invasion of structures in its immediate vicinity. Polyps or tumours not arising in this situation are unlikely to present this pattern of change, which serves to distinguish an angiofibroma from other conditions, for example an antro-choanal polyp, which may present with a nasopharyngeal mass.

The diagnostic role of plain radiography is to identify the antral sign when present. In Holman and Miller's (1965) series this sign was present in 87% of patients. In the series described by Lloyd and Phelps (1986) it was positive in just over 81% (Fig. 11.8). Long-term pressure on the posterior antral wall will cause bowing of the wall anteriorly. Thus any slow-growing tumour, usually but not always of a benign nature, will cause this deformity. The antral sign cannot be considered, therefore, as pathognomonic for angiofibroma. Bowing of the posterior antral wall has been seen in a proven

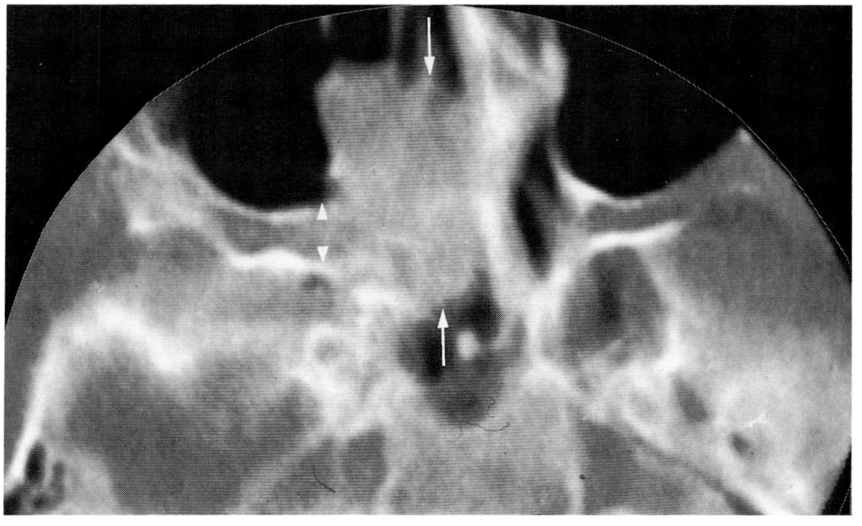

Fig. 11.4. Axial CT scan showing early angiofibroma extending into the nasal cavity and nasopharynx (*arrows*), and widening of the spheno-palatine foramen (*arrowheads*).

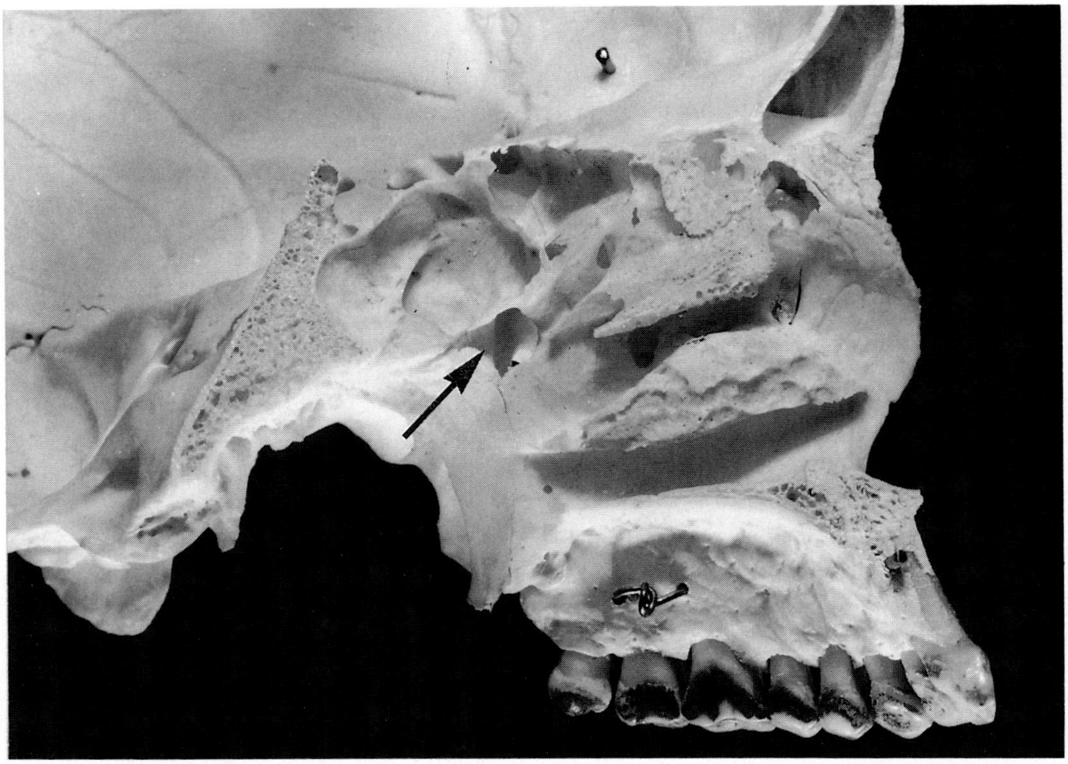

Fig. 11.5. Dried skull showing the site of origin of angiofibroma within the spheno-palatine foramen (*arrow*).

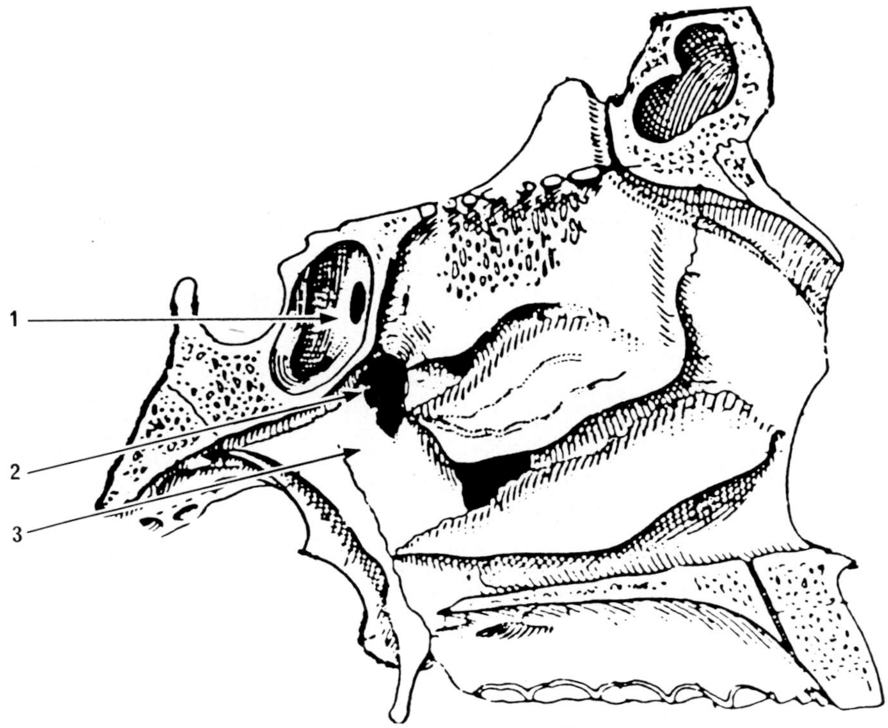

Fig. 11.6. Line drawing showing the close relationship of the sphenoid sinus (*1*), the spheno-palatine foramen (*2*) and the base of the medial pterygoid plate (*3*).

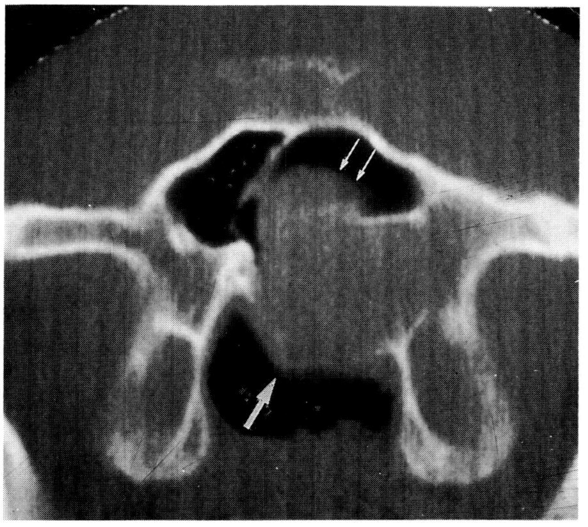

Fig. 11.7. Coronal CT scan showing an angiofibroma projecting into the air space at the posterior nares (*thick arrow*), invading the sphenoid sinus (*thin arrows*) and eroding the base of the medial pterygoid plate.

neurofibroma (Fig. 11.11) and also in a single example of haemangiopericytoma. Schaffer et al. (1978) have also described two cases of tumours other than angiofibromata which gave a positive antral sign. A more important and constant diagnostic feature is the tomographic demonstration of deformity of the base of the medial pterygoid lamina. This was present in 100% of 28 patients reviewed by Lloyd and Phelps (1986) and was demonstrated either by conventional tomography or coronal CT scan. This change does not occur in simple nasopharyngeal polyps; when there is no such bone deformity the radiologist can reassure the surgeon that it is safe to take a biopsy without the danger of severe haemorrhage from an angiofibroma.

The next duty of the radiologist is to map the extent of the tumour prior to surgery. Most important in this respect is to demonstrate whether there is lateral spread into the infratemporal fossa, which may determine the surgical approach used. With no spread into the infratemporal fossa a transpalatal surgical approach may be feasible, but with any degree of involvement of the infratemporal fossa a lateral rhinotomy or transmaxillary approach is required. The extent of the tumour and in particular the degree of involvement of the infratemporal fossa may be demonstrated by enhanced CT scan in the immediate post-injection phase (Fig. 11.12), but is now better achieved by magnetic resonance (Fig. 11.13).

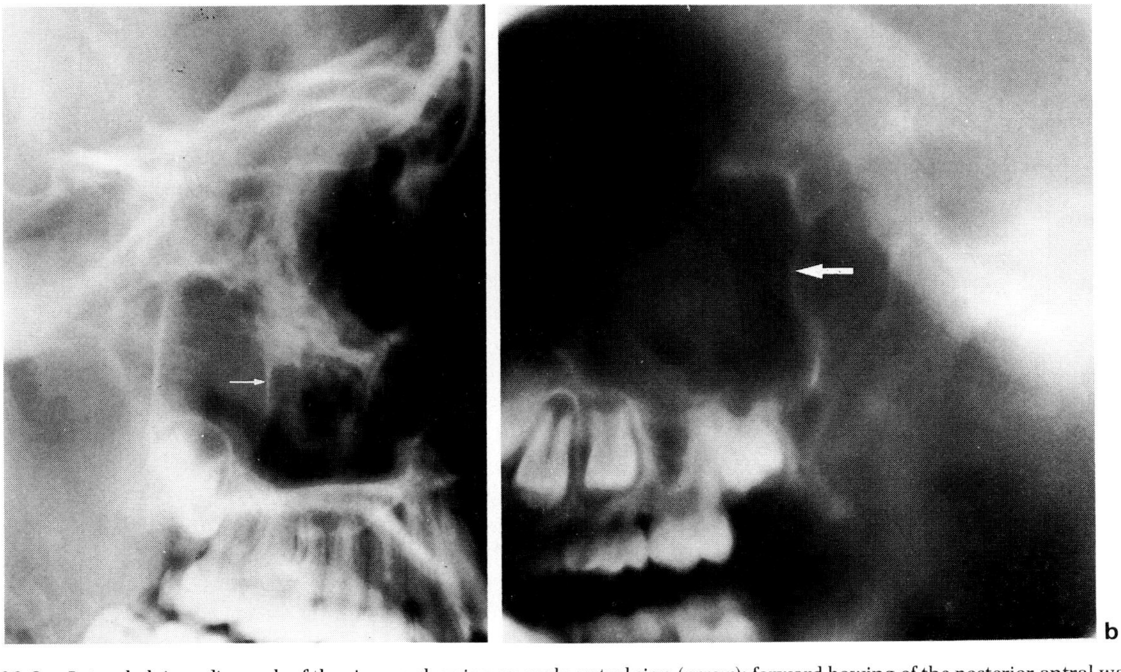

Fig. 11.8. a Lateral plain radiograph of the sinuses showing an early antral sign (*arrow*): forward bowing of the posterior antral wall. **b** Lateral hypocycloidal tomogram showing enlargement of the spheno-palatine fossa with an early antral sign (*arrow*).

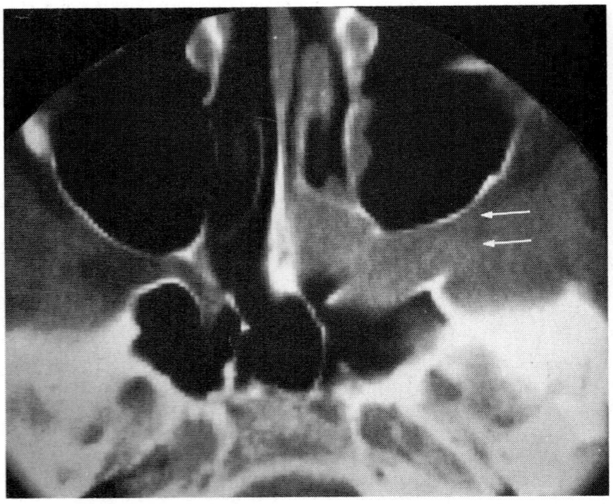

Fig. 11.9. Axial CT scan showing early extension of an angiofibroma into the infratemporal fossa (*arrows*). Note the widening of the pterygo-palatine fossa and the pterygo-maxillary fossa by the tumour.

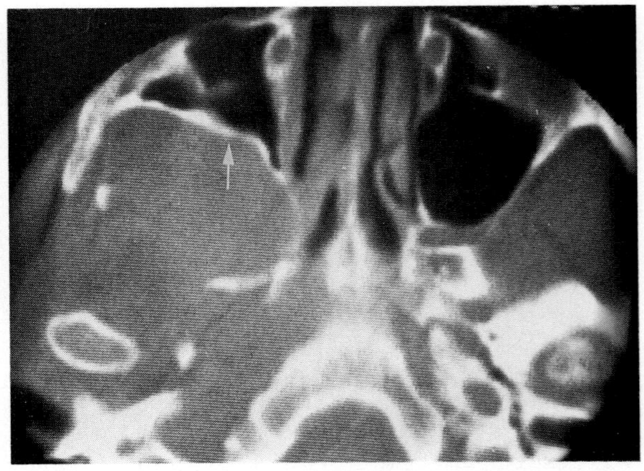

Fig. 11.11. Axial CT scan showing forward bowing of the posterior wall of the maxillary antrum (*arrow*) by a proven neurofibroma.

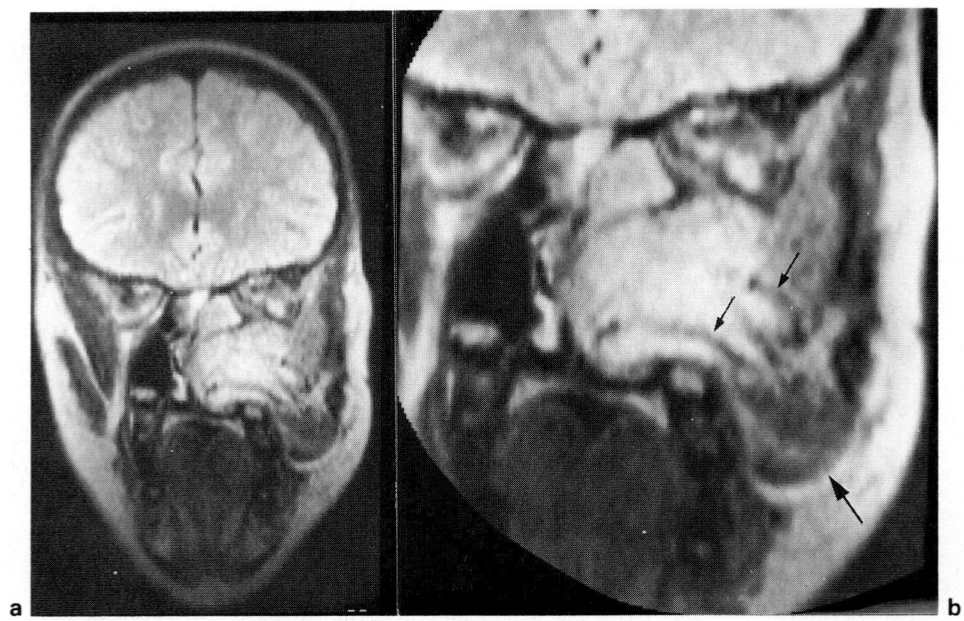

Fig. 11.10. a Coronal magnetic resonance scan showing an angiofibroma. **b** Enlargement scan of the angiofibroma showing large vessels within the tumour (*small arrows*) and the tumour bulge beneath the skin of the cheek (*large arrow*).

Magnetic resonance has several advantages over CT. It better demonstrates the exact extent of the tumour and its superior density resolution shows the edge of the tumour more clearly in relation to other soft tissue structures in the infratemporal fossa. Three-plane imaging and good sagittal sections are an added advantage, and the distinction between tumour invasion of a sinus and secondary mucocoele formation is clearly shown (Fig. 11.14). The vascular nature of angiofibroma as demonstrated by negative signal from the vessels within the tumour is also clearly revealed by magnetic resonance (Figs. 11.10, 11.15). A nasopharyngeal mass in an adolescent male showing this appearance is clearly diagnostic of angiofibroma. Finally, all this is achieved without the use of ionising radiation or intravenous contrast injection.

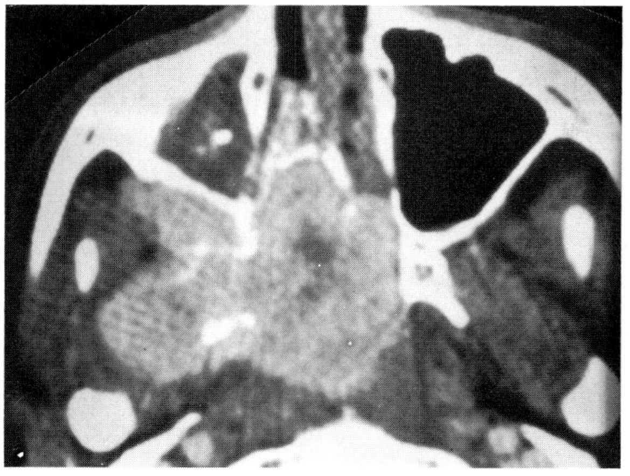

Fig. 11.12. Axial CT scan after contrast injection showing extension of an angiofibroma into the infratemporal fossa.

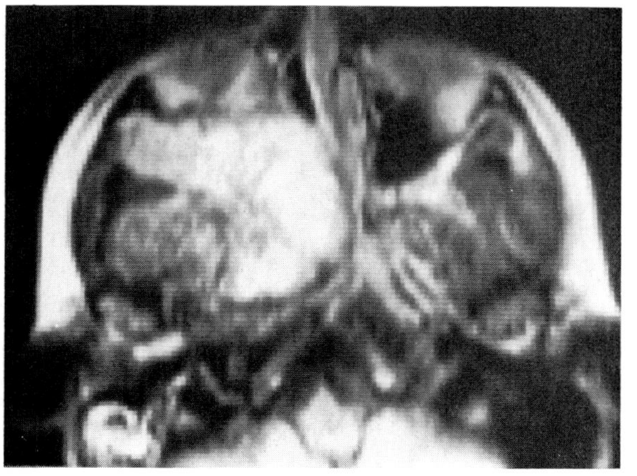

Fig. 11.13. Axial magnetic resonance scan showing extension of an angiofibroma into the infratemporal fossa.

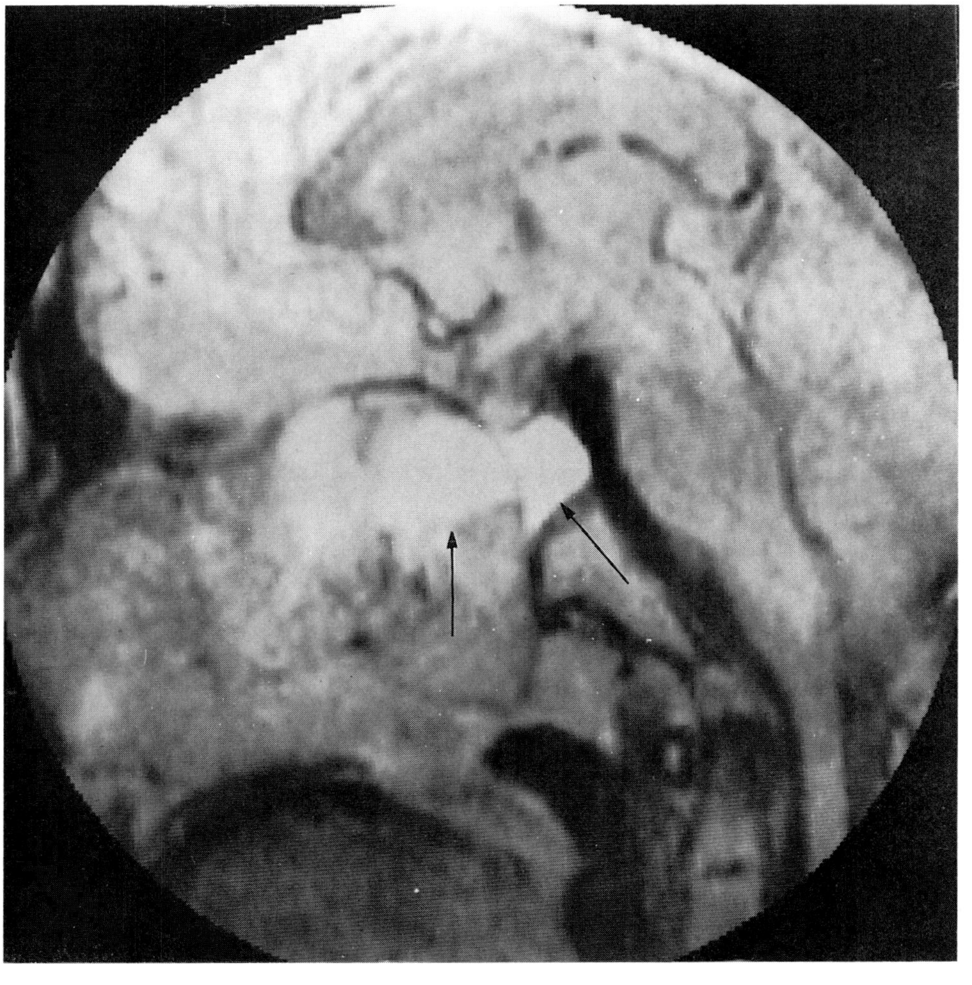

Fig. 11.14. Sagittal magnetic resonance scan of a patient with a large angiofibroma. The white areas (*arrows*) giving a strong signal indicate retained secretion and mucocoele formation in the posterior ethmoids and sphenoid sinus.

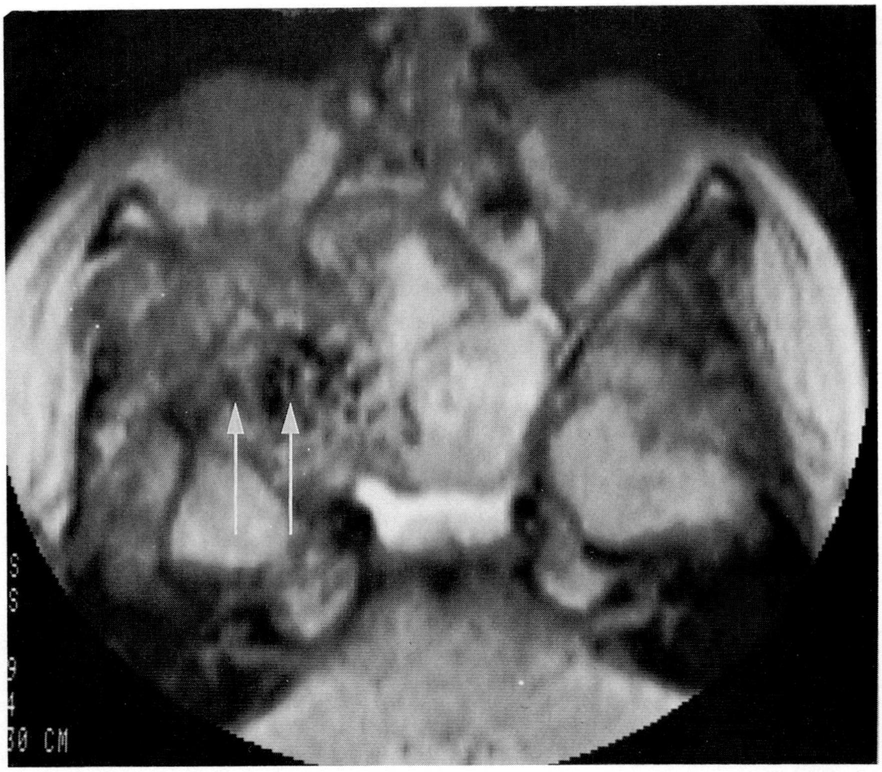

Fig. 11.15. Axial magnetic resonance scan showing the vascularity in an angiofibroma. The vessels are shown as dark areas of negative signal (*arrows*).

Conclusions

1. Angiofibroma originates at the spheno-palatine foramen. It enlarges the foramen and erodes bone locally at the base of the medial pterygoid plate, the floor of the sphenoid sinus and the posterior wall of the maxillary antrum. Further extension leads to invasion of the infratemporal fossa, orbit and middle cranial fossa.

2. The antral sign on plain radiography is not completely reliable. When in doubt coronal tomography, or preferably CT scan, should be undertaken to show signs of early bone erosion of the medial pterygoid lamina, enlargement of the spheno-palatine foramen and invasion of the sphenoid sinus.

3. In the presence of a positive antral sign, three-plane magnetic resonance is the investigative method of choice. It best demonstrates the extent of the tumour; it uses non-ionising radiation; and it will show the vascular nature of the angiofibroma and confirm the diagnosis.

4. Angiography should only be performed if embolisation is deemed necessary prior to surgery. In fact most of these tumours do not require embolisation; and in the series reported by Lloyd and Phelps (1986) there was some evidence to suggest that the recurrence rate was increased and the rate of recurrence in individual patients accelerated by embolisation. The reasons for this are speculative, but it is possible that excessive shrinkage of the tumour after embolisation may lead to minute fragments being overlooked at surgery; these may then quickly re-vascularise in the post-operative period.

Haemangiopericytoma

Haemangiopericytoma is an uncommon vascular tumour believed to derive from the pericyte, a cell with processes which encircle the endothelial cells of capillaries. The tumour was first described by Stout and Murray (1942). These authors believed that the blood vessels were an integral part of the neoplasm and not part of its framework, and because each vessel in the tumour was lined by normal endothelial cells they suggested that the tumour consisted essentially of pericytes. Since that time many cases of vascular neoplasms with these features have been published, and the concept of haemangiopericytoma is now generally accepted. Twenty-three examples of this tumour occurring in the nose and paranasal sinuses have been reported by Compagno (1978). His patients presented with epistaxis and nasal obstruction, most commonly in the sixth and seventh decades of life; the lesions mimicked allergic polyps clinically.

Eleven patients with haemangiopericytoma affecting the facial skeleton have been investigated: in eight the tumour was located in the orbit and in three it was in the nose and sinuses. It is said that haemangiopericytoma of the nose and sinuses has a reduced recurrence rate when compared with those found at other sites in the body (Michaels 1987). An equally benign course has been found in the eight tumours arising in the orbit. In only one patient was there a history of recurrence of the tumour (Fig. 11.16). This followed several previous attempts at surgical removal elsewhere. The evidence is that no recurrence is to be expected if the tumour is removed in toto with intact capsule at first surgery.

Radiological Features

Of the three patients who presented with the tumour in the nose or sinuses, all showed a mass on one side of the nasal cavity. In one the tumour had extended massively into the infratemporal fossa via the pterygo-palatine fossa and pterygo-maxillary fossa. From there it had extended upwards through the inferior orbital fissure to invade the orbital apex (Fig. 11.17). The second tumour expanded one side of the nasal fossa and encroached upon the antral cavity (Fig. 11.18), while the third was essentially naso-ethmoidal in location.

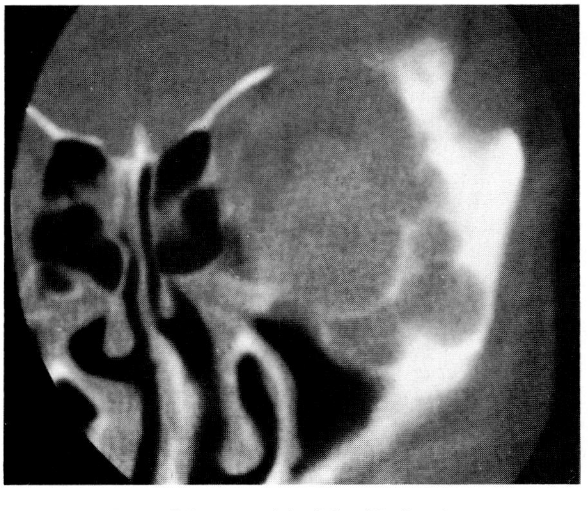

Fig. 11.16. Coronal CT scan of the left orbit showing a recurrent haemangiopericytoma invading the lateral wall of the orbit and maxillary antrum.

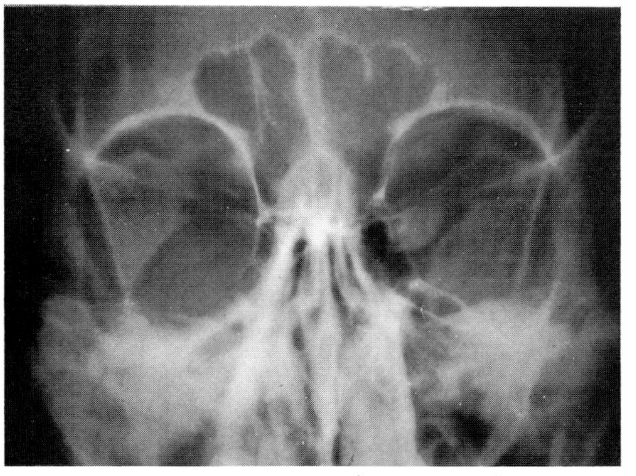

Fig. 11.17. Postero-anterior radiograph showing bone erosion in the orbital apex due to a haemangiopericytoma. See also Fig. 11.19.

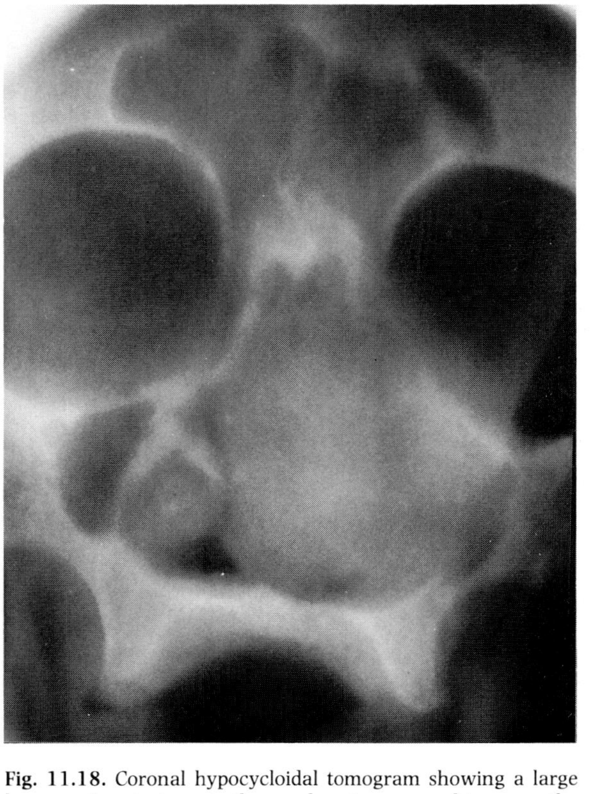

Fig. 11.18. Coronal hypocycloidal tomogram showing a large haemangiopericytoma in the nasal cavity, encroaching upon the left maxillary antrum.

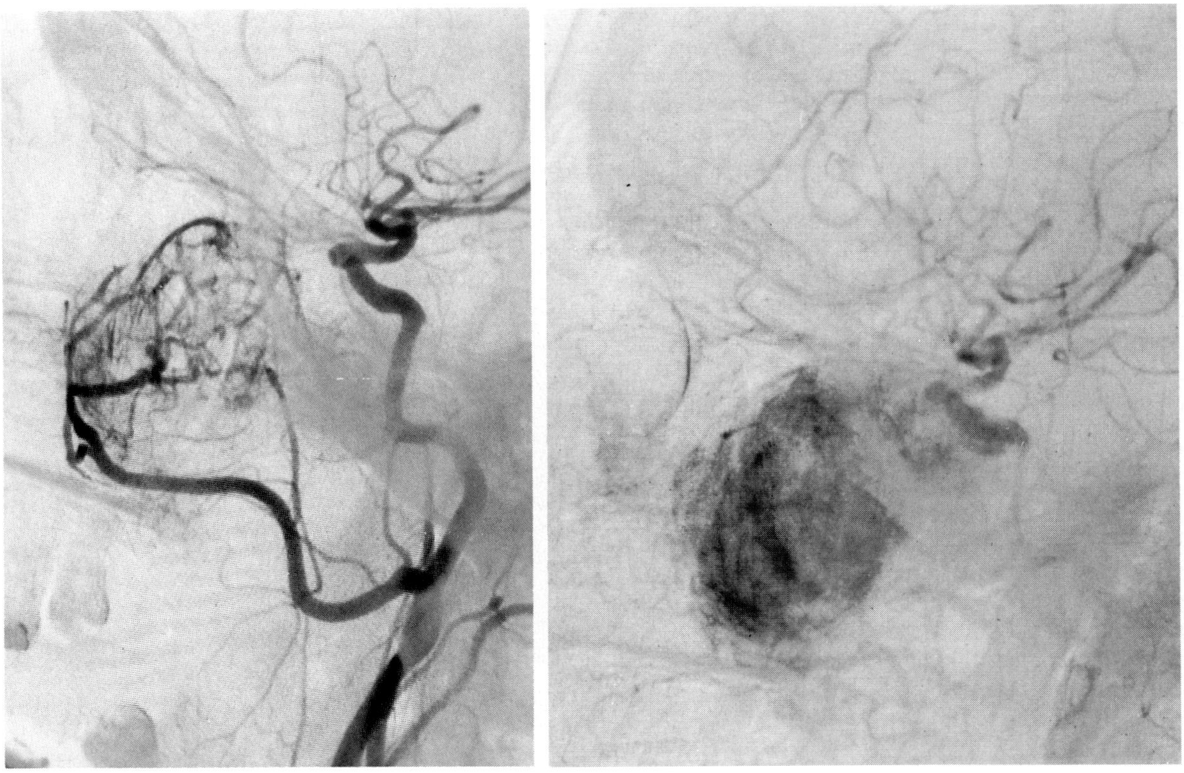

a b

Fig. 11.19a,b. Same patient as Fig. 11.17. **a** Hypervascularity in a haemangiopericytoma which arose in the infratemporal fossa and invaded the posterior wall of the maxillary antrum and the orbital apex. **b** Typical strong tumour blush in the late arterial phase of the arteriogram.

The diagnostic feature of haemangiopericytoma is its high vascularity. All tumours examined by carotid angiography, whether in the nose, sinuses or orbit, have invariably shown a strong tumour blush (Fig. 11.19). This can now be demonstrated adequately by digital subtraction angiography using intravenous contrast, and in this way arterial puncture may be avoided.

Angiosarcoma (Haemangio-endothelioma)

Angiosarcoma is the malignant counterpart of haemangioma and is a very rare tumour in the nasal region. Fu and Perzin (1974) recorded two cases comprising 2% of malignant non-epithelial tumours of the nose and paranasal sinuses; Friedmann and Osborn (1982) encountered only one case representing 0.1% of all tumours of this region.

The clinical features show a broadly based age distribution and no sex predilection. Nasal bleeding and obstruction is the common form of presentation, with swelling of the cheek when the antrum is involved and proptosis when the orbit is invaded. To the naked eye the tumour is seen as highly vascular masses of extremely friable consistency. Microscopically the changes vary according to the degree of differentiation of cell pattern, but the essential feature is the presence of vascular channels lined by malignant endothelial cells.

Two patients with this tumour have been investigated. The first was not investigated ab initio but was referred for CT scan prior to cranio-facial resection for a recurrence of the neoplasm. The second patient was a 50-year-old male who presented with a swelling of the dental alveolus, nasal obstruction and epistaxis. CT showed a massive tumour arising from the floor of the maxillary antrum. The changes were non-specific but were those of a highly malignant neoplasm which had destroyed the alveolar margin, extended into the nasal cavity, and had eroded the posterior wall of the antrum with invasion of the infratemporal fossa (Fig. 11.20).

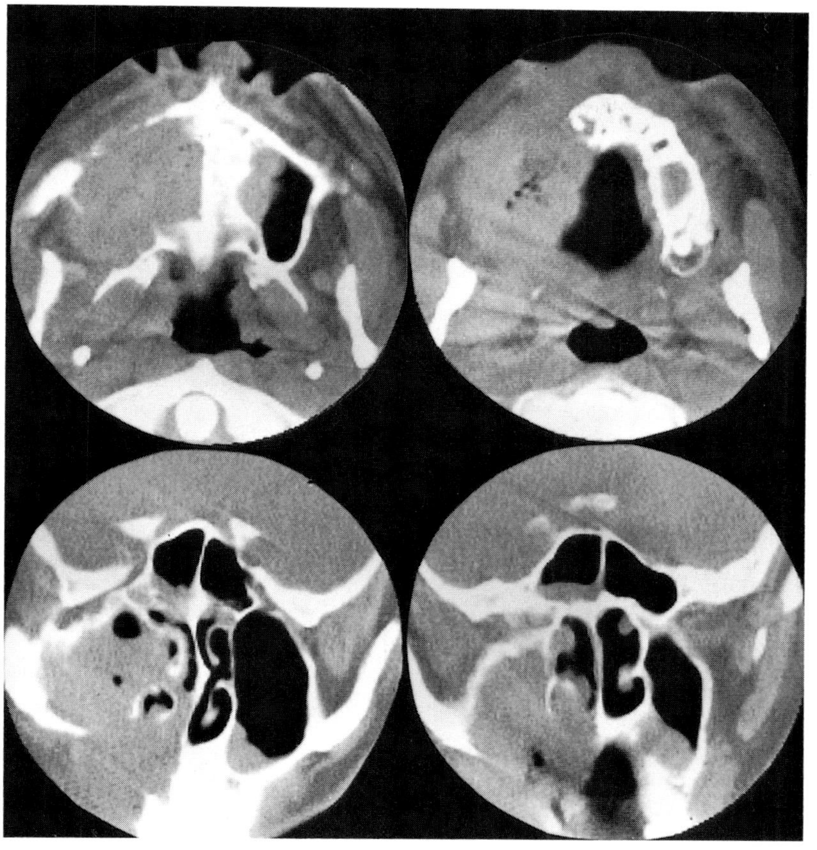

Fig. 11.20. Axial (*above*) and coronal (*below*) CT scans showing an angiosarcoma of the right maxilla. The large tumour has caused massive bone destruction in the alveolus and antral walls.

Cavernous Haemangioma

Cavernous haemangioma is a very rare tumour in the nose and sinuses, but in the adjacent orbit it is the commonest benign neoplasm. It is made up of thin-walled blood-filled channels lined by a single layer of endothelial cells and is found either in the marrow spaces of the bone or as a vascular swelling beneath the mucosa (Michaels 1987). It arises most often from the lateral wall of the nose but may also involve the sinus cavities. Clinically when a tumour arises in these situations it presents with nasal bleeding or obstruction, and a tender swelling of the cheek if the antrum is involved (Fordham 1978). Some tumours may bleed following dental extraction (Broderick and Round 1933). Radiologically most reports have shown an opaque sinus, often with adjacent bone erosion and a nasal mass. Tsutsumiuchi et al. (1982) recorded a solitary phlebolith shown on CT in a cavernous haemangioma located within the maxillary antrum and extending to the nose and ethmoid cells.

Haemangiomata may also be confined to the bone of the sinus wall or to the adjacent skull. Here they produce a well-demarcated osteolytic defect with a fine honeycomb appearance resulting from the formation of numerous thin trabeculae of new bone between the vascular sinuses (Fig. 11.21). In a tangential view or in profile it may be possible to see fine spicules radiating out from the bony surface, producing the so-called sunray or sunburst appearance.

Venous Malformations

Under the title venous haemangioma Fu and Perzin (1974) described three patients who presented with small mass lesions in the nasal vestibule. Microscopically these lesions were composed of small, thick-walled venous channels. The vessels often extended into the adjacent striated muscle and normal cartilage. Although rare in the nose and

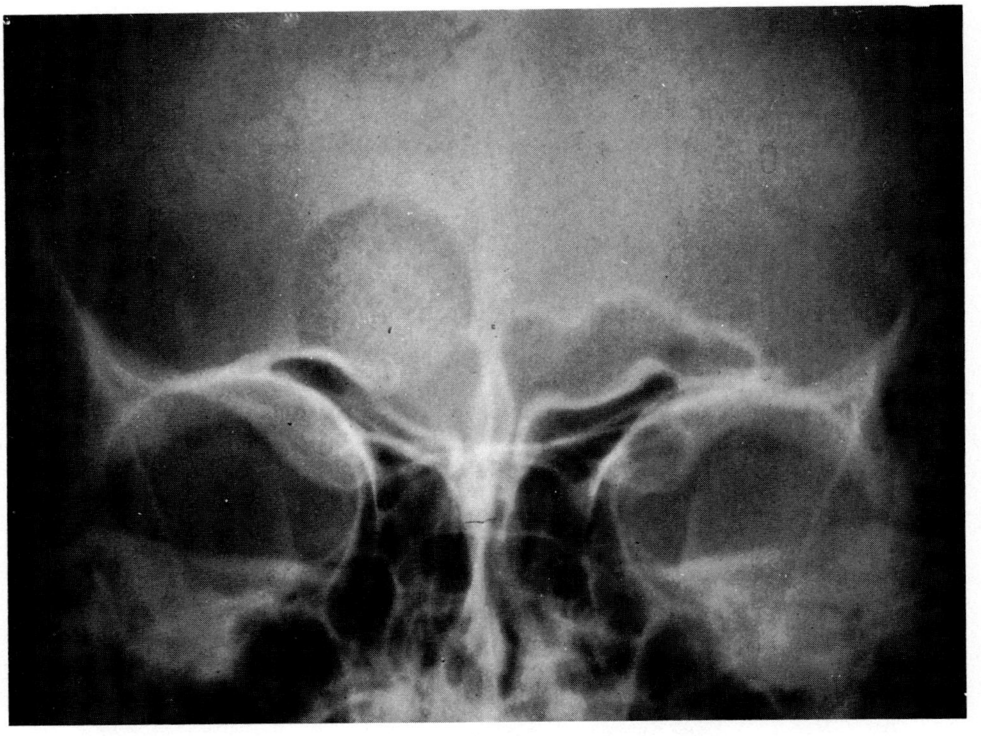

Fig. 11.21. Haemangioma in the frontal bone and frontal sinus. There is a clearly demarcated area of osteolysis in the bone within which is a fine trabecular pattern.

sinuses venous malformations are relatively common in the adjacent orbit. They present radiologically with a classical triad of signs: an enlarged orbit, the presence of phleboliths, and associated venous "lakes" in the frontal bone over the affected orbit.

Congenital venous malformations are best demonstrated by orbital venography (Fig. 11.22). They may replace the normal venous system with a network of dilated venous channels, and in the most extensive malformations the adjacent sinuses may be involved (Fig. 11.23). In the orbit phleboliths invariably indicate the presence of abnormal veins (Lloyd 1982). They may be encountered in a sinus cavity (Fig. 11.24) and presumably have the same diagnostic implications.

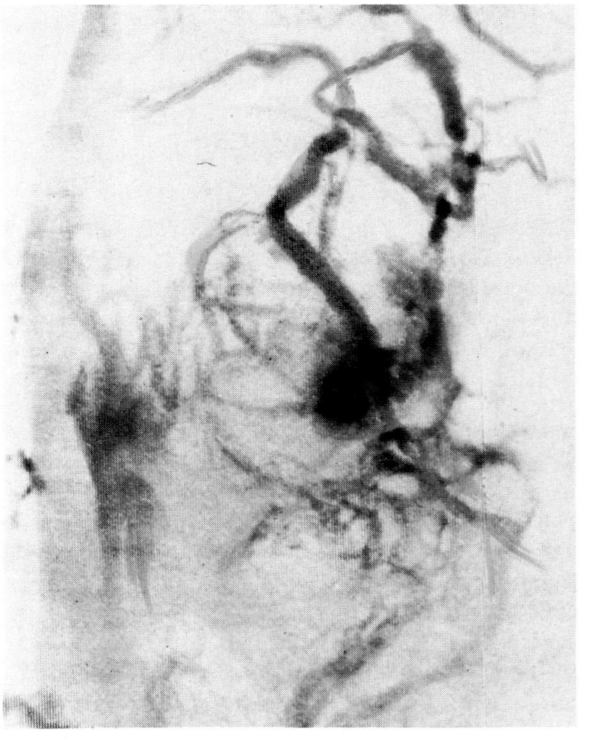

Fig. 11.22. Orbital venogram. Subtraction study showing a venous malformation in the orbital apex.

Angiolymphoid Hyperplasia with Eosinophilia (Kimura's Disease)

In 1948 Kimura et al. described a peculiar disease in which there was an unusual granulomatous lesion combined with hyperplastic changes of lymphatic

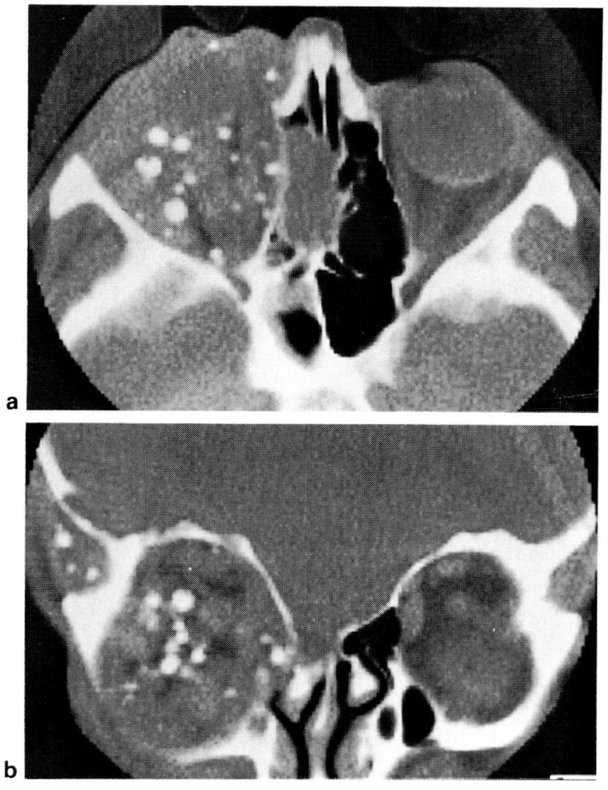

Fig. 11.23a,b. Axial (**a**) and coronal (**b**) CT scans showing a huge venous malformation in the orbit with multiple phleboliths. The venous malformation also involved the adjacent ethmoid cells.

tissue. In 1965 Kawada et al. reported four similar cases which were characterised by flat or dome-shaped swellings and by a blood eosinophilia. The lesions occurred almost exclusively on the cheek, axillary, cubital or inguinal regions. They were composed of hyperplastic lymphoid tissue which contained well-developed lymph follicles and infiltrates of eosinophils. In 1969 Wells and Whimster reported nine similar cases under the heading of angiolymphoid hyperplasia with eosinophilia. The subcutaneous nodules were composed of unencapsulated masses of lymphoid tissue admixed with newly formed vessels and infiltrates of eosinophils. Reed and Terazakis (1972) recorded six further examples of this peculiar angioblastic lesion. The lesions in their patients were either solitary or multiple and presented as tumour-like swellings in the subcutaneous tissue of the head and neck.

The true nature of this lesion has been unclear for a long time, but it is now generally thought to be a neoplasm consisting of proliferating endothelial cells (Welch et al. 1987). It may occur in the orbit and extend from there into the sinuses. Three patients with orbital involvement have been inves-

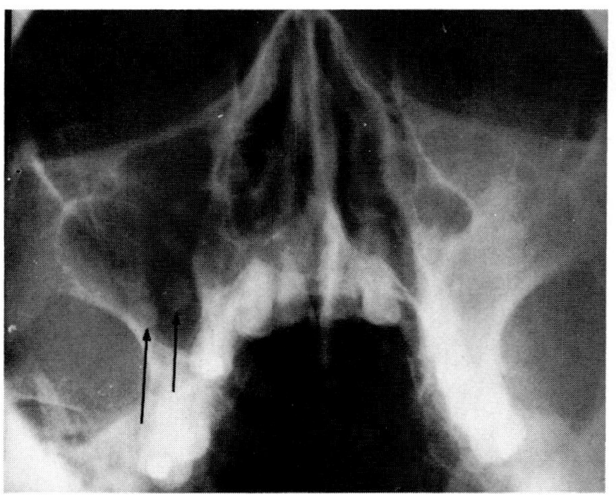

Fig. 11.24. Phleboliths (*arrows*) shown in the right maxillary antrum on routine sinus radiography.

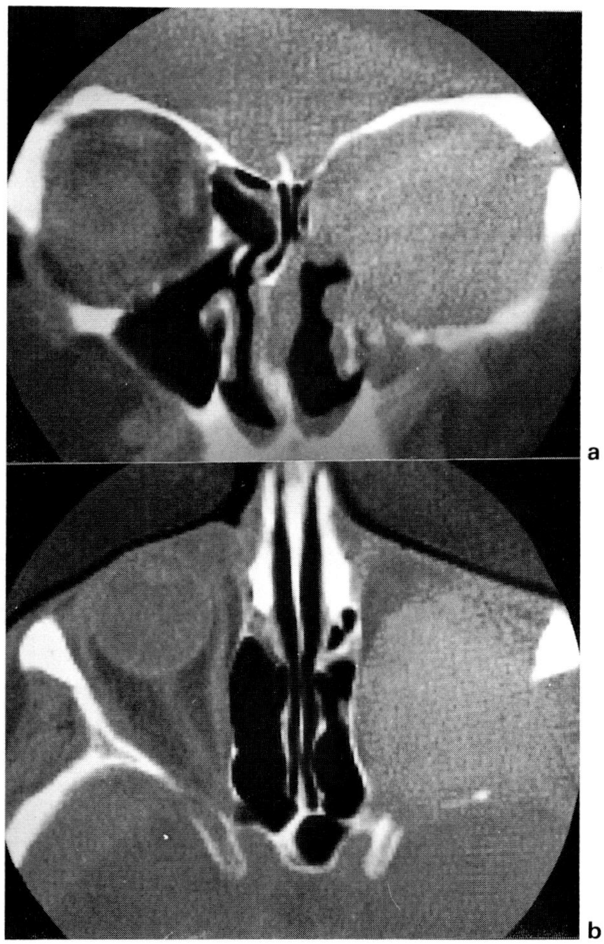

Fig. 11.25a,b. Angiolymphoid hyperplasia with eosinophilia. Coronal (**a**) and axial (**b**) CT scans showing massive invasion of the orbit with extension to the antrum, ethmoids and middle cranial fossa.

tigated during the past 8 years. In two the orbital soft tissue structures were totally replaced by the angiolymphomatous mass, which expanded and eroded the bony walls and invaded the sinuses (Fig. 11.25). The disease has proved difficult to control and in one patient has necessitated cranio-facial resection.

References

Broderick RA, Round H (1933) Cavernous angioma of the maxilla: a fatal haemorrhage after teeth extraction with notes of a similar, non-fatal case. Lancet II:13–15
Chaveau C (1906) Histoire des maladies du pharynx. Baillière, Paris
Chelius MJ (1847) A system of surgery, vol 2. Renshaw, London
Compagno J (1978) Haemangiopericytoma-like tumours of the nasal cavity: a comparison with haemangiopericytoma of soft tissues. Laryngoscope 88:460–469
Fordham SD (1978) Haemangioma of the maxillary sinus. Ear Nose Throat J 57:333–335
Friedberg SA (1940) Nasopharyngeal fibroma. Arch Otolaryngol 31:313–326
Friedmann I, Osborn DA (1982) Pathology of granulomas and neoplasms of the nose and paranasal sinuses. Churchill Livingstone, Edinburgh
Fu Y, Perzin KH (1974) Non-epithelial tumours of the nasal cavity, paranasal sinuses and nasopharynx. I. General features and vascular tumours. Cancer 33:1275–1288
Gosselin L (1873) Fibrome ou polype fibreux nasopharyngien suffocant et rebelle. Clinique Chirugicale de L'hopital de la Charite (Paris) 1:92–116
Holman CB, Miller WE (1965) Juvenile nasopharyngeal fibroma. AJR 94:292–298
Hora JF, Weller WA (1961) Extranasopharyngeal juvenile angiofibroma. Ann Otol Rhinol Laryngol 70:164–170
Kawada AK, Takahashi H, Anzai T (1965) Eosinophilic folliculosis of the skin (Kimura's disease). Jpn J Dermatol 76:61–72
Kimura T, Yoshima S, Ishikawa E (1948) Unusual granulation combined with hyperplastic change of lymphoid tissue. Trans Jpn Pathol Soc 37:179
Lloyd GAS (1982) Vascular anomalies in the orbit. Orbit 1:45–54
Lloyd GAS, Phelps PD (1986) Juvenile angiofibroma: imaging by magnetic resonance, CT and conventional techniques. Clin Otolaryngol 11:247–259
Michaels L (1987) Ear, nose and throat histopathology. Springer, Berlin Heidelberg New York
Reed RJ, Terazakis M (1972) Subcutaneous angioblastic lymphoid hyperplasia with eosinophilia (Kimuras's disease). Cancer 29:489–497
Schaffer K, Victor M, Farley H, Friedman J (1978) Pitfalls in the radiographic diagnosis of angiofibroma. Radiology 127:425–428
Stout AP, Murray MR (1942) Haemangiopericytoma. Ann Surg 116:26–33
Tapia Acuna R (1956) The nasopharyngeal fibroma and its treatment. Arch Otolaryngol 64:451–455
Tsutsumiuchi K, Hasegawa M, Okuno H, Watanabe I, Okayasu I, Suzuki S (1982) Haemangioma of the maxillary sinus. ORL 44: 43–50
Welch NT, Hall PA, Sprague DB (1987) Angiolymphoid hyperplasia with eosinophilia. J R Soc Med 80:384–385
Wells GC, Whimster IW (1969) Subcutaneous angiolymphoid hyperplasia with eosinophilia. Br J Dermatol 81:1–15

12 Lymphoreticular Tumours

Lymphoma

Approximately one quarter of malignant lymphomata arise in an extranodal site, of which the nose and paranasal sinus region is one of the less common. The term lymphosarcoma was first applied to tumours in this situation by Schmidt (1897), and Greifenstein (1937) was the first to report reticulum cell sarcoma occurring in the ethmoid region. The tumours are all of the non-Hodgkin's variety and no convincing example of isolated primary Hodgkin's disease in the nasal region has been presented (Friedmann and Osborn 1982).

The elderly are the most often affected, presenting with nasal obstruction and often a facial swelling. Of 37 patients reported by Wang (1971), two thirds were older than 50 years of age and 50% complained of an asymmetric swelling of the face and mouth. Although originally localised, malignant lymphoma may extend to adjacent sinuses or the orbit. In the antrum it may extend forwards to involve the cheek and the palate may be invaded through the antral floor. Regional lymph node involvement occurs in 20% of cases and generalised lymphadenopathy may sometimes follow.

Radiology and Imaging

Thirteen examples of lymphomata taking origin in the nose and paranasal sinuses have been investigated: seven were demonstrated by CT and two by magnetic resonance; the remainder were examined by pluri-direction tomography and plain radiographs. All but one of these tumours arose in the nasal cavity or in the anterior ethmoid cells and at the time of examination both were usually involved to a greater or lesser extent. The anterior location produces the facial or nasal swelling common to these tumours and it should be noted that a similar anterior site is found in lymphomata in the orbit. In a series of 46 orbital lymphomata (Lloyd 1987), two thirds were anteriorly located and the majority of patients presented with a soft tissue swelling around the eye.

Lymphoma may expand the nasal cavity (Figs. 12.1, 12.2) or cause local bone destruction, which is detectable in most patients at initial plain radiographic examination, and usually seen in the ethmoid cells at the antero-medial border of the orbit. The latter was involved ab initio in 46% of the patients seen (Fig. 12.3). These tumours were of soft tissue density and none showed calcification. They present no distinctive attenuation features on CT. Two patients were examined by magnetic resonance, and both produced a signal of only moderate intensity on T_2 -weighted spin echo sequences on pre-contrast scans. One female patient presented with a non-Hodgkin's lymphoma in the maxillary antrum and orbit (Fig. 12.4), and was examined using gadolinium DTPA. There was no enhancement of the intensity of the signal from the tumour on the post-contrast scans, either on T_1-weighted spin echo sequences or inversion recovery sequences, but the tumour was outlined against the enhanced signal from sinus and nasal mucosa (see Figs. 2.4 and 2.5, p. 19), and against the fat in the orbit and cheek (Fig. 12.5).

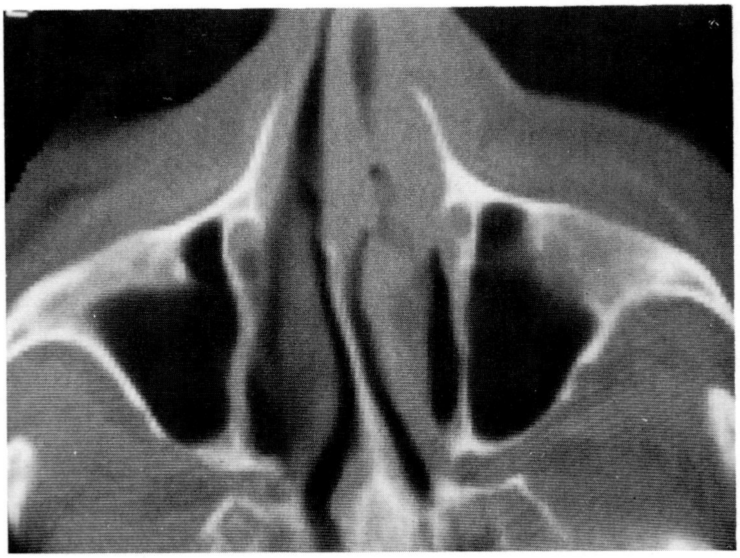

Fig. 12.1. Axial CT scan of a soft tissue mass in the anterior nasal cavity due to lymphoma.

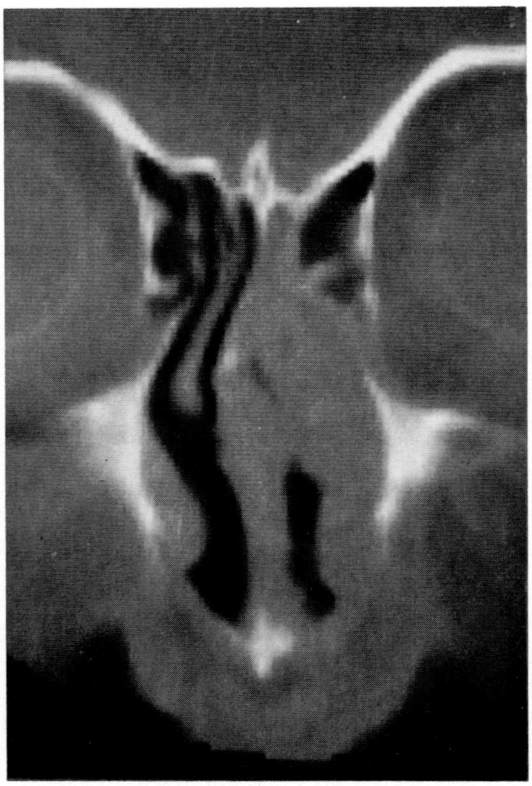

Fig. 12.2. Same patient as Fig. 12.1. Coronal CT scan showing nasal lymphoma.

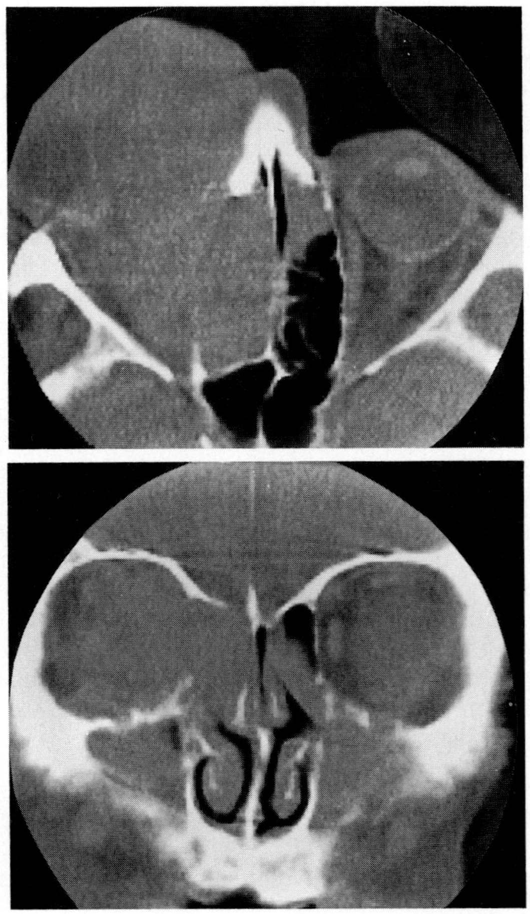

a

b

Fig. 12.3a,b. Axial (**a**) and coronal (**b**) CT scans showing a lymphoma of the ethmoid cells with massive invasion of the orbit.

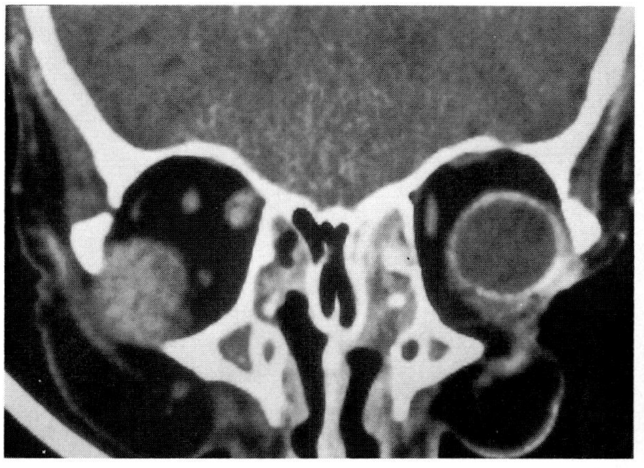

Fig. 12.4. Coronal CT scan showing the orbital extension of a lymphoma of the maxillary antrum.

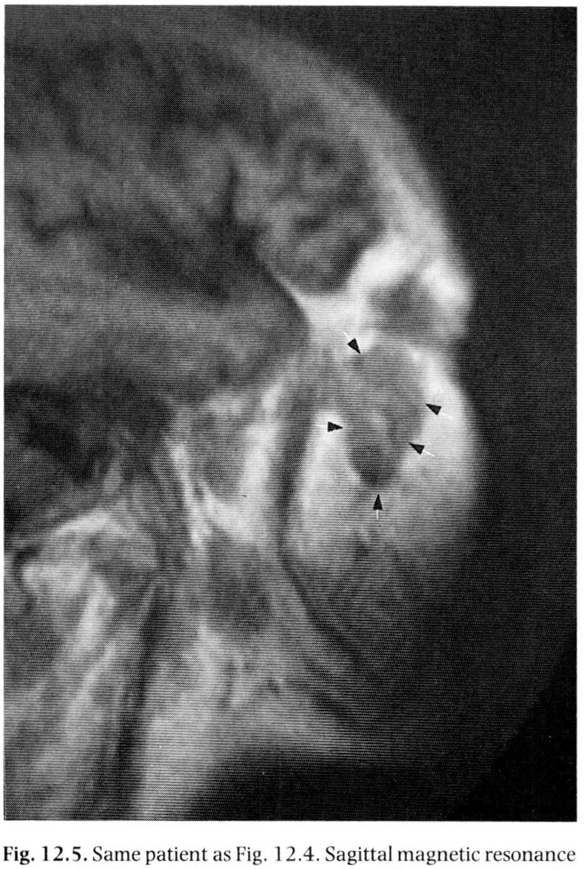

Fig. 12.5. Same patient as Fig. 12.4. Sagittal magnetic resonance scan, using the inversion recovery mode, with intravenous gadolinium DTPA. There is no signal enhancement of the tumour, which has been outlined against the fat of the orbit and cheek (*arrows*).

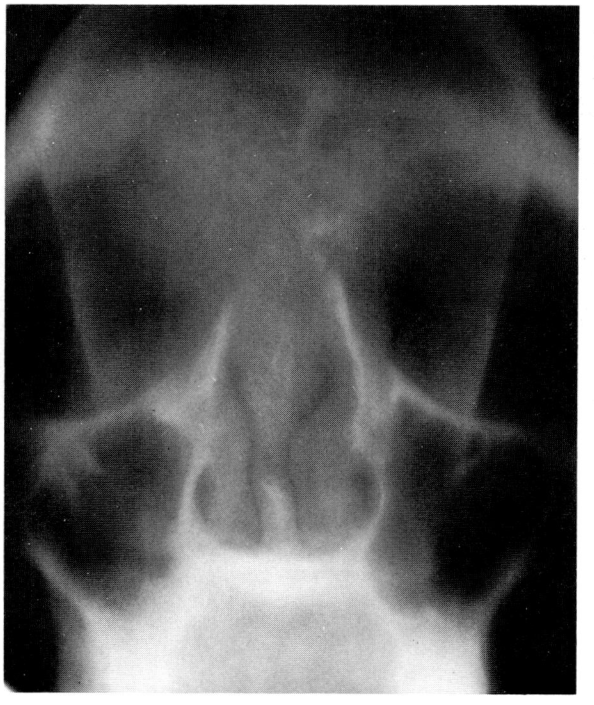

Fig. 12.6. Coronal hypocycloidal tomogram of an elderly female with lymphoma. There is a soft tissue mass over the bridge of the nose, and bone destruction in the nasal bones and adjacent frontal sinus.

In summary, the diagnosis of lymphoma should be considered when an elderly patient presents with a facial swelling in or around the nose that is accompanied by a mass in the anterior ethmoid cells and nasal cavity and associated with demonstrable bone destruction (Fig. 12.6).

Plasmacytoma

Multiple myeloma is a disease of unknown aetiology which almost always starts in the bone marrow. Occasionally it may present as a solitary lesion in bone or as an extramedullary soft tissue tumour. Extramedullary plasmacytoma may arise in any structure containing reticulo-endothelial tissue, and such tumours may either be isolated or form the initial lesion of a generalised disseminated condition (Booth et al. 1973). Extramedullary tumours of either form are infrequent, but the head and neck area – especially the upper respiratory tract, with its abundant lymphatic tissue – is by far the most common site. In the series of 192 plasmacytomata

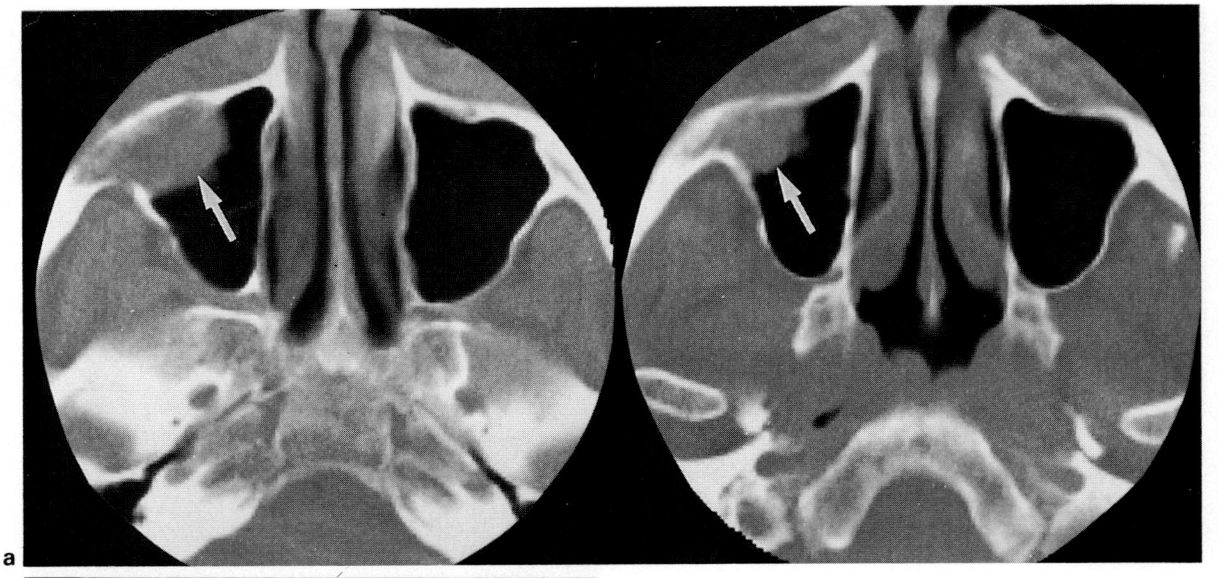

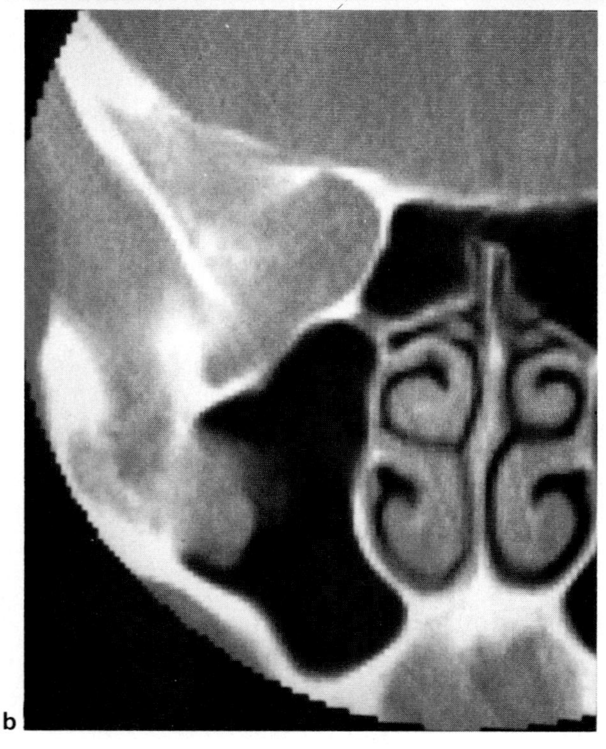

Fig. 12.7a,b. Axial (a) and coronal (b) CT scans of a plasmacytoma, showing a slightly lobulated mass (*arrows*) in the alveolar recess of the right maxillary antrum associated with local bone destruction.

of the head and neck reported by Castro et al. (1973), the nose and paranasal sinuses were the commonest locations and accounted for 37.5% of cases. The nose is the usual site, followed by the maxillary antrum (Heatly 1953). Clinically the symptoms are non-specific: epistaxis and rhinorrhoea followed by nasal obstruction are the common presenting features.

Radiology and Imaging

There are two main tasks for the radiologist in evaluating these patients prior to surgery: the first is to show the extent of the local disease by conventional radiographic and imaging studies; the second is to determine the presence or absence of generalised bone disease by skeletal survey. In some

a

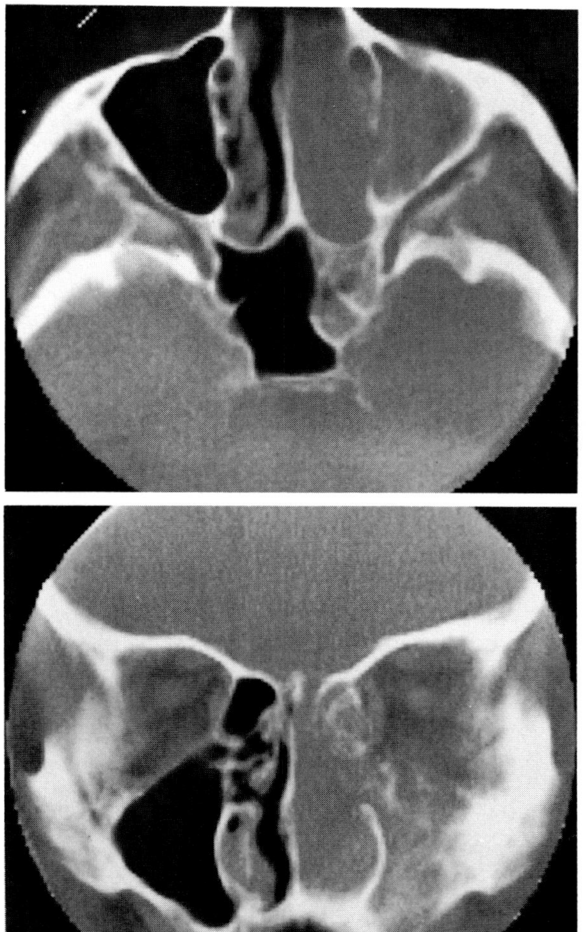

b

Fig. 12.8. Plasmacytoma. Axial (**a**) and coronal (**b**) CT scans showing a mass in the nasal cavity, with bone destruction in the ethmoids and maxillary antrum.

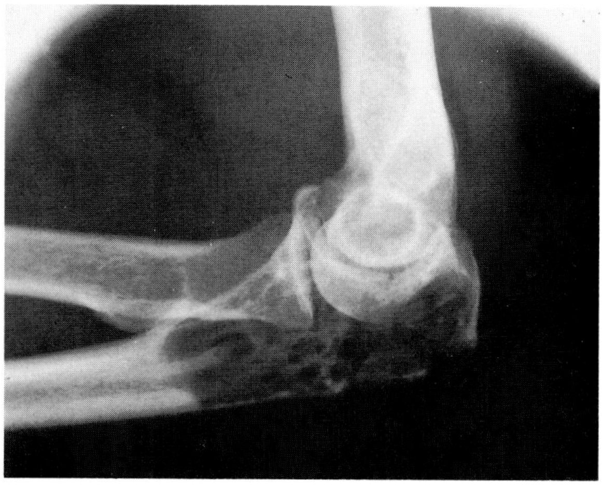

Fig. 12.9. Osteolytic defect in the olecranon, which was associated with an extramedullary plasmacytoma in the nasal cavity.

reported series less than 50% of tumours in the nose and sinuses have shown plain radiographic changes, probably because of the usual location of this tumour in the nasal cavity. Some form of tomography is therefore essential for demonstrating the soft tissue mass. Generally its appearance is nonspecific, but in some cases plasmacytoma produces a polypoidal thickening of the mucosa or lobulated appearance with underlying invasion of the soft tissues or bone (Michaels 1987). This can be shown on CT and is a distinctive combination, especially when localised to an unusual site (Fig. 12.7). The diagnosis is also suggested when characteristic skeletal changes occur in combination with a destructive lesion in the nose and paranasal sinuses (Figs. 12.8, 12.9).

References

Booth JB, Cheesman AD, Vincenti NH (1973) Extramedullary plasmacytoma of the upper respiratory tract. Ann Otol Rhinol Laryngol 82:709–715

Castro EB, Lewis JS, Strong EW (1973) Plasmacytoma of paranasal sinuses and nasal cavity. Arch Otolaryngol 97:326–329

Friedmann I, Osborn DA (1982) Pathology of granulomas and neoplasms of the nose and paranasal sinuses. Churchill Livingstone, Edinburgh

Greifenstein A (1937) Die Klinik der Retothelsarkome, dergestellt auf Grund von 31 eigenen Beobachtungen. Arch Nasen Ohren Kehlkopfheilkunde 143:189–215

Heatly CA (1953) Primary plasma cell tumors of the upper air passages with particular reference to involvement of the maxillary sinus. Ann Otol Rhinol Laryngol 62:289–306

Lloyd GAS (1987) The orbit and eye. In: Sutton D (ed) Textbook of radiology and imaging, vol 2. Churchill Livingstone, Edinburgh, chap. 51

Michaels L (1987) Ear, nose and throat histopathology. Springer, Berlin Heidelberg New York

Schmidt M (1978) Die Neubildungen in den oberen Luftwegen in die Krankheiten der oberen Luftwege. Zweite Auflage Kap 18:620–654

Wang CC (1971) Primary malignant lymphoma of the oral cavity and paranasal sinuses. Radiology 100:151–153

13 Tumours of Neurogenic Origin

Peripheral Nerve Tumours

There are four varieties of peripheral nerve tumours, all of which are derived from the proliferation of Schwann cells:

1. Neurilemmoma
2. Neurofibroma
3. Plexiform neurofibroma associated with neurofibromatosis
4. Malignant schwannoma

Each of these tumours has been identified in the nose and paranasal sinuses.

The typical neurilemmoma is encapsulated and consists histologically of spindle cells arranged in compact bundles with nuclear palisading and the characteristic Antoni type A and B areas. The plexiform neurofibroma and solitary neurofibroma are distinguished from neurilemmoma by the presence of axons: they are essentially fibrous lesions with nerve fibres traversing the tumour. Malignant schwannomata resemble the pattern of the neurilemmoma histologically, but the cells are obviously malignant.

Radiology and Imaging

Solitary tumours, usually neurilemmomata, may occur in the nose and sinuses (Fig. 13.1). One particular form can be readily diagnosed on plain radiography and by soft tissue imaging techniques. This is the infraorbital neurofibroma, which is often seen as a solitary lesion (Figs. 13.2, 13.3) but may be part of more generalised neurofibromatosis (Fig. 13.4). In either case it shows as a soft tissue mass in the roof of the maxillary antrum, causing expansion of the infraorbital canal.

Neurofibromatosis affects the orbit and skull generally and the changes which may occur are striking and characteristic: enlargement of the orbit is

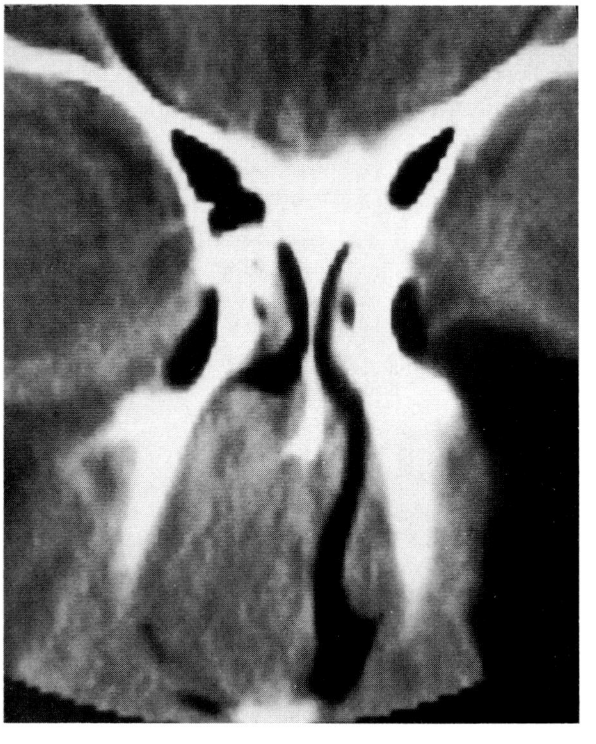

Fig. 13.1. Coronal CT scan showing a solitary neurilemmoma in the right side of the nasal cavity.

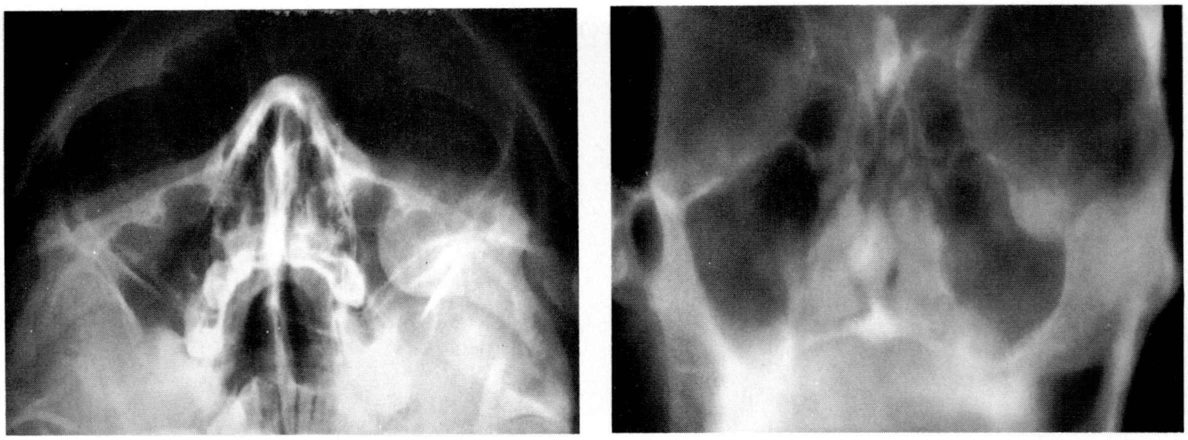

a b

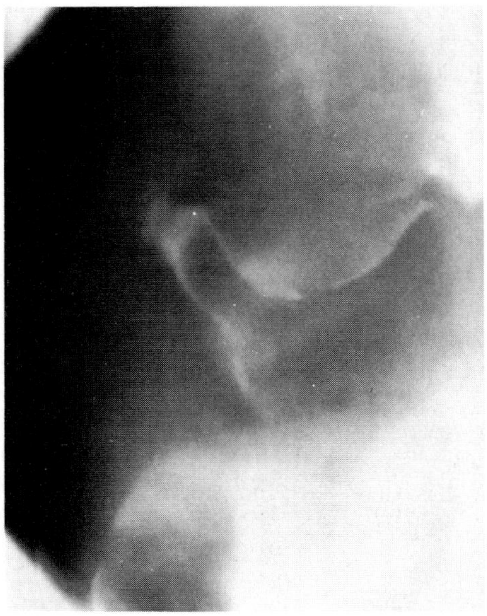

c

◀ **Fig. 13.2a–c.** Neurilemmoma of the infraorbital nerve shown on plain radiography (**a**) and on coronal hypocycloidal tomography (**b**). The sagittal tomogram (**c**) demonstrates enlargement of the infraorbital canal.

Fig. 13.3. Soft tissue mass associated with enlargement of the infraorbital canal (*arrow*). At surgery this was shown to be a malignant schwannoma. ▼

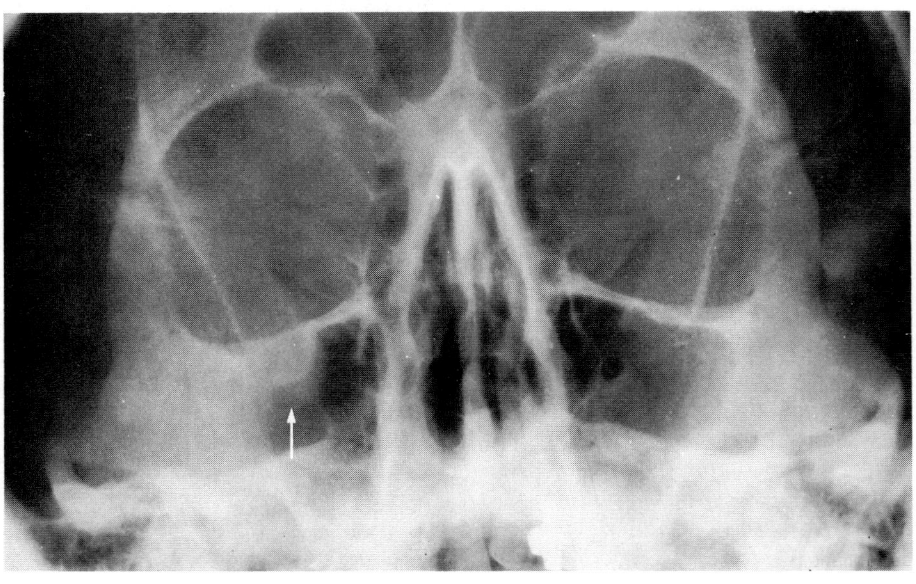

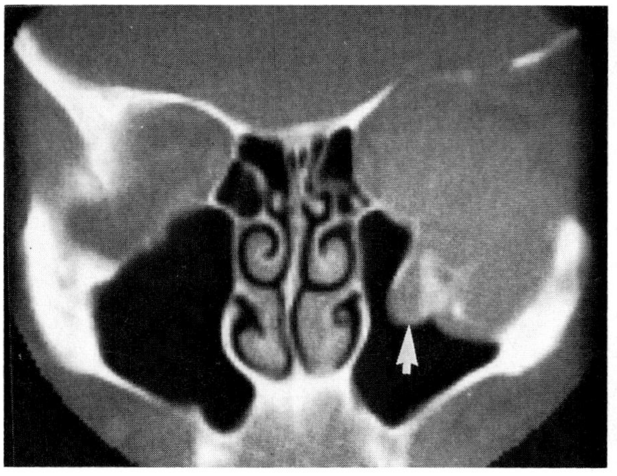

Fig. 13.4. Coronal CT scan showing orbital neurofibromatosis with involvement of the infraorbital nerve (*arrow*).

accompanied by elevation of the sphenoid ridge to the level of the orbital roof; there is often a large defect in the greater wing of the sphenoid forming the posterior boundary of the orbit, producing an encephalocoele (Fig. 13.5a). The result of these changes is the so-called empty orbit of neurofibromatosis seen on the postero-anterior radiograph (Lloyd 1975). The paranasal sinuses are usually only secondarily affected, the enlargement of the orbit associated with plexiform neurofibromatosis causing a deformity and underdevelopment of the ipsilateral maxillary antrum (Fig. 13.5b).

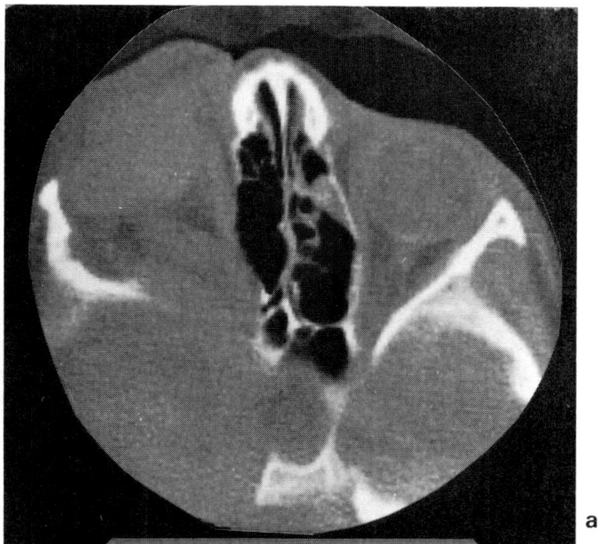

a

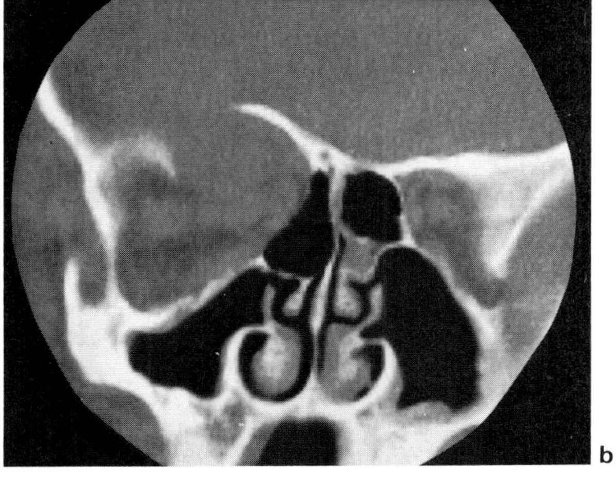

b

Fig. 13.5. a Axial CT scan showing orbital neurofibromatosis with typical posterior encephalocoele. **b** Same patient. The coronal CT scan shows that the orbital neurofibromatosis is associated with a deformity of the antrum and ethmoid cells.

Olfactory Neuroblastoma

Olfactory neuroblastoma is a malignant neoplasm of the olfactory apparatus and is composed of undifferentiated neuroectodermal tissue. The tumour was first described in the literature by Berger et al. in 1924: they called it esthesioneuroepithelioma. Berger and Coutard in 1926 described another intranasal neurogenic tumour differing in histological pattern and named it esthesioneurocytoma. These are now considered to be subgroups of the same tumour type under the generic name of olfactory neuroblastoma; one contains both nervous and epithelial elements, the other is derived from nervous elements only. The tumour arises from neuroectodermal tissue in the cribriform plate, upper third of the nasal septum and along the superior turbinate. It is said to be slightly commoner in women (54%) than in men (46%) (Elkan et al. 1979). These authors, when assimilating the reports of 97 cases, found a bimodal incidence with the majority of cases in the 50–60 year age group but another peak (16%) between 11 and 20 years. The age distribution of the tumour is unlike that of adrenal neuroblastoma, which is almost always a disease of childhood and adolescence.

Histological similarities with sympathetic neuroblastoma do exist however. Well-demarcated lobules of uniform tumour cells with congeries of blood vessels are characteristic features. The tumour cells are small with little cytoplasm. Pseudorosettes occur in 50% of cases. Electron microscopy shows neurofibrils in the lobules and this technique may be helpful in confirming the diagnosis (Michaels 1987). It is recognised that the his-

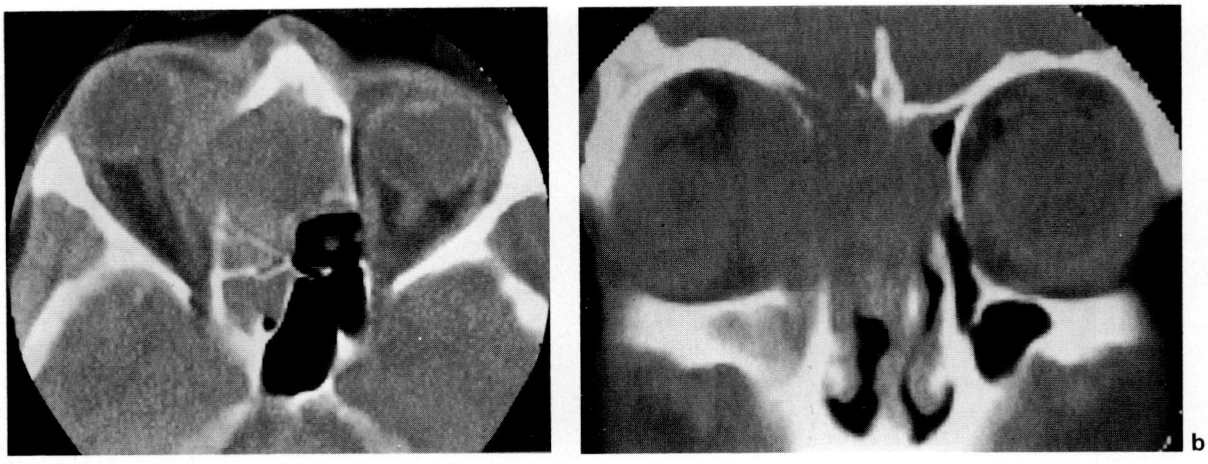

Fig. 13.6. a Axial CT scan showing an olfactory neuroblastoma with early invasion of the orbit. **b** Same patient. The coronal CT scan shows that the olfactory neuroblastoma has also eroded through the frontal sinus and cribriform plate.

tological features do not necessarily indicate the clinical aggressiveness of the tumour (Elkan et al. 1979).

The presenting symptoms are non-specific. Olsen et al. (1983) and Kadish et al. (1976) noted nasal obstruction in 70% of patients, epistaxis in 50%–70%, anosmia, diplopia, epiphora, proptosis and metastatic cervical lymphadenopathy. The tumour usually grows slowly, eroding surrounding bony structures. The site of origin ensures that the cribriform plate is involved early in the disease. Preformed holes in this bone allow rapid penetration intracranially (Harrison 1984). The latter author reviewed the surgical pathology in eight patients and described dural involvement and extension into the anterior cranial fossa in the absence of radiological demonstration of bone erosion.

The diagnosis of olfactory neuroblastoma is established by biopsy, but some characteristic features may be demonstrated radiologically. It presents as a mass in the nose and ethmoid air cells. Initially there is unilateral involvement but later the tumour extends to both sides of the ethmoid labyrinth. Bone erosion is evident in most patients at initial examination and the orbit is often invaded. The feature of the orbital extension is a displacement of the orbital contents rather than infiltration by tumour (Fig. 13.6). Characteristically the anterior cranial fossa is invaded through the cribriform plate on the affected side. This may be demonstrated by CT or magnetic resonance.

Radiology and Imaging

Twenty-four patients with this tumour have been investigated. Eighteen were examined by CT and eight of these also by magnetic resonance, four using the paramagnetic contrast agent gadolinium DTPA. The patients ranged in age from 9 to 67 years; ten were female and fourteen male. The results of plain film radiography were non-specific, the typical changes being those of a soft tissue mass in the nose with clouding of the frontal sinus and ethmoid cells on the affected side. Bone erosion in the region of the cribriform plate is not easy to assess on plain radiographs. Before the introduction of CT it was usually adequately shown by coronal hypocycloidal tomography, but this technique has been largely superseded by high-resolution CT.

The olfactory epithelium covers both superior turbinates and the upper part of the nasal septum, and it is probable that olfactory neuroblastoma

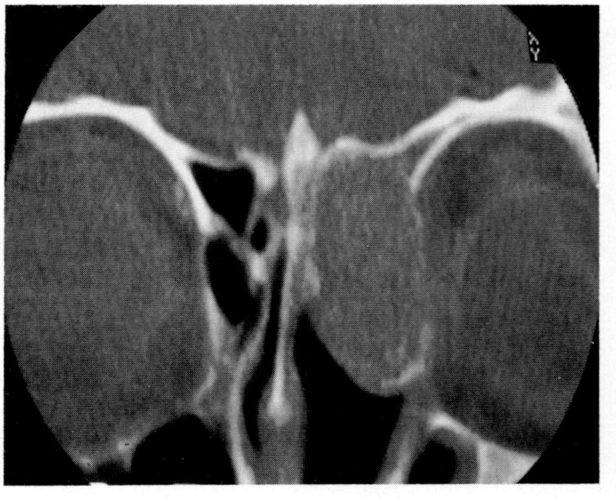

Fig. 13.7. Olfactory neuroblastoma. In this patient the tumour is shown on coronal CT scan to be invading the horizontal section of the frontal sinus.

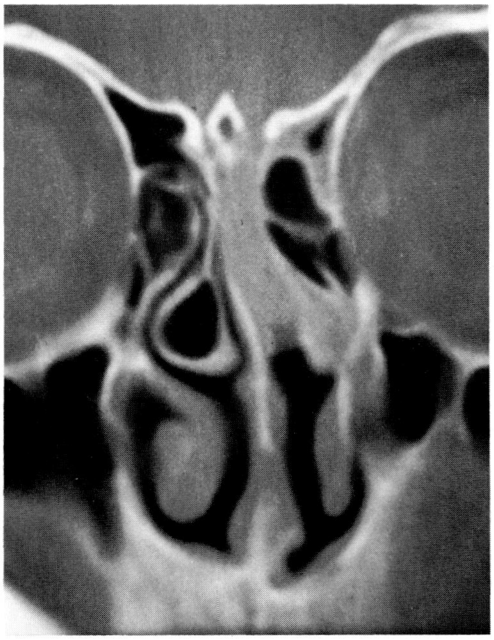

Fig. 13.8. Coronal CT scan of an early olfactory neuroblastoma. In this patient the tumour has already eroded the cribriform plate from the upper part of the nasal cavity.

takes origin in these areas. Tumour growing in the superior nasal meatus readily gains access to the frontal sinus and ethmoids by upward extension (Figs. 13.6b, 13.7). The thin bone of the fronto-ethmoidal cells in this area presents little barrier to invasion of the anterior cranial fossa. The other route of invasion is directly via the cribriform plate as described by Harrison (1984) (Fig. 13.8).

The principal task of the radiologist is to map the extent of the tumour pre-operatively. Although olfactory neuroblastoma is radiosensitive, it is not curable by radiotherapy and the correct treatment is primary cranio-facial resection. In these patients the most important area of involvement is the anterior cranial fossa. The surgical exposure required is to a large degree determined by tumour extension above the cribriform plate. The other important area is the orbit. The presence and degree of orbital involvement will influence the decision to preserve or exenterate the orbital contents.

In olfactory neuroblastoma the level of enhancement on CT following intravenous contrast is variable. Rosengran et al. (1979) reported no enhancement, while Manelfe et al. (1978) noted an increase in attenuation from 30–40 Hounsfield units to 66–80 after intravenous contrast. Contrast enhancement is of little value in establishing the diagnosis but is important for showing the extent

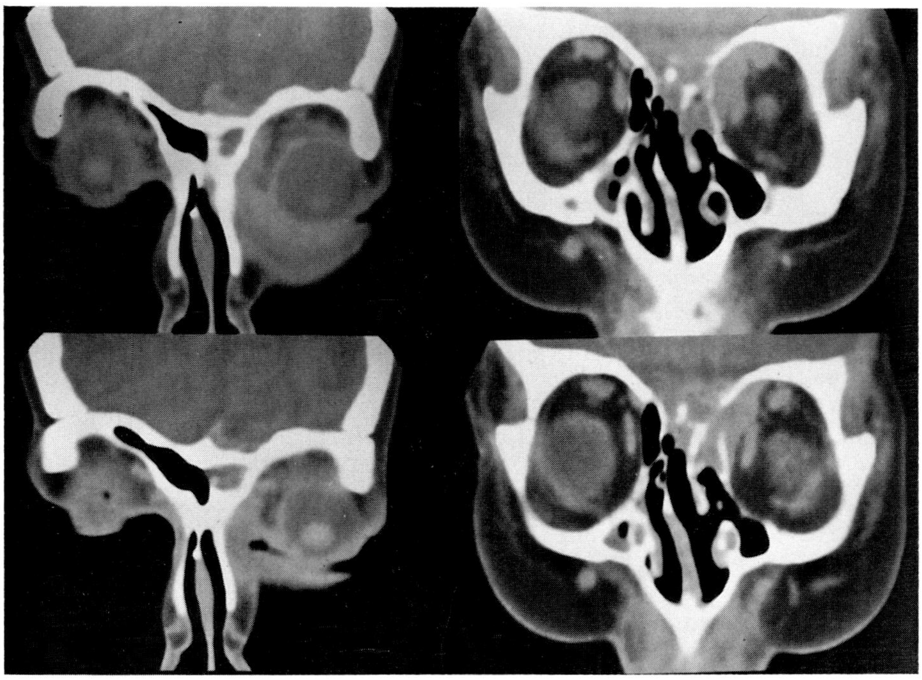

Fig. 13.9. Coronal CT scans of an olfactory neuroblastoma taken after a bolus injection of intravenous contrast, showing the extension of enhanced tumour above the cribriform plate.

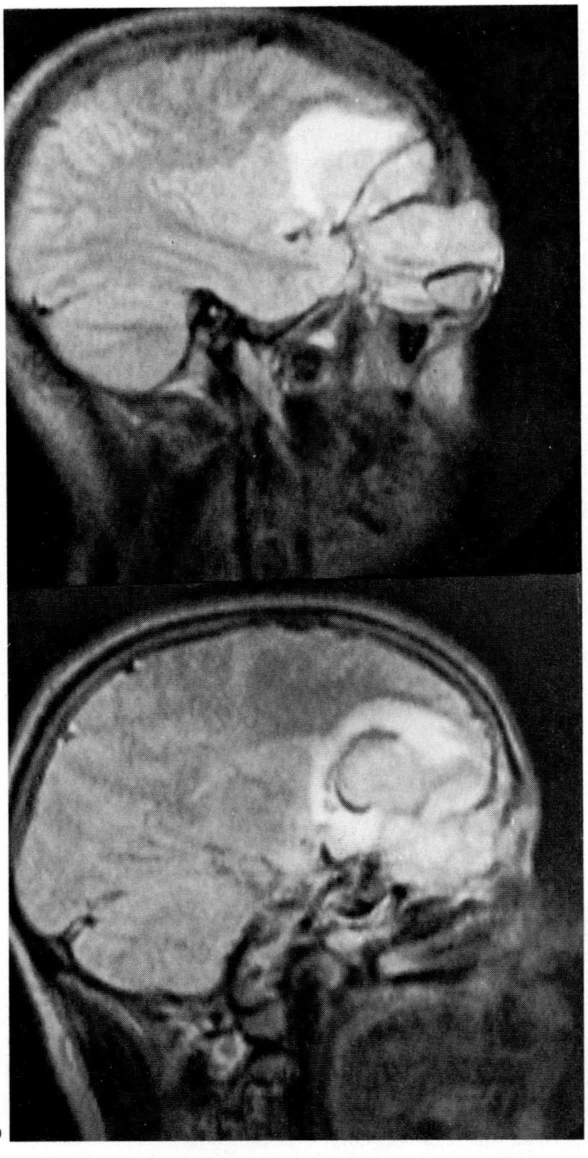

a

b

Fig. 13.10a,b. Massive extension of olfactory neuroblastoma into the anterior cranial fossa shown on sagittal magnetic resonance scans.

of tumour invasion of the anterior cranial fossa. The normal procedure has been to give a bolus of contrast immediately prior to direct coronal CT scanning (Lund et al. 1983), so that the enhanced tumour is outlined against the non-enhanced brain tissue (Fig. 13.9).

An alternative and superior way of demonstrating the extent of these tumours is by magnetic resonance (Fig. 13.10). Like most tumours of the paranasal sinuses olfactory neuroblastoma produces the strongest signal on T_2 -weighted spin echo sequences. The signal intensity varied considerably in the eight patients examined and in some the tumour produced very high signal. The effectiveness of this modality has been increased by the use of gadolinium DTPA. The most emphatic demonstration of tumour above the cribriform plate was shown on inversion recovery sequences following intravenous gadolinium. The nasal mucosa also enhances with this paramagnetic contrast agent (Carr and Gadian 1985), so that poor differentiation in the nose between enhanced tumour and normal mucosa may lead to an overestimate of tumour size. However in the sinus cavities this disadvantage is offset by superior discrimination between tumour and fluid or mucosal thickening. Tumours producing high signal on T_2-weighted spin echo sequences may be difficult to distinguish from retained secretion in the sinus cavities. It is, however, possible to make this distinction following intravenous gadolinium. The degree of enhancement of tumour by gadolinium is most marked on T_1-weighted images and is particularly well shown on inversion recovery sequences (Fig. 13.11). After intravenous gadolinium discrimination between tumour and retained secretion or concomitant inflammatory changes becomes more obvious because retained secretion in the sinuses is not affected by the gadolinium. This is clearly demonstrated on the pre- and post-contrast T_1-weighted spin echo sequences shown in Fig. 13.12.

The magnetic resonance signal characteristics of olfactory neuroblastoma are almost certainly related to the highly vascular nature of this tumour. In the eight patients examined by this technique the typical features were those of an intense or moderately intense signal on pre-contrast T_2-weighted spin echo sequences, and strong signal enhancement after intravenous gadolinium on T_1-weighted sequences. A characteristic feature of the response to gadolinium was an enhancement of tumour higher than that of turbinate mucosa on inversion recovery and less than that of the mucosa when T_1-weighted spin echo sequences were employed.

Meningioma

A meningioma in the nose or paranasal sinuses may occur as a primary tumour or as the secondary extension of an intracranial growth.

Primary Meningioma

Meningiomata are thought to arise from "cap" cells or meningocytes located in clusters at the tips of

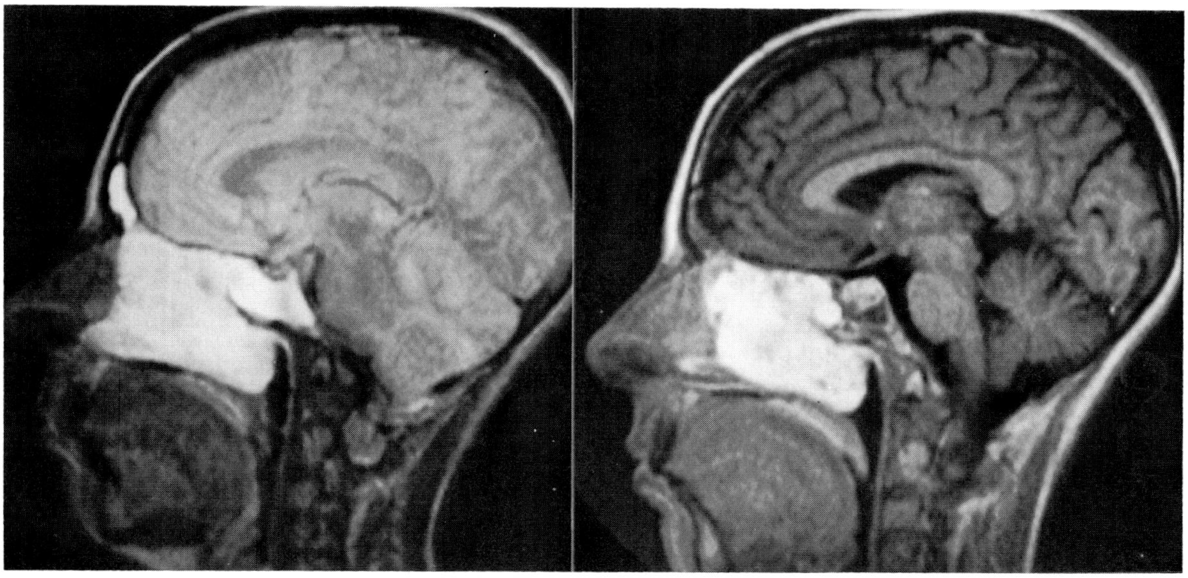

a b

Fig. 13.11a,b. Olfactory neuroblastoma demonstrated by magnetic resonance. **a** Sagittal scan using a T_2-weighted spin echo sequence made prior to intravenous gadolinium DTPA. There is poor discrimination between tumour and retained secretion in the frontal and sphenoid sinuses. **b** Inversion recovery sequence made after intravenous gadolinium. Enhancement of the tumour now allows discrimination between tumour and retained secretion, and the tumour extent is optimally demonstrated.

the arachnoid villi (Kernohan and Sayre 1952). In primary extracranial tumours the meningocyte is believed to arise from arachnoid cells along cranial nerve sheaths or from embryonal ectopic arachnoid cells. Another theory of origin is that of Shuangshoti and Panyathanya (1973), who believed that meningiomata can arise directly from multipotential mesenchymal cells wherever situated. Meningiomata arising primarily in the sinuses are rare. Willen et al. (1979) could find only nine examples

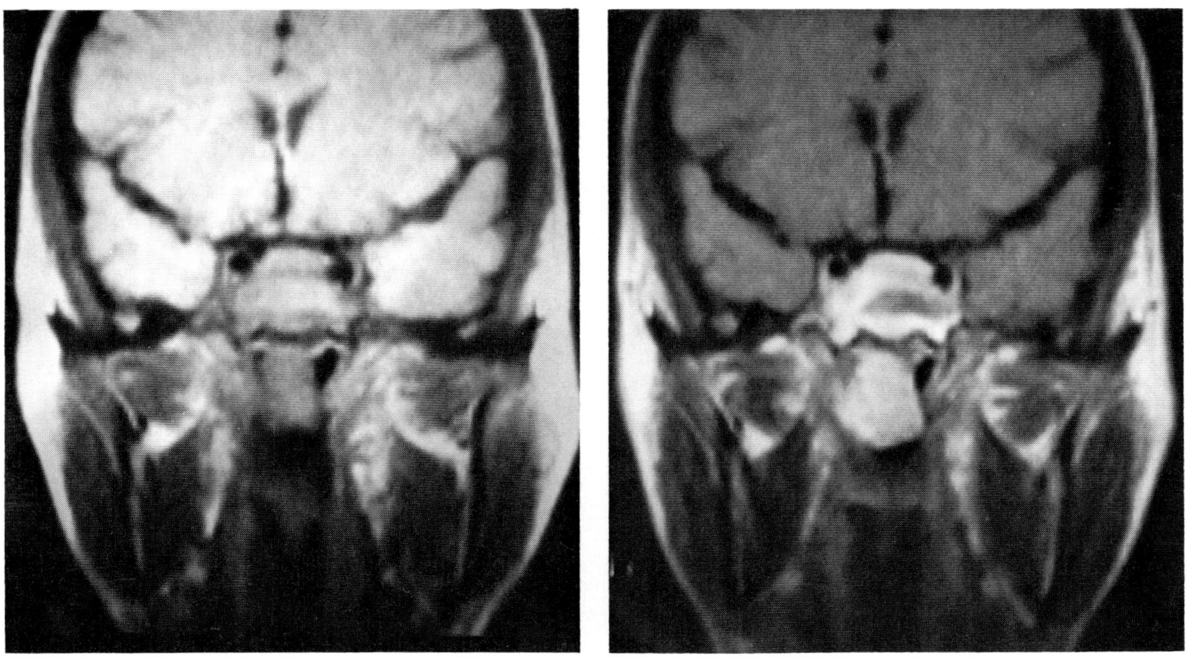

a b

Fig. 13.12a,b. T_1-weighted spin echo sequences before (**a**) and after (**b**) intravenous gadolinium. Coronal sections through the sphenoid sinus. The thickened mucosa enhances, the retained fluid centrally does not. This was correctly interpreted as inflamed mucosa with no tumour present. Note enhancement of tumour tissue below the sphenoid in the nasopharynx.

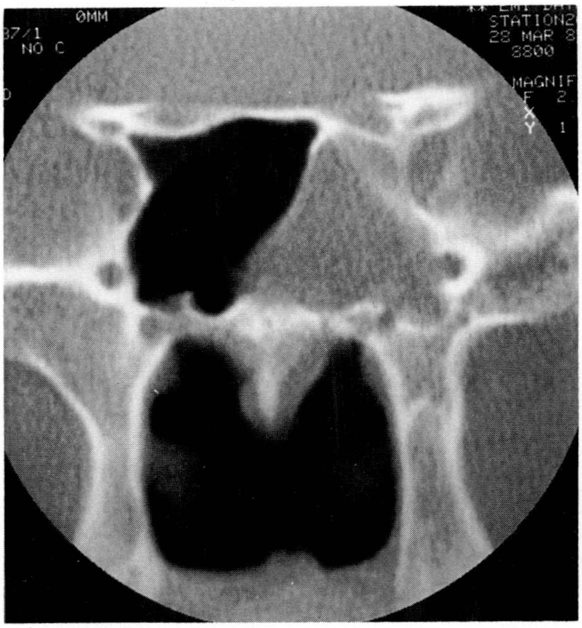

Fig. 13.13. Recurrent meningioma in the sphenoid sinus invading the pterygoid canal and foramen rotundum.

in the literature, and seven meningiomata were seen in a 20-year period at the Institute of Laryngology in London (Papavasiliou et al. 1982). They have been reported in the nose (Lindstrom and Lindstrom 1969; Kjeldsberg and Minckler 1972; Willen et al. 1979), where they may present as a nasal polyp. Tumours have also been reported in the maxillary antrum (Hill 1962), in the frontal sinus (New and Devine 1947; Majoros 1970) and in the sphenoid sinus (Sardar et al. 1979).

Histologically meningiomata in this situation show a meningothelial or fibroblastic pattern, in which tumour cells of epithelial or fibroblastic appearance are arranged concentrically around small blood vessels, sometimes with psammoma bodies (Michaels 1987). Most of them are benign in that metastases are very rare; but they spread locally invading foramina (Fig. 13.13), whilst pressure erosion may result in spread from one cavity to another. Recurrence after surgery is common.

Radiology and Imaging

Six patients with histologically proven meningiomata arising primarily in the paranasal sinuses have been investigated. In addition to conventional X-ray studies all six patients had CT scans and two were examined by magnetic resonance. The average age of the patients was 45 years and the age range 11–75, with two patients in the second decade and three in the seventh. Five of these patients had known tumour recurrence, some multiple; one patient was lost to follow-up.

The tumours showed initially as a soft tissue mass in the sinus involved, demonstrated either on a plain radiograph or by CT scan (Fig. 13.14). On CT the tumour usually enhanced strongly after contrast. Calcification was shown in the tumour mass in four patients. Hyperostosis was present in two patients ab initio and was shown eventually in varying degree in all patients following repeat scans for tumour recurrence. In four patients this involved the cribriform plate and anterior fossa (Fig. 13.15).

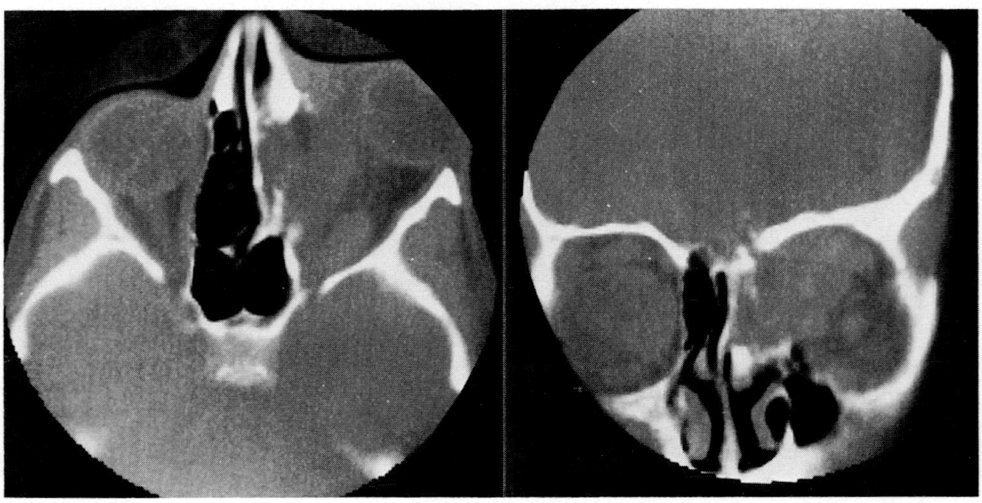

a b

Fig. 13.14a,b. Axial (**a**) and coronal (**b**) CT scans showing meningioma of the ethmoids invading the orbit. Note the hyperostosis in the cribriform plate area and in the roof of the orbit.

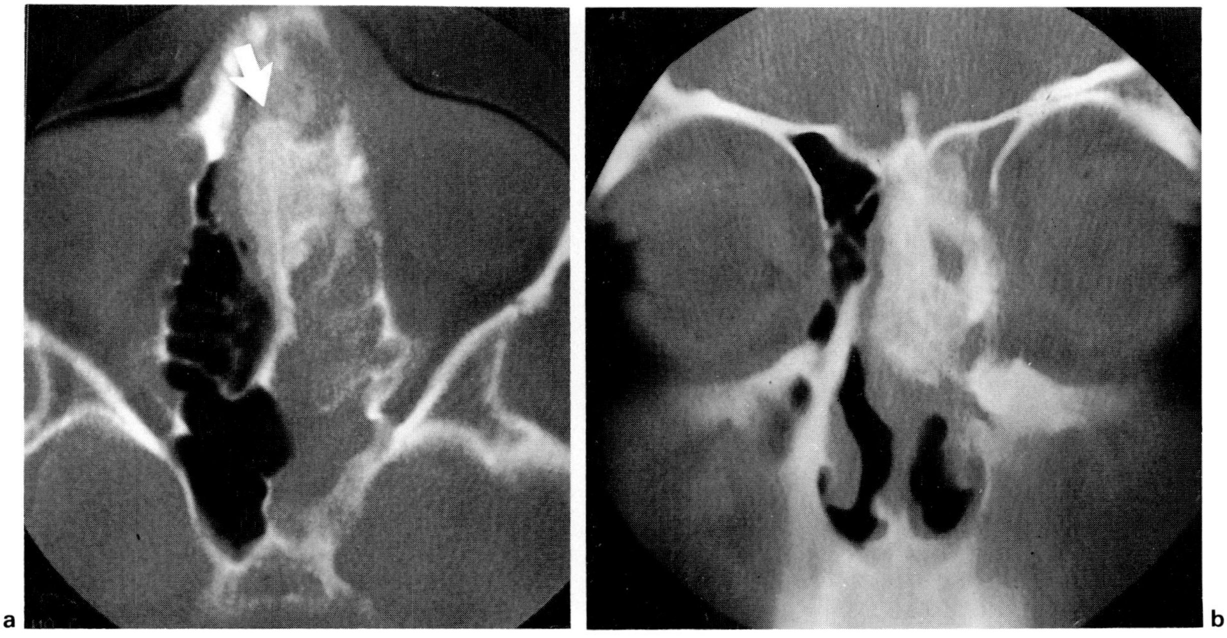

Fig. 13.15a,b. Recurrent meningioma in the ethmoid cells. **a** Axial CT scan showing calcification and hyperostosis (*arrow*). **b** Coronal section showing the calcification and hyperostosis.

Two of the recurrent extradural meningiomata were examined by magnetic resonance. Failure to demonstrate hyperostosis and calcification is a drawback of this technique and generally it was of less diagnostic value than CT. However, in one patient the tumour showed a low-intensity rim (Fig. 13.16), a sign which has been described for intracranial meningiomata (Zimmerman et al. 1985; Mawhinney et al. 1986).

Another change characteristic of meningioma may occur in the sinuses: this is the condition known as pneumosinus dilatans (see Chap. 6). It may be provoked in the frontal sinuses by an intracranial subfrontal meningioma, or the sphenoid sinuses may be affected by a meningioma of the tuberculum sellae or planum sphenoidale. It may also occur as the secondary effect of an extradural meningioma in the orbit (Lloyd 1985) (see Fig. 6.42, p. 68).

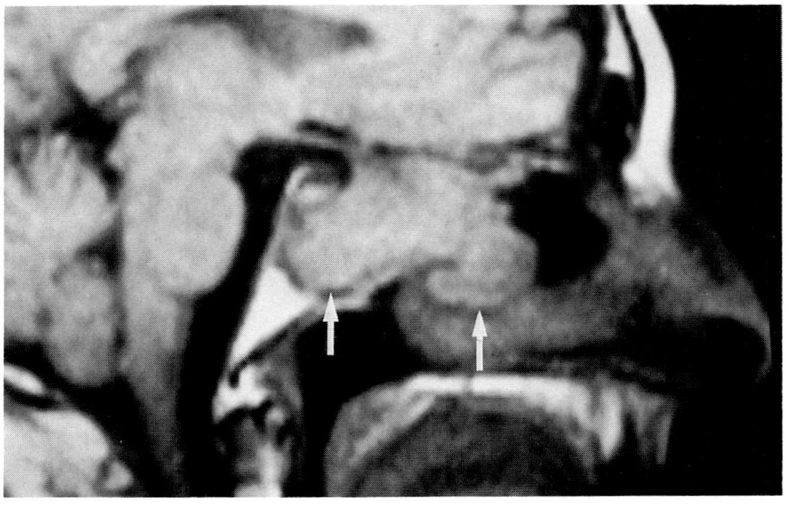

Fig. 13.16. Sagittal magnetic resonance scan of a recurrent meningioma in the nose and sphenoid sinus. The tumour shows a hypointense border (*arrows*).

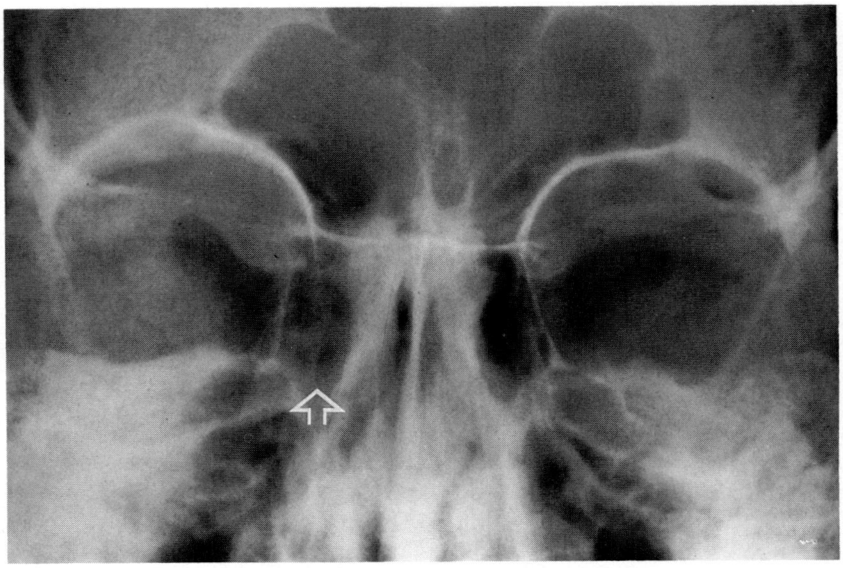

Fig. 13.17. Occipito-frontal view showing hyperostosis on the greater and lesser wings of the sphenoid (*right*) due to a sphenoidal ridge meningioma. Note that there is loss of translucence in the right ethmoids and sphenoid sinuses (*arrow*).

Secondary Meningioma in the Sinuses

Secondary involvement of the sinuses by a primary intracranial meningioma is well recognised and usually occurs as a recurrence, following surgery to the intracranial growth (Kendall 1973). Kendall described five patients with post-operative involvement of the sinuses. In contrast a pre-operative survey of 80 patients with sphenoidal ridge meningioma revealed only one patient showing invasion of the sinuses. He concluded that surgery rather than chronicity of the lesion favours the extracranial spread of meningioma.

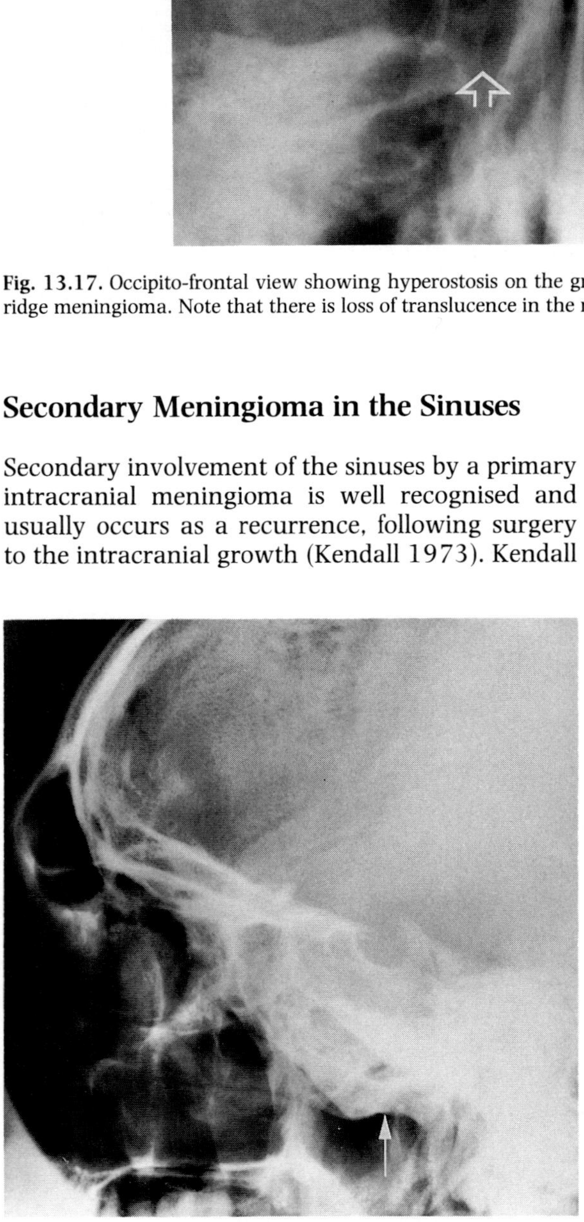

Fig. 13.18. Same patient as Fig. 13.17. Lateral skull radiograph showing an opacity in the sphenoid sinus and nasopharynx. The meningioma presented as a mass in the nasopharynx (*arrow*) and the diagnosis was confirmed at biopsy.

Two examples of sphenoid ridge meningioma invading the sphenoid sinus have been seen prior to any surgery. One patient presented with proptosis and a mass in the nasopharynx; biopsy of the latter confirmed the diagnosis of meningioma (Figs. 13.17, 13.18).

References

Berger L, Coutard H (1926) L'esthesioneurocytome olfactif. Bulletin de l'Association de France d'etude. Cancer 15:404–421

Berger L, Luc, Richard (1924) L'esthesioneuroepitheliome olfactif. Bulletin de l'Association de France d'etude. Cancer 13: 410–421

Carr DM, Gadian DG (1985). Contrast agents in magnetic resonance imaging. Clin Radiol 136:561–568

Elkan D, Hightower S, Lim Meng, Cantrell RW, Constable WC (1979) Esthesioneuroblastoma. Cancer 44:1087–1094

Harrison D (1984) Surgical pathology of olfactory neuroblastoma. Head Neck Surg 7:60–64

Hill CL (1962) Meningioma of the maxillary sinus. Arch Otolaryngol 76:547–549

Kadish S, Goodman M, Wang CC (1976) Olfactory neuroblastoma: a clinical analysis of 17 cases. Cancer 137:1571–1576

Kendall B (1973) Invasion of the facial bones by basal meningiomas. Br J Radiol 46:239–244
Kernohan JW, Sayre GP (1952) Tumours of the central nervous system. Armed Forces Institute of Pathology, Washington DC
Kjeldsberg CR, Minckler I (1972) Meningioma presenting as nasal polyps. Cancer 29:153–156
Lindstrom CG, Lindstrom DW (1969) On extracranial meningioma. Acta Otolaryngol (Stockh) 68:451–456
Lloyd GAS (1975) Radiology of the orbit. Saunders, London and Philadelphia
Lloyd GAS (1985) Orbital pneumosinus dilatans. Clin Radiol 36: 381–386
Lund VJ, Howard DJ, Lloyd GAS (1983) CT evaluation of paranasal sinus tumours for craniofacial resection. Br J Radiol 56: 439–446
Majoros M (1970) Meningioma of the paranasal sinuses. Laryngoscope 80:640–645
Manelfe C, Bonafe A, Fabre P, Pessey JJ (1978) CT in olfactory neuroblastoma. J Comput Assist Tomogr 2:412–420
Mawhinney RR, Buckley JH, Worthington BS (1986) Magnetic resonance imaging of the cerebello-pontine angle. Br J Radiol 59:961–969
New GB, Devine MD (1947) Neurogenic tumour of nose and throat. Arch Otolaryngol 46:163–179
Michaels L (1987) Ear, nose and throat histopathology. Springer, Berlin Heidelberg New York
Olsen K, De Santo L (1983) Olfactory neuroblastoma. Arch Otolaryngol 109:797–802
Papavasiliou A, Sawyer R, Lund V (1982) Effects of meningiomas on the facial skeleton. Arch Otolaryngol 108:255–257
Rosengran J, Bao Shan J, Wallace S, Danzinger J (1979) Radiographic features of olfactory neuroblastoma. AJR 132:945–948
Sadar ES, Conomy JP, Benjamin SP, Levine HL (1979) Meningiomas of the paranasal sinuses, benign and malignant. Neurosurgery 4:227–231
Shuangshoti S, Panyathanya R (1973) Ectopic meningiomas. Arch Otolaryngol 98:102–105
Willen R, Gad A, Willen M, Qvarnstrom O, Stahle J (1979) Extracranial meningioma presenting as a nasal polyp. ORL 41:234–239
Zimmerman RD, Fleming CA, Saint-Louis LA, Lee BCP, Manning JJ, Deck MDF (1985) Magnetic resonance imaging of meningiomas. Am J Neuroradiol 6:149–157

14 Tumours of Muscle Origin

Tumours of muscle origin are divided into skeletal muscle tumours (rhabdomyoma, rhabdomyosarcoma) and smooth muscle tumours (leiomyoma, leiomyoblastoma and leiomyosarcoma).

Skeletal Muscle Tumours

Although benign rhabdomyoma has been found in the nasopharynx, no example of this tumour has been reported as occurring in the nose and sinuses. In contrast rhabdomyosarcoma (which represents 8% of all malignant disease in children) shows a predilection for the head and neck region especially the orbit, but the nose and paranasal sinuses may also be primarily involved. Histologically the tumour is derived from the malignant rhabdomyoblast, and has been classified by Horn and Enterline (1958) into pleomorphic, embryonal, alveolar and botryoid sarcoma. The presence of cross-striation within the cell structure is the most characteristic histological feature.

The fact that the nose and paranasal sinuses do not normally contain skeletal muscle poses a problem of derivation of these tumours. Some authors (Cooper 1934) have thought that their origin might be from the medial pterygoid muscle with secondary invasion of the sinuses, but the most commonly accepted view is that the rhabdomyosarcomata of children and adolescents are derived from embryonic tissue, either immature muscular tissue or undifferentiated mesenchymal tissue with a potency for aberrant differentiation into muscle fibres.

Rhabdomyosarcoma in the sinuses is a disease of the adolescent and young adult. Of 10 patients investigated 60% were in the second or third decade; the average age was 26.7 years and the age range 10–56 years. The common form of presentation is facial swelling with or without pain. Other clinical features include epistaxis, nasal obstruction, epiphora, and proptosis when the orbit is involved (90% of patients). Spread to the sphenoid sinus and skull base may give rise to cranial nerve palsies and the tumour may invade the neck directly via the nasopharynx or by lymph node metastases. Fu and Perzin (1976) reported 50% cervical node involvement in their series of 16 rhabdomyosarcomata. This is in marked contrast to malignant epithelial sinus tumours, where neck metastases are unusual. The overall clinical picture is that of widespread tumour involvement at initial examination, subsequently confirmed by CT and magnetic resonance. This widespread extent of the tumour at presentation would account for the poor prognosis of rhabdomyosarcoma in the nose and sinuses compared with similar tumours of orbital origin. For example Sutow et al. (1970) recorded a 75% 5-year survival for orbital rhabdomyosarcoma as opposed to 21% survival for patients with rhabdomyosarcoma elsewhere in the head and neck. Orbital rhabdomyosarcoma characteristically involves the anterior orbit, produces early clinical signs and does not as a rule invade bone until a late stage of the disease.

Smooth Muscle Tumours

In the nose and sinuses smooth muscle tumours, either benign or malignant, are extremely rare. Leiomyoma has so far only been reported in the nasal cavity. Leiomyosarcoma is also found in the nose but may in addition involve the paranasal sinuses. The only smooth muscle normally present here is in the vasculature, and some believe that this tissue constitutes the origin of leiomyomatous tumours, although their derivation from multipotent mesenchymal cells is at least an equally valid possibility. Histologically the tumours form interlacing bundles of spindle cells in which "myofibrils" may be demonstrated by appropriate staining. This feature is constantly found in leiomyoma and in the greater proportion of leiomyosarcomata (Stout and Hill 1958). The sarcomatous smooth muscle tumours have a very poor prognosis, spreading to involve more than one cavity in the sinuses, often with secondary spread to the orbit, and producing systemic metastases typically in the lung fields. Spread to cervical lymph nodes may also occur.

Radiology and Imaging

In a period of 7 years 12 patients with histologically proven muscle-derived tumours of the nose and paranasal sinuses have been investigated. Nine had skeletal muscle tumours, all of which were rhabdomyosarcomata. Of the three tumours of smooth muscle origin one was a leiomyoblastoma and two leiomyosarcomata. The age range of the patients with skeletal muscle tumours was 11–56 years and the average age 28.5 with a peak distribution in the second decade. The two patients with a leiomyosarcoma were aged 47 and 56 respectively and the single patient with a leiomyoblastoma was 5 years old (Papavasiliou and Michaels 1981). All patients were examined by conventional radiography and eleven by CT. In addition three patients had magnetic resonance scanning.

The results of these investigations can be summarised as follows:

1. All patients were shown to have a soft tissue mass in the nose or paranasal sinuses by plain radiography. This was confirmed by tomography – whether conventional, computerised or magnetic resonance tomography.

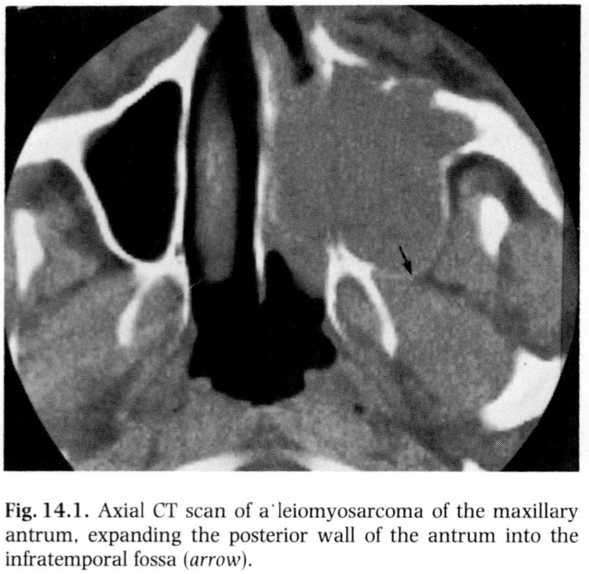

Fig. 14.1. Axial CT scan of a leiomyosarcoma of the maxillary antrum, expanding the posterior wall of the antrum into the infratemporal fossa (*arrow*).

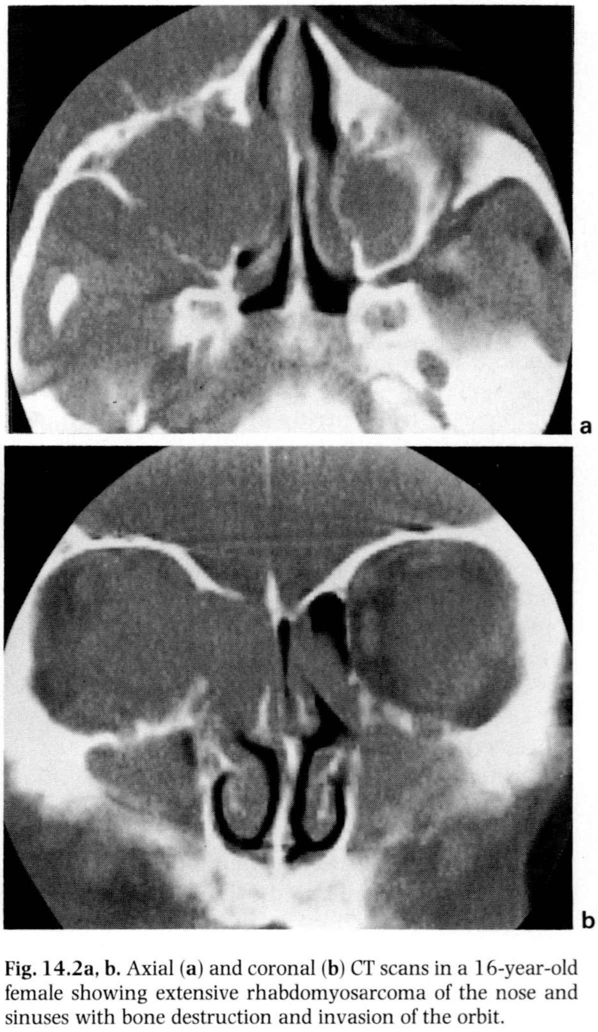

a

b

Fig. 14.2a, b. Axial (**a**) and coronal (**b**) CT scans in a 16-year-old female showing extensive rhabdomyosarcoma of the nose and sinuses with bone destruction and invasion of the orbit.

2. In all patients the tumour involved the nasal cavity and adjacent maxillary antrum.

3. Of the nine patients with rhabdomyosarcoma, in eight (89%) the ethmoids were also involved and in five (55%) the sphenoids or frontal sinuses were involved by tumour.

4. Tumour calcification was not observed. In some patients displaced fragments of sinus wall were incorporated within the tumour mass but no new bone formation or ectopic calcification was demonstrable.

5. At initial examination the orbit was found to be involved in eight (89%) of the patients with rhabdomyosarcoma and in both patients with leiomyosarcoma, in addition to infratemporal fossa extension (Figs. 14.1, 14.2, 14.3 and 14.4). Only the nasal cavity was involved in the child with the leiomyoblastoma (Fig. 14.5).

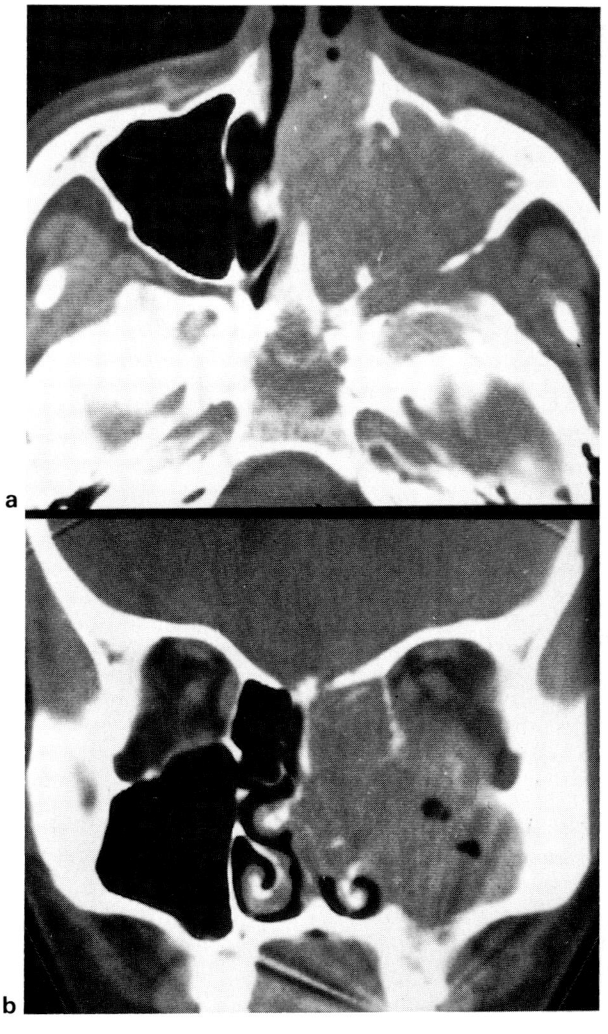

Fig. 14.3a, b. Axial (**a**) and coronal (**b**) CT scans of a 21-year-old male with a rhabdomyosarcoma of the maxillary antrum. There is invasion of the orbit and pterygo-palatine fossa.

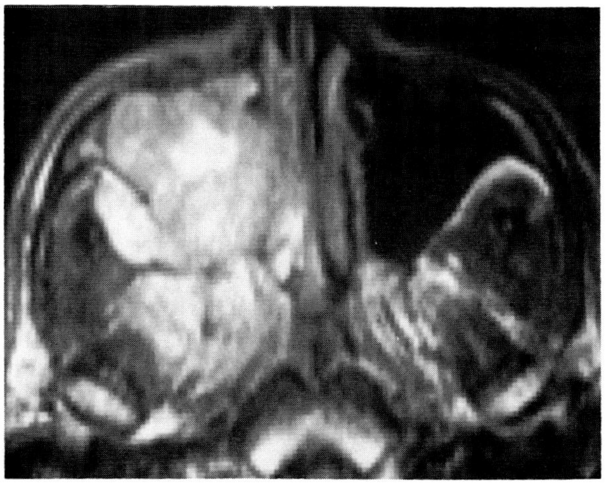

Fig. 14.4. Axial magnetic resonance scan showing a very large rhabdomyosarcoma of the maxillary antrum in a 41-year-old male. There is massive invasion of the infratemporal fossa.

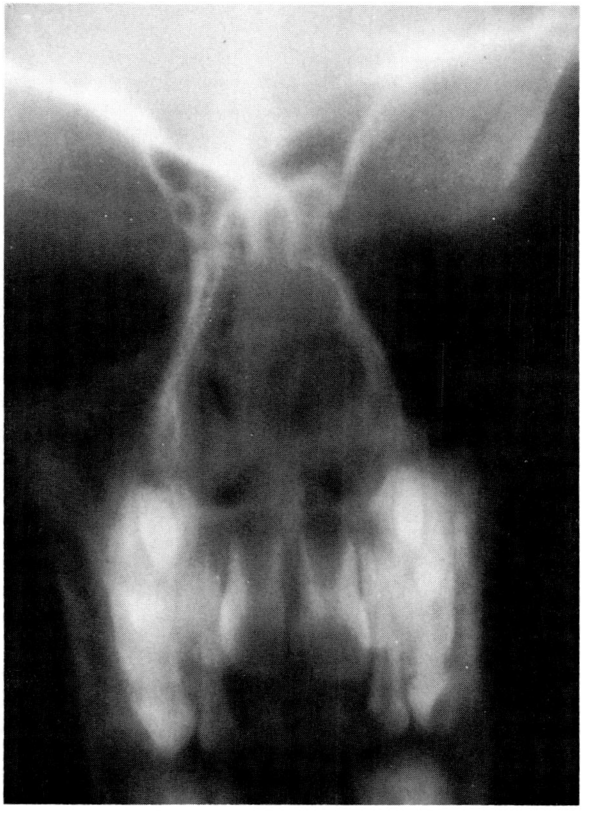

Fig. 14.5. Coronal hypocycloidal tomogram showing bone erosion and expansion of the nasal cavity in a 5-year-old. This was histologically a leiomyoblastoma or epithelioid leiomyoma.

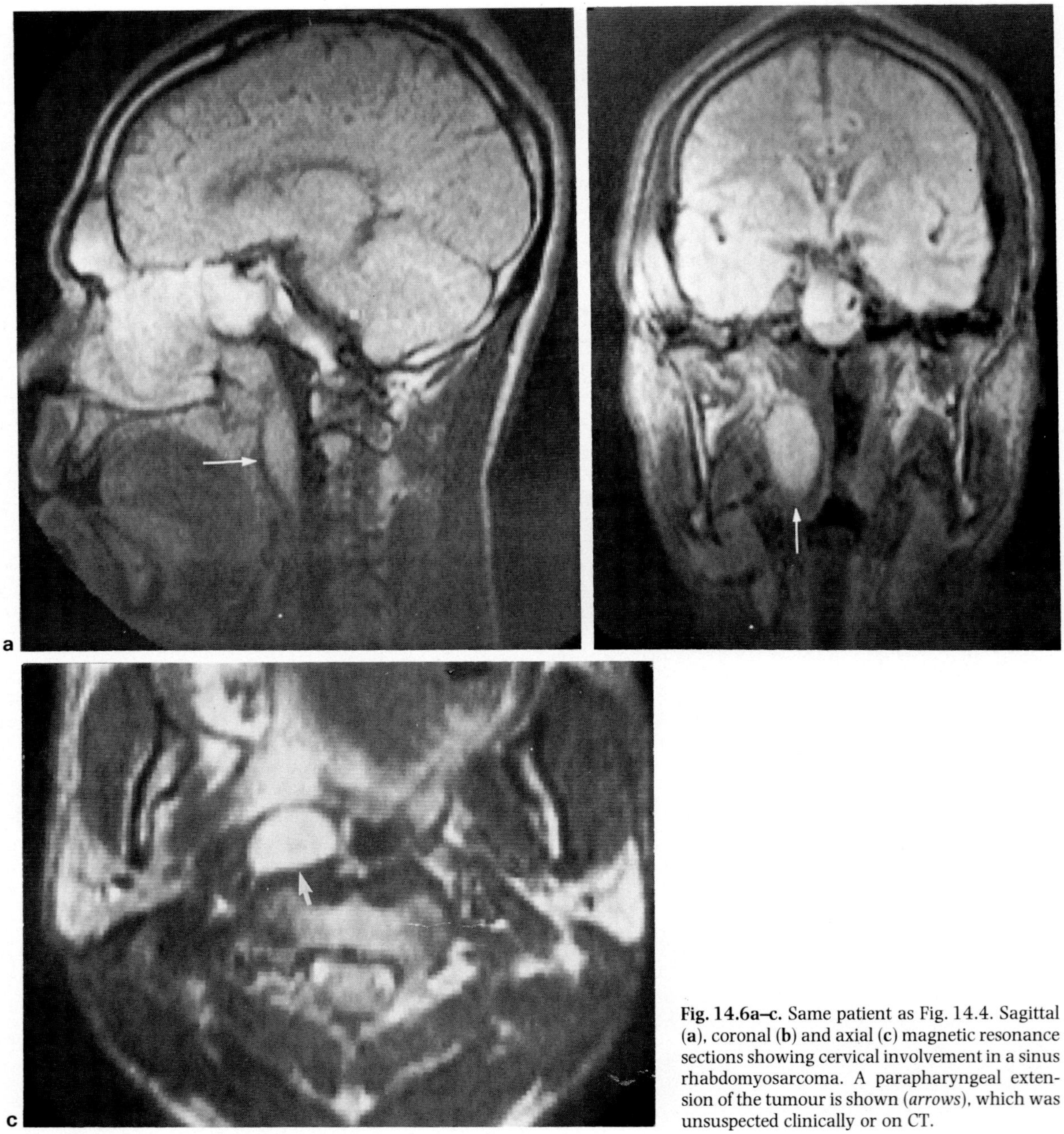

a b c

Fig. 14.6a–c. Same patient as Fig. 14.4. Sagittal (**a**), coronal (**b**) and axial (**c**) magnetic resonance sections showing cervical involvement in a sinus rhabdomyosarcoma. A parapharyngeal extension of the tumour is shown (*arrows*), which was unsuspected clinically or on CT.

6. Magnetic resonance studies were carried out in three of the cases of rhabdomyosarcoma. In all there was a high-intensity signal from the tumour on T_2-weighted spin echo sequences. At presentation these tumours were all well advanced. The high incidence of neck involvement by this tumur is well recognised (Fu and Perzin 1976), and two of these patients had cervical involvement when scanned. One had direct extension of the tumour from the maxillary antrum and pterygo-palatine fossa to the parapharyngeal region. This was unsuspected clinically or by CT scan and only recognised on the magnetic resonance scans (Fig. 14.6). The demonstration of the full extent of the tumour was a result of the wide coverage achieved by three-plane multislice imaging using a head coil, which allowed total scanning of the head and neck. This is a major advantage of the method and in this respect it has revolutionised the imaging of head and neck tumours.

References

Cooper KG (1934) Plasmacytoma and rhabdomyoma of the paranasal sinuses. Arch Otolaryngol 20:329–339

Enzinger FM, Weiss SW (1983) Soft tissue tumours. Mosby, St Louis

Fu Y, Perzin KH (1976) Non-epithelial tumours of the nasal cavity, paranasal sinuses and nasopharynx. V. Skeletal muscle tumours (rhabdomyoma and rhabdomyosarcoma). Cancer 37:364–376

Horn RL, Enterline HT (1958) Rhabdomyosarcoma: a clinico-pathological study and classification of 39 cases. Cancer 11:181–199

Papavasiliou A, Michaels L (1981) Unusual leiomyoma of the nose (leiomyoblastoma): report of a case. J Laryngol Otol 95:1281–1286

Stout AP, Hill WT (1958) Leiomyosarcoma of the superficial soft tissues. Cancer 11:844–854

Sutow WW, Sullivan MP, Ried HL, Taylor HG, Griffith KM (1970) Prognosis in childhood rhabdomyosarcoma. Cancer 25:1384–1390

15 Fibro-osseous Disease

The principal conditions included under the heading of fibro-osseous disease are fibrous dysplasia, ossifying fibroma and benign osteoblastoma. In the nose and sinuses accurate differentiation of these lesions on histological grounds can be difficult. The criteria used for the different entities show some overlap and the problem is further compounded by the variable histology which is often found in different parts of the same lesion (Michaels 1987). In particular the dividing line between fibrous dysplasia and ossifying fibroma is unclear. Some authors regard the latter as a variety of monostotic fibrous dysplasia, while others maintain that it is a separate entity.

Fibrous dysplasia is a condition in which normal bone is replaced by fibrous tissue of spindle cells and poorly formed trabeculae of woven bone with an irregular shape and distribution. Ossifying fibroma may be considered as a localised form of this condition but differs in that there is the presence of osteoid, which is not often found in fibrous dysplasia. Unlike fibrous dysplasia, in ossifying fibroma the lesion is well delimited from the surrounding bone and at the periphery the abnormal bone may show a lamellar structure frequently rimmed by osteoblasts. A more distinctive histological pattern is often seen in these lesions. This was referred to as calcific spherulation by Sherman and Sternberg (1948) and more recently Margo et al. (1985) have used the term "psammomatoid ossifying fibroma". Juvenile ossifying fibroma and cementifying fibroma are names that have also been given to this change. The histological pattern is that of small spherular masses composed of calcified material, osteoid or bone and sometimes referred to as psammoma-like bodies because of their similarity to the psammomata of meningioma.

Osteoblastoma is a benign neoplasm which consists microscopically of a vascular fibrous stroma containing irregular trabeculae of bone and osteoid surrounded by proliferating osteoblasts (see Chap. 18).

Fibrous Dysplasia

The aetiology of this disorder of bone formation is unknown. In the past some authors have attributed the initiation of fibrous dysplasia to trauma, but the most widely held view is that it is a developmental defect derived from embryological faults, since the lesions arise in childhood, enlarge during the period of body growth, and cease to grow after puberty.

There are monostotic and polyostotic forms of the disease. The monostotic type may involve any of the facial bones but the maxilla is the most commonly affected. The polyostotic variety of the condition can occur with or without systemic changes. In the female the association of polyostotic fibrous dysplasia with sexual precocity and café au lait pigmentation of the skin is known as Albright's syndrome.

Irrespective of the site of origin of the disease it is predominantly one of young subjects and since its common location is in the maxilla it usually presents with swelling and deformity of the cheek, sometimes associated with nasal obstruction, proptosis and disturbances of vision. Some cases are

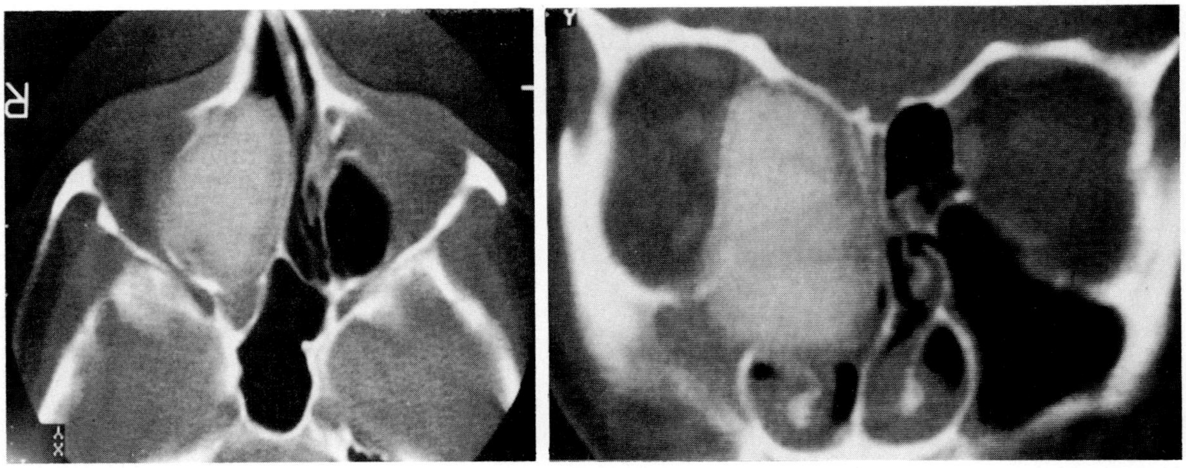

Fig. 15.1a, b. Axial (a) and coronal (b) CT scans of a 21-year-old female with an ossifying fibroma of the ethmoids. There is typical "ground glass" opacity and very clear demarcation of the mass from normal structures.

asymptomatic, the lesion being identified as an incidental finding.

Ossifying Fibroma

The name ossifying fibroma was originally used by Montgomery (1927) to describe fibro-osseous lesions of the jaws, and since the introduction of this term there has been much argument concerning the histological classification of the condition: whether it constitutes a separate entity – a benign tumour composed of cellular fibrous tissue containing bone or osteoid – or simply a variant of fibrous dysplasia.

The radiological evidence for ossifying fibroma was first presented by Sherman and Sternberg (1948). These authors reported the radiographic changes in 12 patients with the condition in the maxilla and mandible, and they described what they believed to be characteristic appearances. The picture was that of: (1) a unilocular osteolytic lesion, oval or spherical in shape; (2) a distinct boundary to the lesion described as "egg shell' in character; (3) progressive enlargement with the formation of spherical densities in the mass; and (4) a unique growth change in the maxilla: the dissolution of adjacent bone without pressure displacement.

Some of these tumours were said to be outstanding microscopically in that they showed extensive spherulation, a change synonymous with that described subsequently as "psammomatoid" by Margo et al. (1985). Sherman and Sternberg (1948) suggested that the dissimilarity in the radiographic appearances of ossifying fibroma and fibrous dysplasia is that the former is a unilocular lesion while fibrous dysplasia may be multilocular, not confined to one part of the bone and often shows diffuse hyperostosis. Later Sherman and Glauser (1958) described the radiological changes in 17 histologically proven cases of fibrous dysplasia of the jaws, and described three types radiologically: one showing diffuse homogeneous sclerosis tending generally to follow the contour of the bone; a second type having a multiloculated osteolytic appearance, oval in shape, with septa and frequently calcific strands or flecks; and a third type presenting a unilocular pattern, which could not always be distinguished from ossifying fibroma.

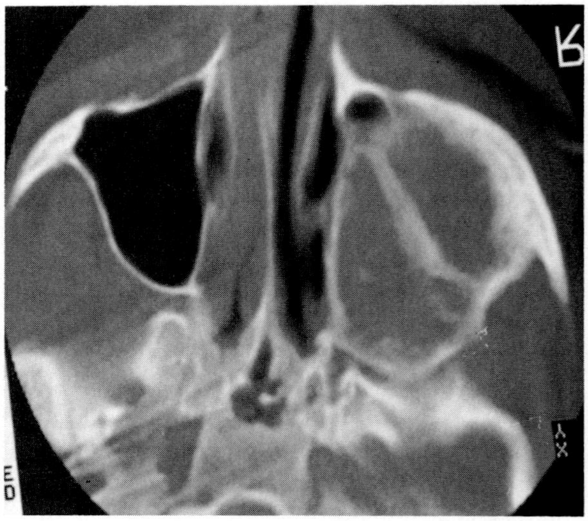

Fig. 15.2. Axial CT scan showing expansion of the left antrum due to an ossifying fibroma.

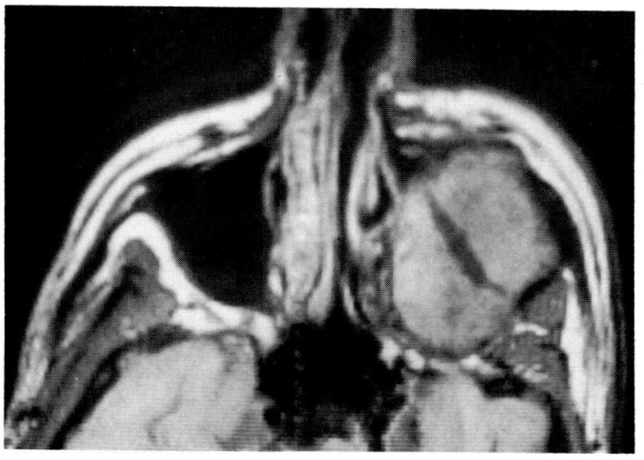

Fig. 15.3. Same patient as Fig. 15.2. Axial magnetic resonance scan. The diagonal bar of calcification is shown as an area of signal void.

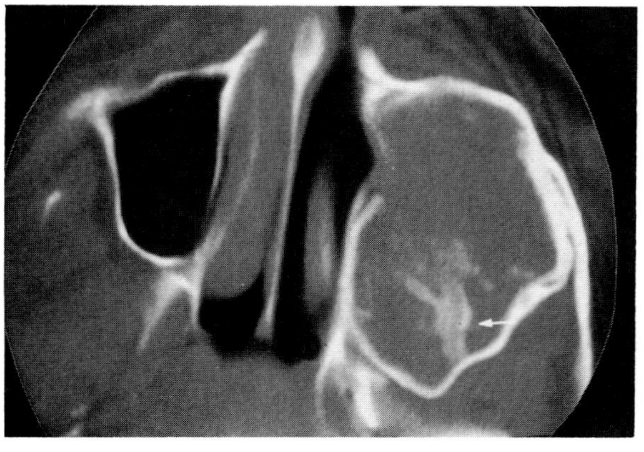

Fig. 15.4. Same patient as Fig. 15.3. Axial CT scan showing nodule of calcification in the posterior part of the mass (*arrow*).

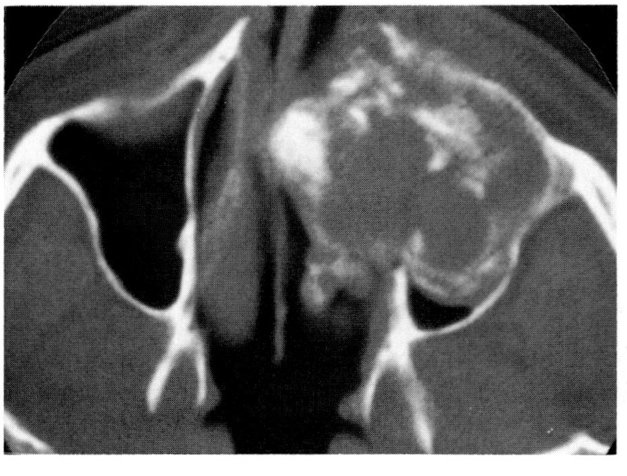

Fig. 15.5. Axial CT scan of an ossifying fibroma in a 38-year-old female.

Radiology and Imaging

Since the papers described above there has been a revolution in imaging techniques which has resulted in much new information concerning fibro-osseous disease affecting the nose and paranasal sinuses. Thirty-one patients with this disease have been investigated: 16 with fibrous dysplasia and 15 with ossifying fibroma.

The diagnosis was based on clinical and radiological evidence initially, but histological evidence was obtained in a third of the cases classified as fibrous dysplasia and all those of ossifying fibroma. All patients were examined by plain radiography and the majority had conventional tomography. Fourteen had CT scans (7 in each category) and 3 had magnetic resonance studies. The histology was that of fibro-osseous disease, and the majority of the cases diagnosed as ossifying fibroma showed spherulation and psammomatoid appearances. In a few patients the latter change was present in some sections whilst others showed the trabecular pattern usually associated with fibrous dysplasia. There was therefore no clear-cut histological distinction between the two conditions.

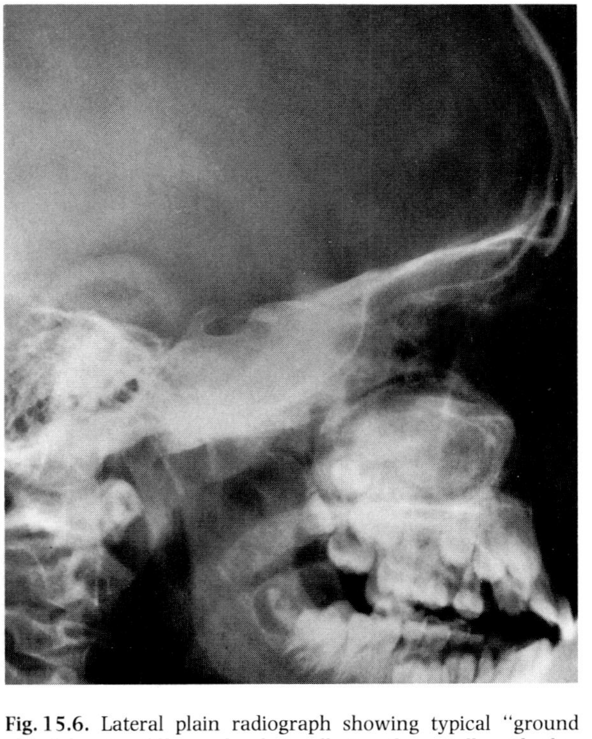

Fig. 15.6. Lateral plain radiograph showing typical "ground glass" density of fibrous dysplasia affecting the maxilla and sphenoid bone.

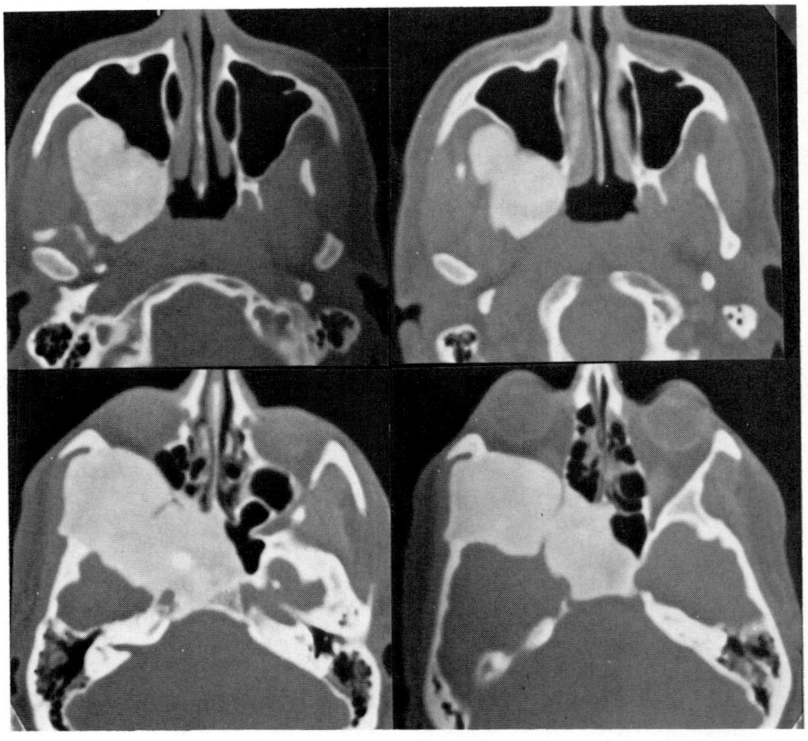

Fig. 15.7. Four axial CT scans showing the dense bone changes of fibrous dysplasia of the maxilla and sphenoid bone. The sphenoid is extensively affected and the enlarged bone is encroaching on the orbit, and on the middle and temporal fossae of the skull.

The imaging changes in this series were of two sorts: either a dense, "ground glass" appearance similar to that seen in the long bones in fibro-osseous disease (Fig. 15.1) or a more osteolytic process with expansion of the bone and strands or nodules of dense calcification within the lesion (Figs. 15.2–15.5). In some patients a mixture of these two processes was apparent, but the changes were essentially the same whether the lesion was classified as fibrous dysplasia or ossifying fibroma. However in fibrous dysplasia the lesion was more diffuse with widespread involvement of one or

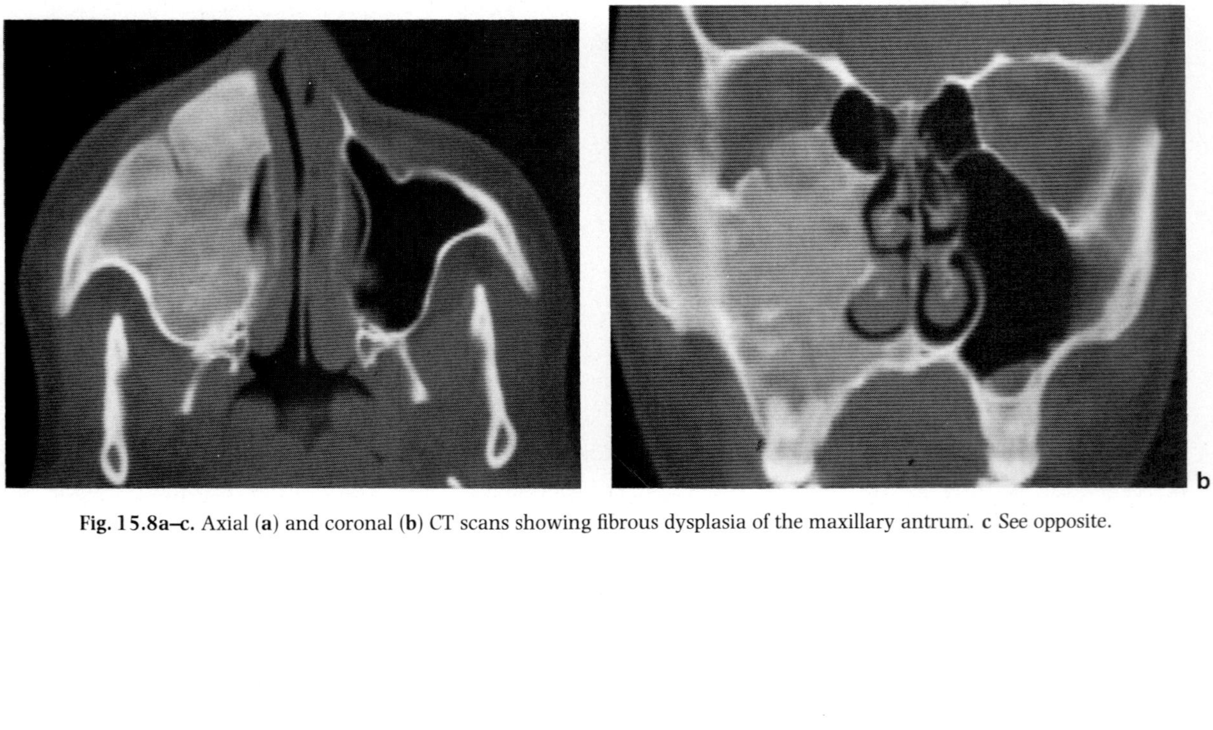

a b

Fig. 15.8a–c. Axial (**a**) and coronal (**b**) CT scans showing fibrous dysplasia of the maxillary antrum. **c** See opposite.

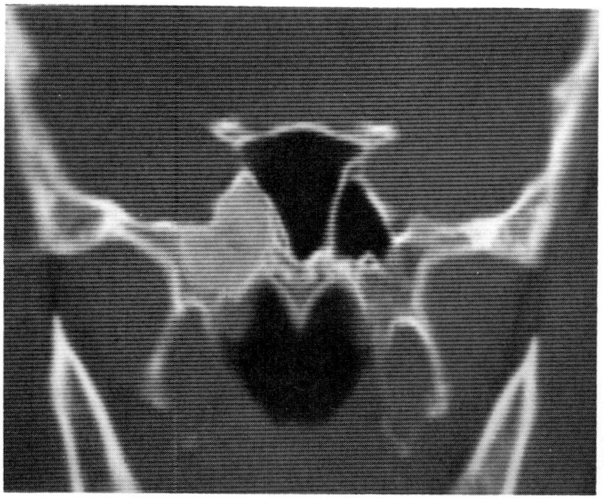

c

Fig. 15.8. (*continued*). **c** A more posterior coronal scan shows that the sphenoid is also affected.

occasionally two of the bones forming the facial skeleton (Figs. 15.6, 15.7 and 15.8), while in ossifying fibroma the lesion was discrete, usually round (Fig. 15.9) or oval in shape, confined to a part of one bone and frequently appeared encapsulated (Fig. 15.1). One particular feature of fibrous dysplasia affecting the maxilla is its typical situation in the lateral wall of the antrum and zygoma, with encroachment of the dense bone on the antral cavity (Figs. 15.10, 15.11).

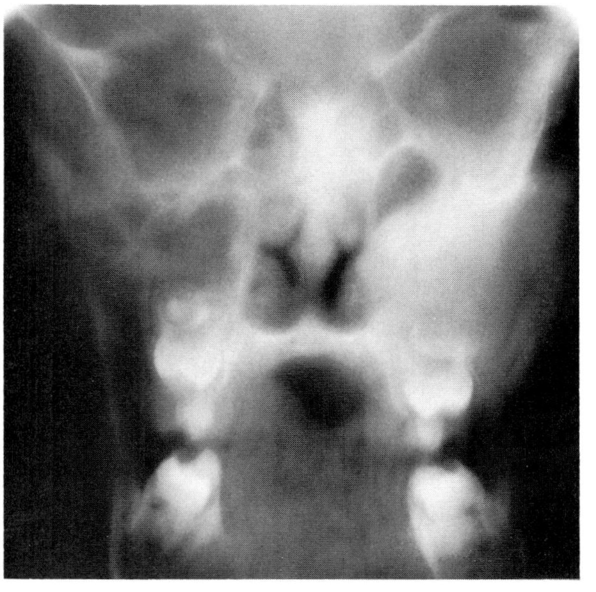

Fig. 15.10. Coronal hypocycloidal tomogram showing fibrous dysplasia of the maxilla. There is typical encroachment by the dense bone on the lumen of the maxillary antrum.

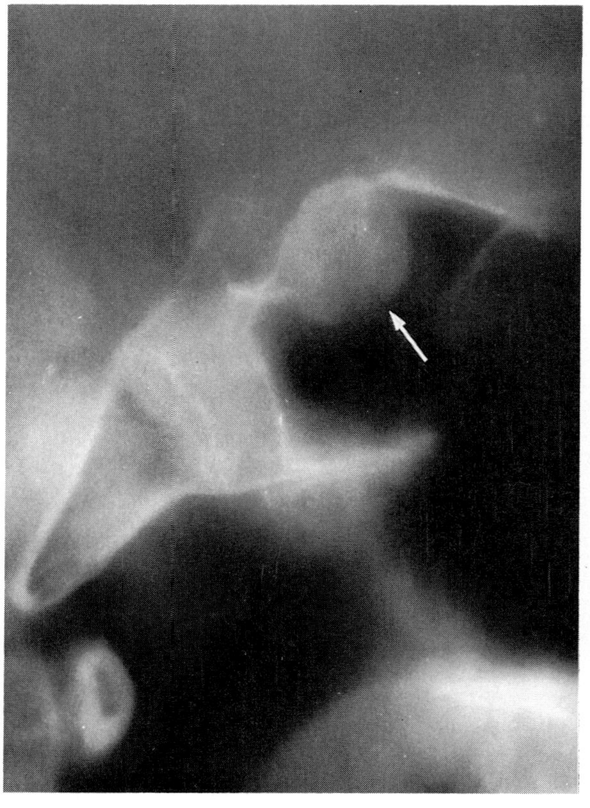

Fig. 15.9. Lateral hypocycloidal tomogram of an ossifying fibroma in the sphenoid sinus of a 14-year-old male. It appears as a discrete rounded mass (*arrow*) arising from the pituitary floor.

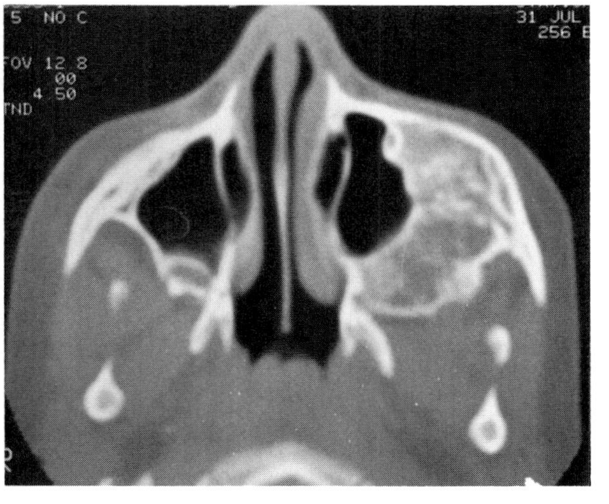

Fig. 15.11. Axial CT scan of a child with fibrous dysplasia of the maxilla, showing bone expansion with encroachment on the antral cavity.

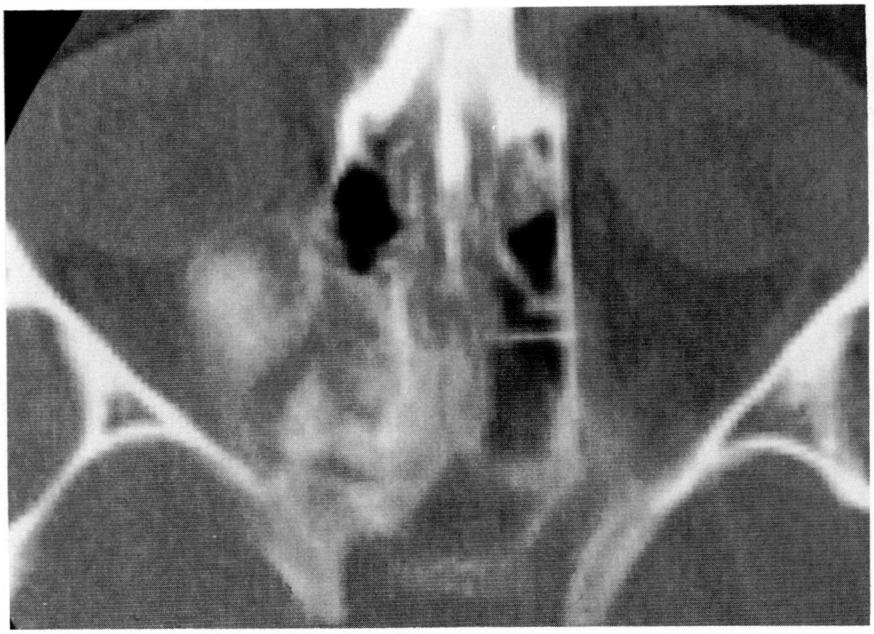

Fig. 15.12. Ossifying fibroma shown on axial CT scan. This was associated with pneumosinus dilatans (see Chap. 6).

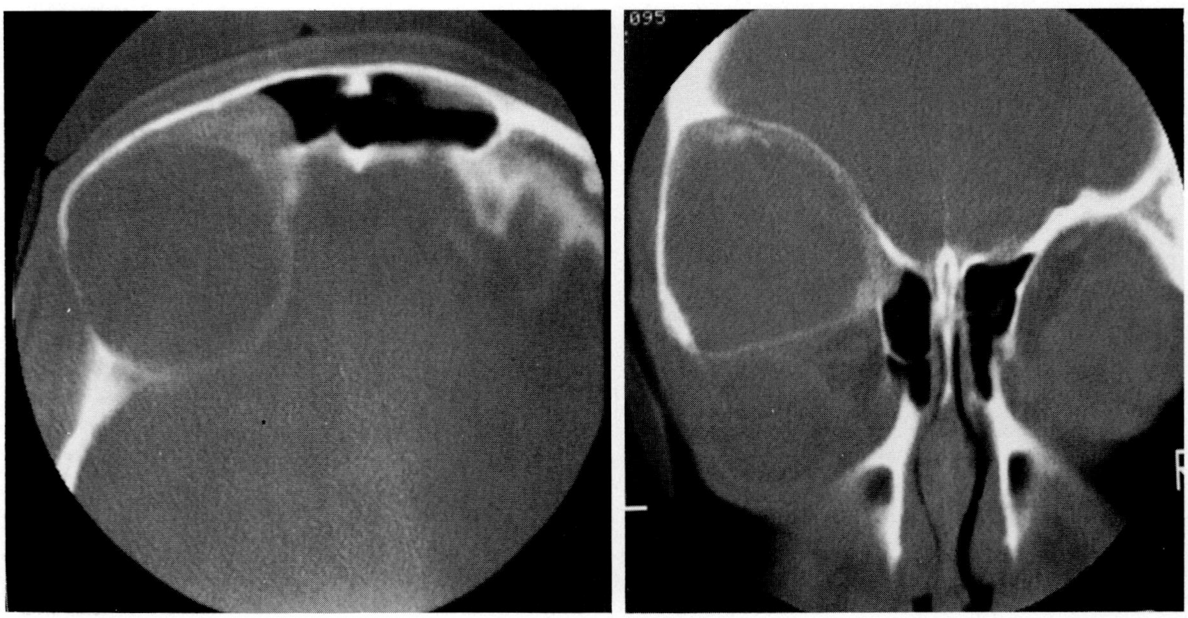

a b

Fig. 15.13a,b. Axial (**a**) and coronal (**b**) CT scans using wide window settings showing a ballooned expansion of the orbital roof. The lesion was partially cystic at surgery. Note the solid tissue within the lumen of the frontal sinus. This showed typical "psammomatoid" ossifying fibroma on microscopy.

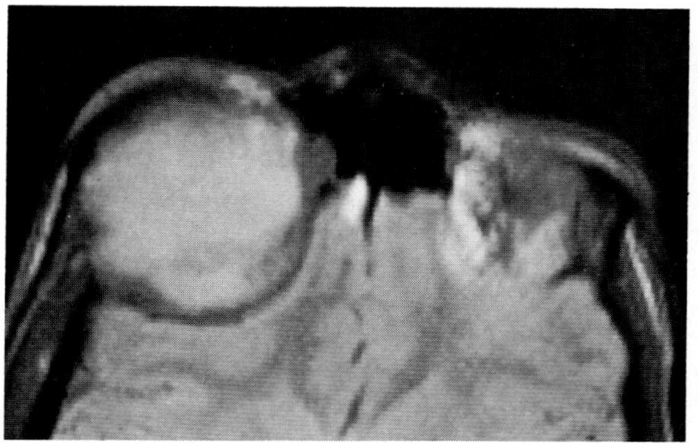

Fig. 15.14. Same patient as Fig. 15.13. Axial magnetic resonance scan. The T_2-weighted image shows high signal from the area of expansion in the orbital roof.

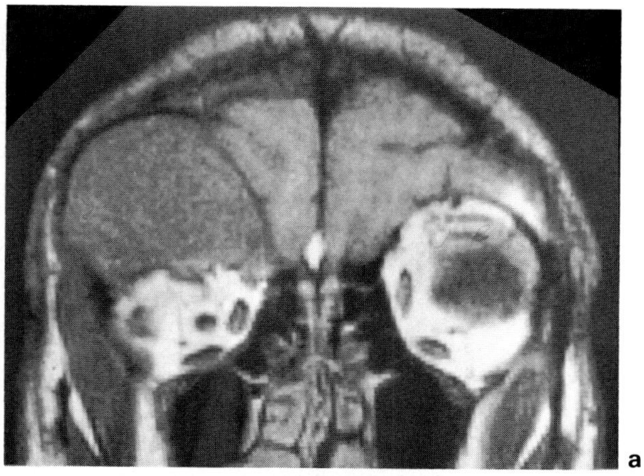

a

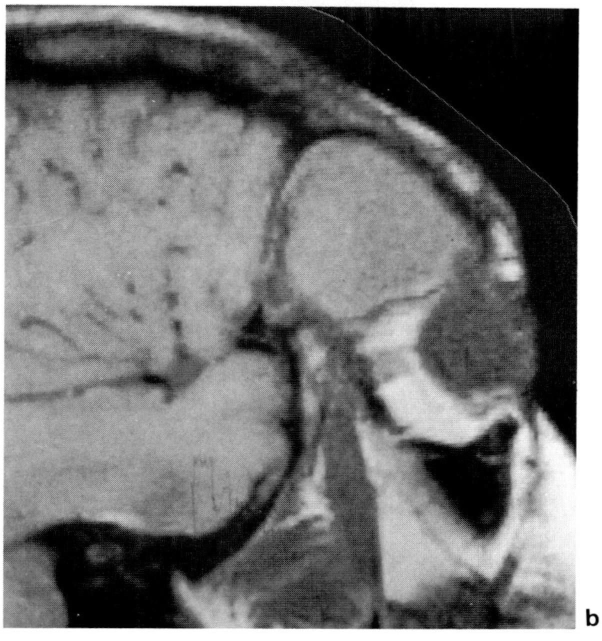

b

Fig. 15.15a,b. Same patient as Fig. 15.14. Coronal (**a**) and sagittal (**b**) magnetic resonance scans of the ossifying fibroma expanding the orbital roof.

Two further changes may be seen when the facial bones and sinuses are affected by fibro-osseous disease:

1. Three patients in this series showed abnormal dilatation of the sinus cavities when their bony walls were affected by fibro-osseous disease; this was the condition known as pneumosinus dilatans. Two of these patients were females with bilateral fibrous dysplasia of the maxillae. In both the antra were enlarged and the orbits encroached upon (Figs. 6.43 and 6.44, pp. 69 and 70). The third patient was a 24-year-old male who was shown to have an ossifying fibroma growing into the orbit from the ethmoid labyrinth. This was accompanied by dilatation of the adjacent fronto-ethmoidal air cells (Fig. 15.12) (see Chapter 6).

2. Three patients presented radiologically with a ballooned expansion of the orbital roof, and were shown histologically to have ossifying fibroma. Margo et al. (1985), in reviewing a large series of patients presenting in this manner, regarded this change as unique to the orbit and diagnostic of psammomatoid ossifying fibroma (POF). There was, however, evidence in two of the three patients we investigated that the horizontal part of the frontal sinus was involved and it may well be that this expansion of the diploë of the orbital roof stems from frontal sinus pathology (Figs. 15.13, 15.14 and 15.15). Other authors have described frontal sinus involvement by ossifying fibroma (Corbet 1951; Thomas and Kaspar 1966); and Lehrer (1969) reported that these lesions in the orbital roof could be mistaken for a frontal sinus mucocoele. A cystic form of ossifying fibroma has also been seen in the other paranasal sinuses (Figs. 15.16, 15.17).

In conclusion it would seem from a practical point of view that there is justification in distinguishing between fibrous dysplasia and ossifying fibroma even though there may be no clear demarcation of pathogenesis (Friedmann and Osborn 1982). The fact that ossifying fibroma may enlarge after the cessation of skeletal growth (Williams and Faccini 1973), and that very large examples of this encapsulated lesion can be removed without recurrence (Lund 1982), makes the distinction important in tems of clinical management and assessment for surgery.

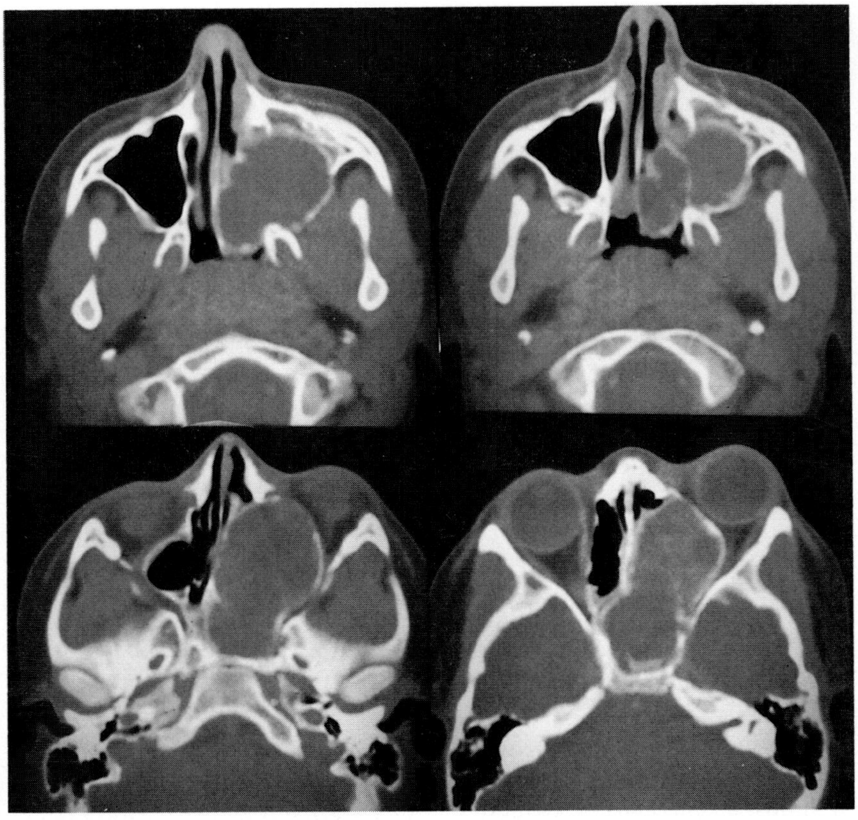

Fig. 15.16. Cystic ossifying fibroma in a 9-year-old male. Axial CT scans showing expansion of the sphenoid, ethmoid and maxillary sinuses with low attenuation.

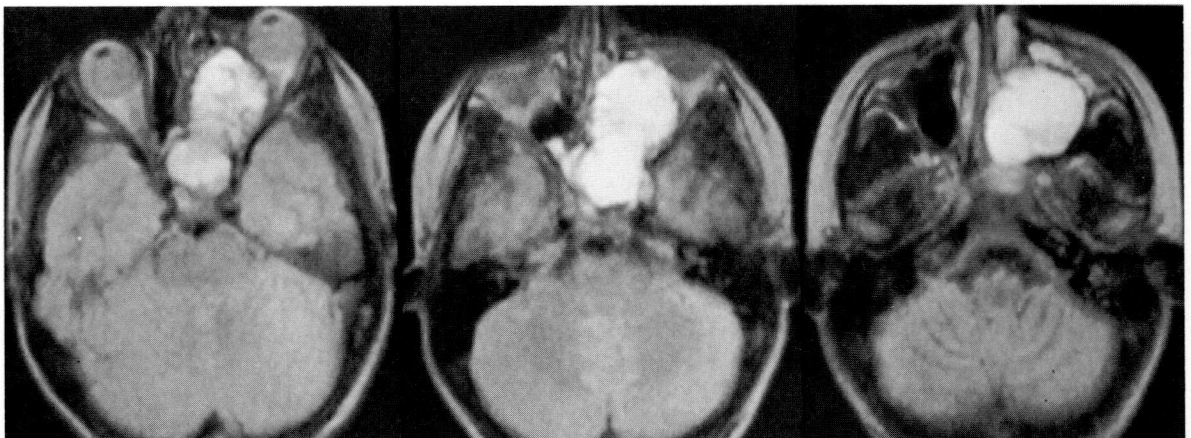

Fig. 15.17. Same patient as Fig. 15.16. T_2-weighted axial magnetic resonance scans showing high signal from the expanded areas. The lesion gave no signal on the inversion recovery sequences, and the spin characteristics were typical of a cystic mass or mucocoele in the sinuses.

References

Corbet CC (1951) Calcifying or ossifying fibroma of frontal sinus. J Laryngol Otol 65:607–608

Friedmann I, Osborn EA (1982) Pathology of granulomas and neoplasms of the nose and paranasal sinuses. Churchill Livingstone, Edinburgh

Lehrer HZ (1969) Ossifying fibroma of the orbital roof: its distinction from "blistering" or "intraosseous meningioma". Arch Neurol 20:536–541

Lund VJ (1982) Ossifying fibroma. J Laryngol Otol 96:1141–1147

Margo CE, Ragsdale BD, Pernan KI, Zimmerman LE, Sweet DE (1985) Psammomatoid (juvenile) ossifying fibroma of the orbit. Ophthalmology 92:150–159

Michaels L (1987) Ear, nose and throat histopathology. Springer, Berlin Heidelberg New York

Montgomery AH (1927) Ossifying fibromas of the jaw. Arch Surg 15:30–44

Sherman RS, Glauser OJ (1958) Radiological identification of fibrous dysplasia of the jaws. Radiology 71:553–558

Sherman RS, Sternberg WCA (1948) The roentgen appearances of ossifying fibroma of bone. Radiology 50:595–609

Thomas GK, Kaspar KA (1966) Ossifying fibroma of the frontal bone. Arch Otolaryngol 83:43–46

Williams JL, Faccini JM (1973) Fibrous dysplastic lesions of the jaws in Nigerians. Br J Oral Surg 11:118–125

16 Fibrous Tissue Tumours

Fibromatosis

Fibromatosis is the name given to a locally aggressive fibrous tissue tumour which grows slowly and infiltrates locally but does not metastasise. It may cause morbidity and even death due to local infiltration which is difficult to control surgically. The condition is best categorised as a less aggressive form of fibrosarcoma.

The lesion consists of fibrocytes with a benign appearance and associated with abundant collagen. Fu and Perzin (1976) described six patients with this disease with ages ranging from 2 to 61 years. Presenting symptoms were nasal obstruction or a mass in the maxillary area associated with pain and epistaxis. The radiographic changes in these patients consisted of clouding of one or more of the paranasal sinuses, with focal bone destruction in two patients; in one there was increased bone density suggesting fibrous dysplasia.

Two patients with this condition have been seen, both of whom were 2-year-old children who were examined by CT scan. There was expansion of the maxilla involving the antrum (Fig. 16.1) and in one case this was associated with areas of increased density and was indistinguishable from the changes seen in fibrous dysplasia (Fig. 16.2). Overall the appearances were those of a benign expansile mass rather than a radiologically malignant lesion.

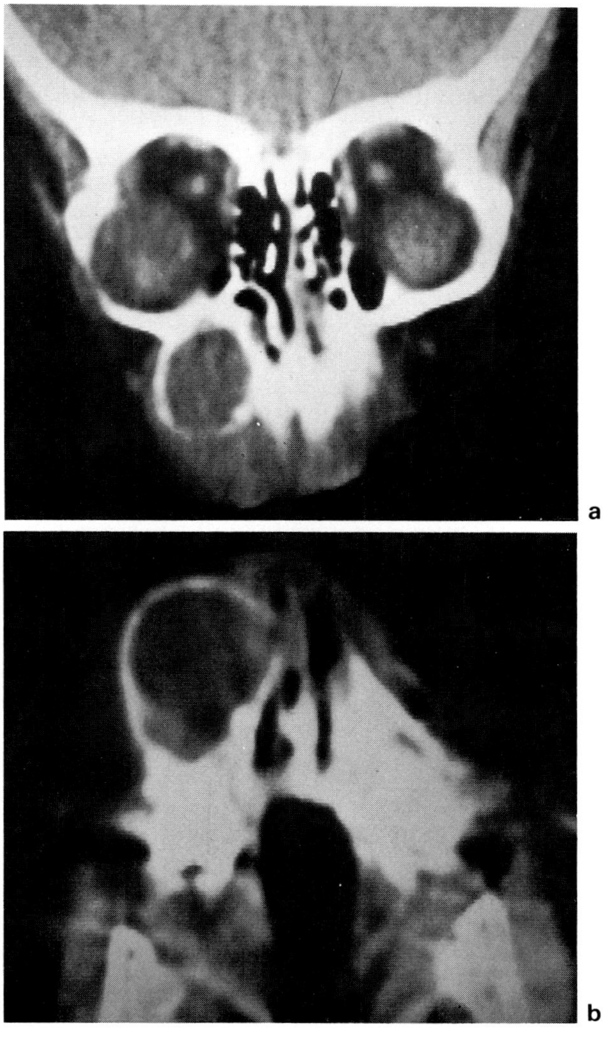
a

b

Fig. 16.1a, b. Fibromatosis in a 2-year-old child. Coronal (**a**) and axial (**b**) CT scans showing a cyst-like expansion of the maxillary antrum.

Fibrosarcoma

In the past a variety of other lesions have been misdiagnosed as fibrosarcoma in the nose and

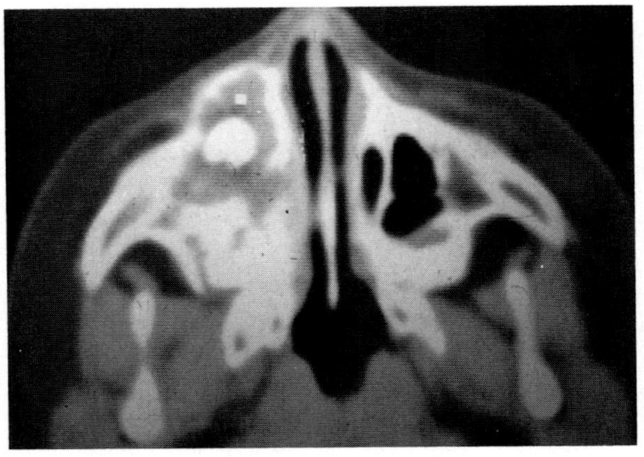

Fig. 16.2. Fibromatosis. Axial CT scan showing expansion of the bone with a soft tissue mass in the lower part of the maxillary antrum and areas of nodular calcification. These appearances could not be distinguished from fibrous dysplasia.

paranasal sinuses and in fact the tumour is rare. Fu and Perzin (1976) reported 13 examples out of 256 non-epithelial tumours, and Jackson et al. (1977) only 3 cases out of 115 malignant neoplasms.

The tissue of origin is uncertain but fibrosarcoma is thought to arise from the periosteum of bone (Hoggins and Brady 1962), and is generally considered to be a tumour of fibroblasts, consisting of fusiform cells separated by collagen. Mitotic figures are always present in variable numbers and are most numerous in the poorly differentiated tumours. The neoplasm does not produce bone or cartilage but new bone formation may be found at the edge of the lesion due to reactive change in the adjacent bone. The tumour has a strong tendency to recur, but metastases are unusual except in the poorly differentiated neoplasms.

Fibrosarcoma has a broadly based age distribution and usually presents clinically as a facial mass with pain and sometimes nasal obstruction. The only known aetiological agent is irradiation, and the literature includes a number of reports of this tumour occurring after radiotherapy (Pettit et al. 1954; Cade 1957).

Malignant Fibrous Histiocytoma

In 1973 Townsend et al. described the first case of fibrous histiocytoma in the nose and paranasal sinuses. Before that date the condition had been described under a variety of other names. Blitzer et al. (1977), in a review of 29 examples of this tumour affecting the deep structures of the head and neck (six in the sinuses), listed 21 synonyms including such names as sclerosing haemangioma, dermatofibroma, xanthofibroma and fibroxanthosarcoma. Some tumours previously diagnosed as fibrosarcoma were in fact examples of this neoplasm (Friedmann and Osborn 1982). Perzin and Fu (1980) have described nine examples of this tumour in the nose, paranasal sinuses and nasopharynx. These authors believe that fibrous histiocytoma is most probably derived from undifferentiated mesenchymal stem cells that have the ability to differentiate into two different pathways – one fibroblastic, the other histiocytic – the proportion of these two elements varying in different lesions.

The age distribution of fibrous histiocytoma is broadly based, ranging from early childhood to the ninth decade; the sexes are affected equally. The clinical picture has no specific features, the common form of presentation being nasal obstruction, epistaxis and proptosis when the orbit is affected. The maxillary sinus is most often involved but the nasal cavity and the ethmoids may also be affected. Histologically the tumour consists of spindle cells arranged in a storiform pattern (meaning "like a rush mat") with histiocytic cells, foam cells and multinucleate giant cells. Mitoses both normal and atypical may be seen. The recurrence rate is high and long-term survival exceptional.

Radiology and Imaging of Fibrosarcoma and Malignant Fibrous Histiocytoma

Three examples of malignant fibrous histiocytoma and four patients with fibrosarcoma have been investigated. The imaging features were similar in both varieties of tumour and were non-specific. They showed bone expansion and destruction in the sinuses (Figs. 16.3, 16.4) without new bone formation, calcification or reactive change in the bone. In both tumour types there was one example of a tumour arising in the infratemporal fossa and involving the antrum secondarily (Figs. 16.5, 16.6). One undifferentiated fibrosarcoma took origin in the frontal sinus and had invaded the anterior cranial fossa at initial examination. A feature of the three malignant fibrous histiocytomata was the relentless advance of the tumour, with many recurrences (Fig. 16.7) necessitating multiple CT investigations and repeated surgery.

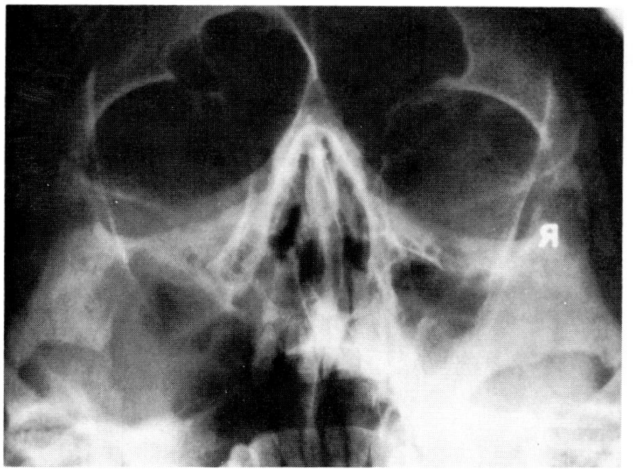

Fig. 16.3. Occipito-mental view showing massive bone destruction in the maxilla by a fibrosarcoma.

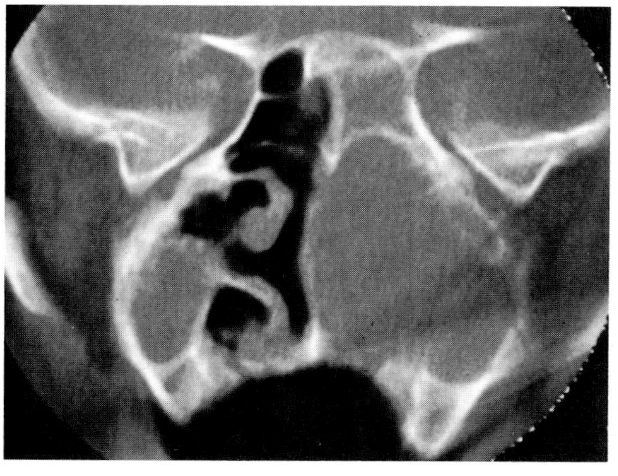

Fig. 16.4. Fibrosarcoma. Coronal CT scan showing expansion of the maxillary antrum and bone destruction.

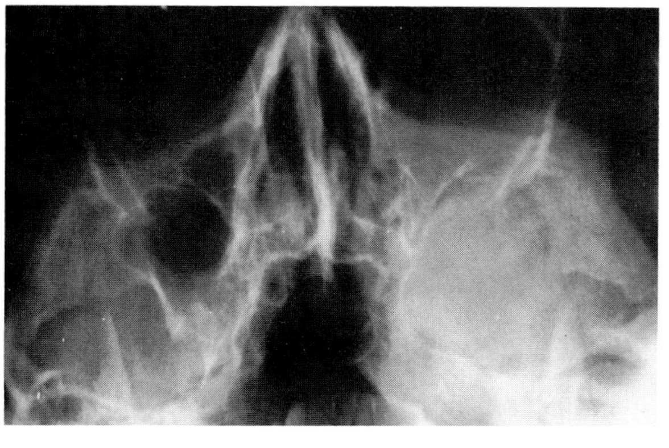

Fig. 16.5. Occipito-mental view showing expansion of the infratemporal fossa and erosion of the maxillary antrum by a large fibrosarcoma.

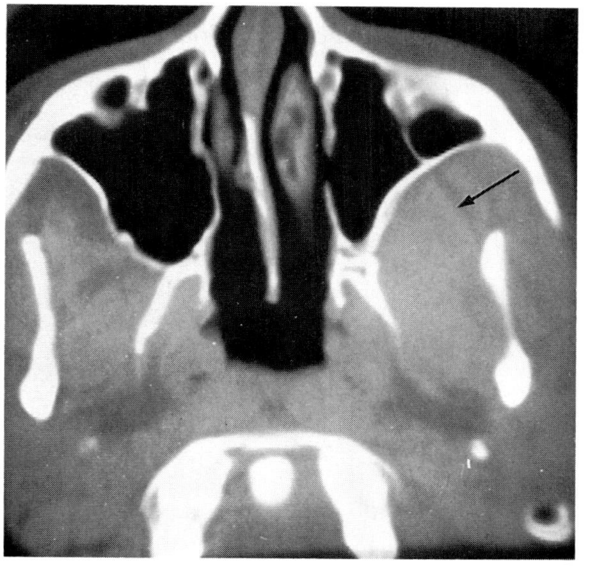

Fig. 16.6. Axial CT scan showing a malignant fibrous histiocytoma arising in the infratemporal fossa (*arrow*) and indenting the posterior wall of the maxillary antrum.

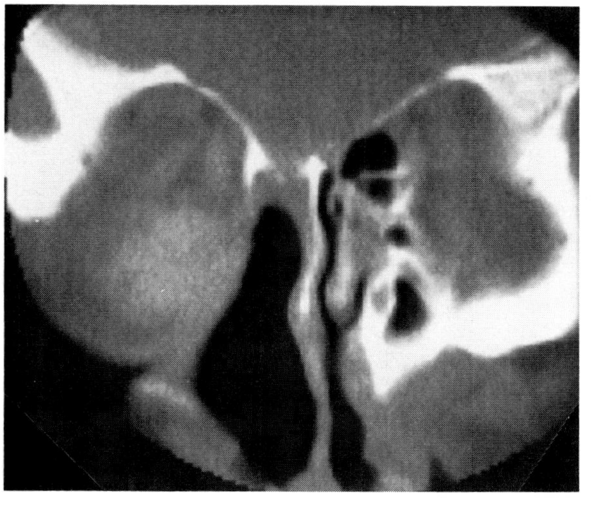

Fig. 16.7. Recurrent malignant fibrous histiocytoma in the inferomedial orbit shown on coronal CT. A maxillectomy had been performed 2 years previously.

References

Blitzer A, Lawson W, Biller HF (1977) Malignant fibrous histiocytoma of the head and neck. Laryngoscope 87:1479–1499

Cade S (1957) Radiation induced cancer in man. Br J Radiol 30: 3393–3402

Friedmann I, Osborn DA (1982) Pathology of granulomas and neoplasms of the nose and paranasal sinuses. Churchill Livingstone, Edinburgh

Fu Y, Perzin KH (1976) Non-epithelial tumours of the nasal cavity, paranasal sinuses and nasopharynx. VI. Fibrous tissue tumours. Cancer 37:2912–2928

Hoggins GS, Brady CL (1962) Fibrosarcoma of the maxilla. Oral Surg Oral Med Oral Pathol 15:34–38

Jackson RT, Fitz-Hugh GS, Constable WC (1977) Malignant neoplasms of the nasal cavities and paranasal sinuses. Laryngoscope 87:726–736

Perzin KH, Fu Y (1980) Non-epithelial tumours of the nasal cavity, paranasal sinuses, and nasopharynx: a clinico-pathological study. XI. Fibrous histiocytomas. Cancer 45:2250–2266

Pettit VD, Chamness JT, Ackerman LV (1954) Fibromatosis and fibrosarcoma irradiation therapy. Cancer 7:149–158

Townsend GL, Neel HB, Weiland LH, Devine KD, McBean JB (1973) Fibrous histiocytoma of the paranasal sinuses. Arch Otolaryngol 98:51–52

17 Cartilaginous Tumours

Chondromata are very rare in the nose and paranasal sinuses. Some arise from the nasal septum and are found incidentally during examination of the nose; others occur in the ethmoid cells. Kilby and Ambegoakar (1977) found on review of the literature that 50% of reported tumours arose from the ethmoids and 17% from the septum. It is generally considered likely that many of the cases that have been described in the literature as chondromata were in fact chondrosarcomata; and malignant features of the cartilage cells may be seen in the illustrations accompanying some of these publications (Michaels 1987).

Chondrosarcomata are less uncommon. They occur in the nasal cavity and have a wide age distribution. According to Lichtenstein and Jaffe (1943) chondrosarcomata arise from mature cartilage. An origin from the nasal septum would explain their presence in the nasal cavity but not in the sinuses, where they are probably derived from cartilaginous cell rests. The prognosis for patients with chondrosarcoma depends upon the resectability of the tumour and the degree of histological differentiation. They may be cured by total surgical excision, but posteriorly located tumours involving the sphenoid and skull base are not totally resectable. Overall 5-year survival rates are recorded as 60% (Fu and Perzin 1974) and 77% (Evans et al. 1977). Clinically they present with nasal obstruction, facial swelling, proptosis and visual disturbances when the orbit is involved.

Radiology and Imaging

Seventeen patients with histologically proven chondrosarcomata have been investigated. Sixteen of these were studied by CT and three also had magnetic resonance scans, one using the paramagnetic contrast agent gadolinium DTPA. The average age of the patients in the series was 48.2 years with an age range of 12–71 and a bimodal distribution peaking in the third and seventh decades. The distinctive feature of these tumours is the presence of calcification within the soft tissue mass shown on plain radiographs (Fig. 17.1) and conventional or

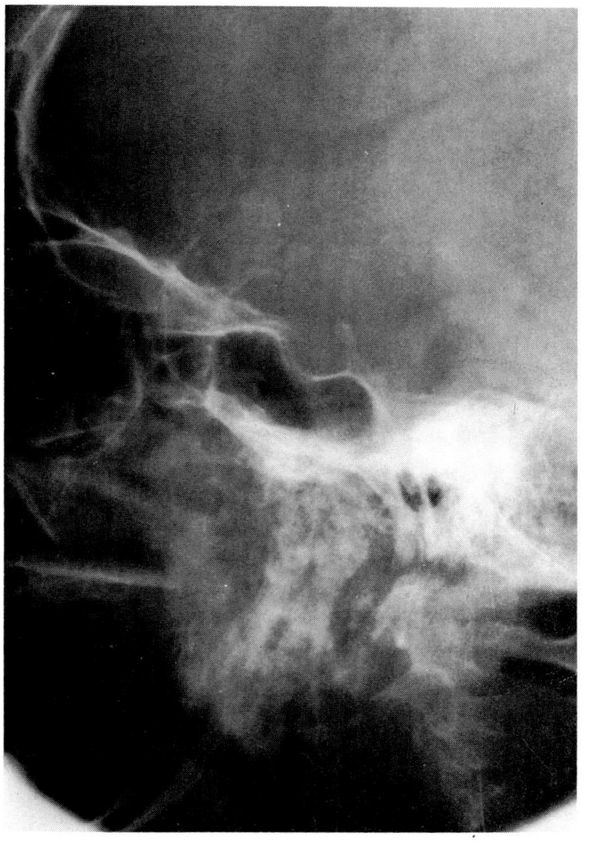

Fig. 17.1. Plain lateral radiograph showing calcification in a large chondrosarcoma of the posterior maxilla and nasopharynx.

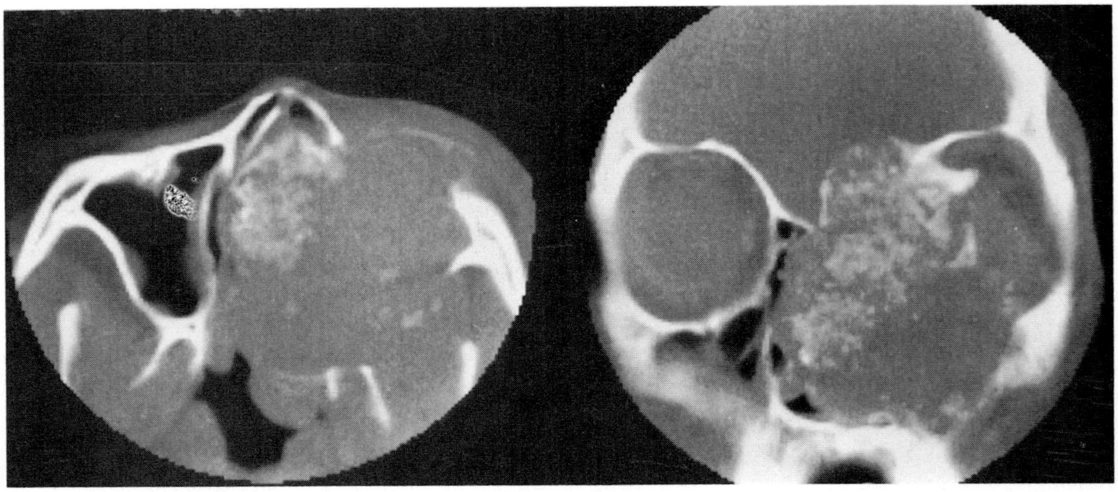

a b

Fig. 17.2a,b. Axial (a) and coronal (b) CT scans of a chondrosarcoma showing multiple confluent calcifications.

computerised tomography. Calcification was present in 73% of the tumours: in some it consisted of one or two punctate areas, extending in some instances to multiple confluent calcifications (Fig. 17.2). However the most typical changes seen were dense irregular plaques of calcification scattered throughout the mass. When present they are diagnostic of chondrosarcoma (Figs. 17.3, 17.4).

The location of these tumours is predominantly naso-ethmoidal (93%), the antra and sphenoids being less frequently involved. Extension to the orbit was present in 60% of cases and to the anterior and middle fossae of the skull in 66%. Nearly 50% of the tumours were centrally located in the nose and sinuses, some entirely confined to the nasal cavity

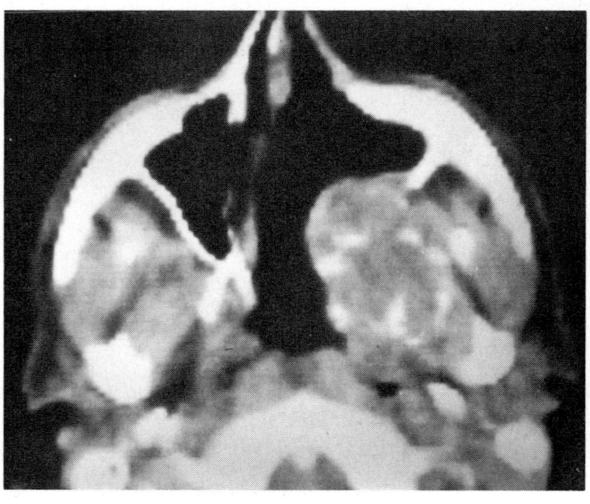

Fig. 17.3. Axial CT scan showing dense punctate calcification in a chondrosarcoma.

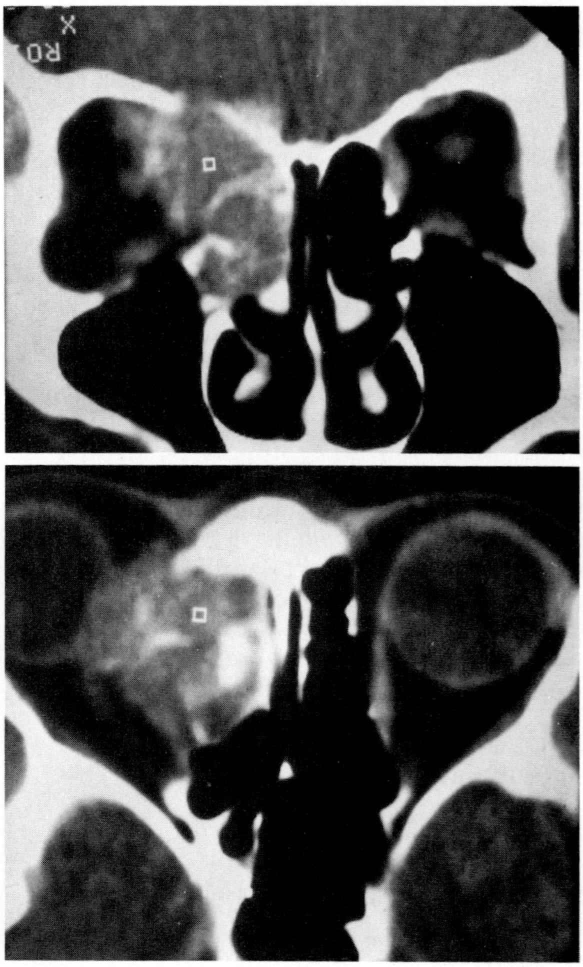

a b

Fig. 17.4a,b. Coronal (a) and axial (b) CT scans showing irregular plaques of calcification in a chondrosarcoma of the ethmoids invading the orbit.

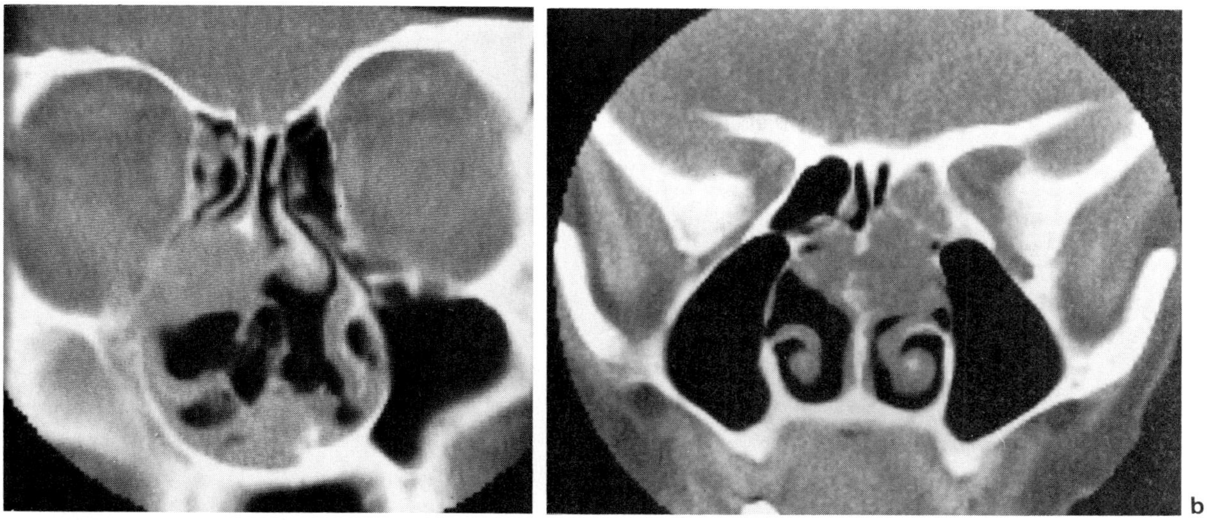

a b

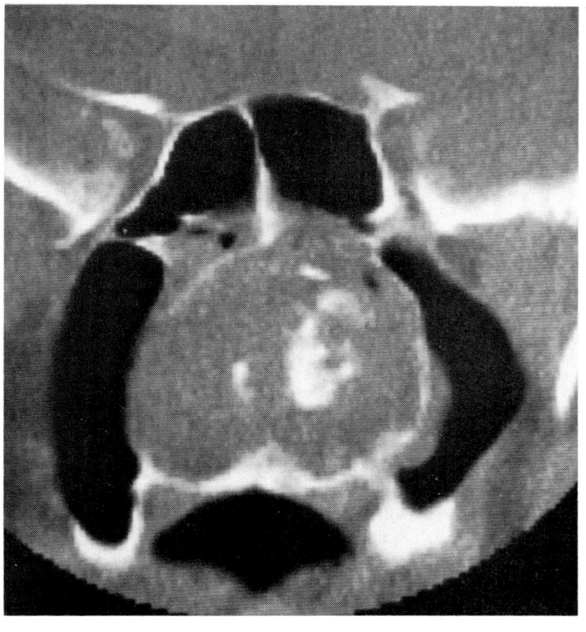

c

Fig 17.5a–c. Three examples of chondrosarcoma of the nasal septum shown on coronal CT scans, **a** Shows central necrosis in the mass. **c** Shows a typical dense plaque of calcification within the tumour.

and clearly taking origin from the septum (Fig. 17.5). The septal chondrosarcomata were more frequently seen in the younger age group but not exclusively so. These tumours gave relatively low signal on all magnetic resonance sequences in the three patients examined. This allows good differentiation of tumour from retained secretion on the T_2-weighted spin echo sequences, however, and the extent of the tumour is still optimally demonstrated, especially in sagittal section (Fig. 17.6). The tumour in the patient receiving intravenous gadolinium DTPA showed differential enhancement between the peripheral and central areas of the mass. The outer, more cellular layers of tumour showed strong enhancement, while the chondromatous core was unaffected and gave low signal (Figs. 17.7, 17.8).

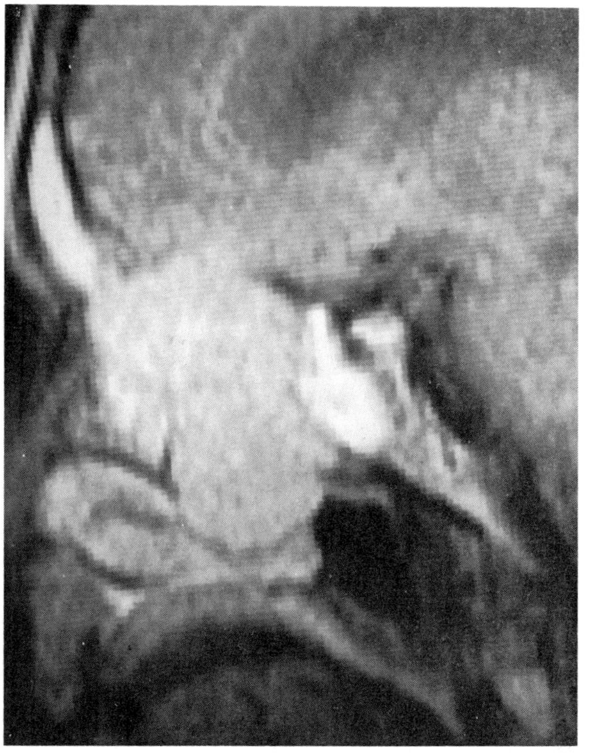

Fig. 17.6. Sagittal magnetic resonance scan of an extensive chondrosarcoma of the nose and sinuses. The tumour shows relatively low signal on a T_2-weighted spin echo sequence, but there is good differentiation of tumour from retained secretion in the frontal and sphenoid sinuses.

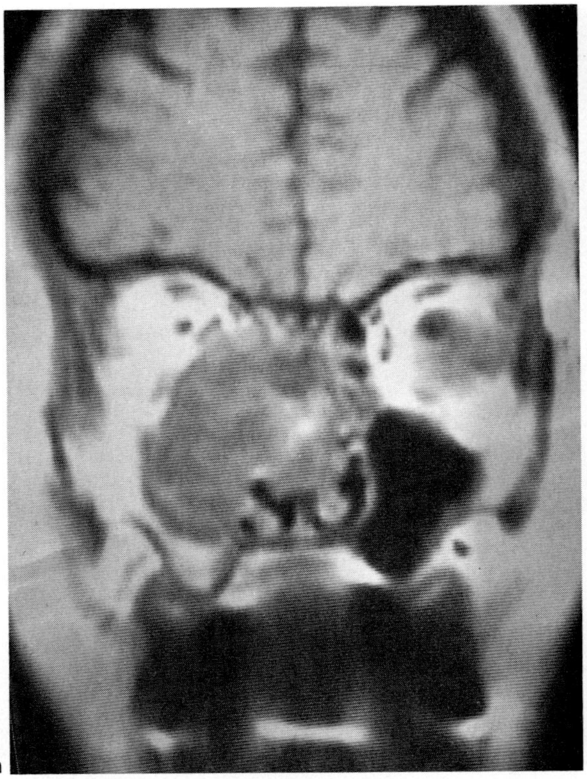

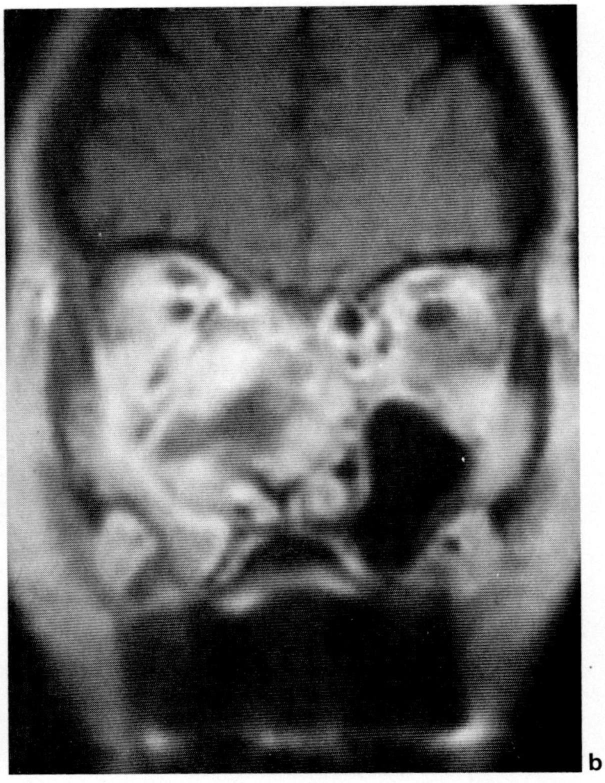

a b

Fig. 17.7a,b. Coronal magnetic resonance scans of a chondrosarcoma. **a** T_1-weighted spin echo sequence before contrast. **b** Same sequence using intravenous gadolinium DTPA as contrast agent. There has been differential enhancement of the tumour, the more peripheral and microscopically cellular part showing enhanced signal. The central, chondromatous part of the tumour does not enhance.

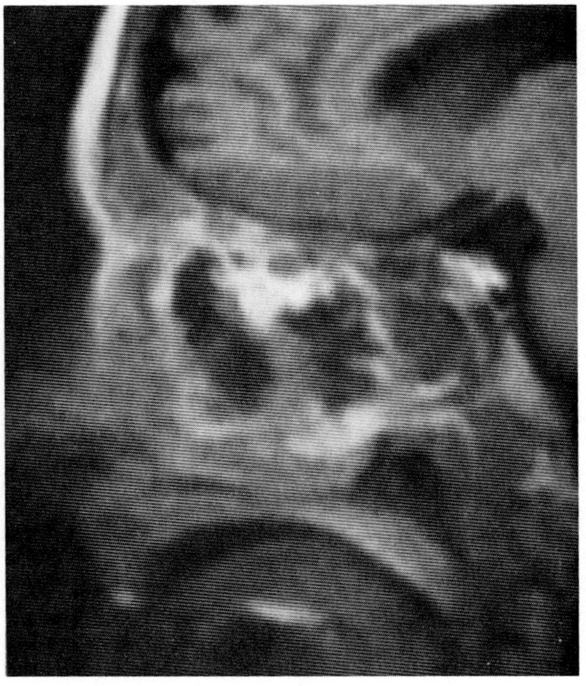

Fig. 17.8. Same patient as Fig. 17.7. Sagittal inversion recovery magnetic resonance scan after 12 cc of intravenous gadolinium DTPA. The more cellular peripheral part of the tumour shows enhanced signal. The central chondromatous part is unaffected by the paramagnetic agent.

References

Evans HL, Ayola AG, Romsdahl MM (1977) Prognostic factors in chondrosarcoma of bone. Cancer 40:818–831

Fu Y, Perzin KH (1974) Non-epithelial tumours of the nasal cavity, paranasal sinuses and nasopharynx. III. Cartilaginous tumours. Cancer 34:453–463

Kilby D, Ambegoakar AG (1977) The nasal chondroma. Two case reports and a survey of the literature. J Laryngol Otol 91: 415–426

Lichtenstein L, Jaffe HL (1943) Chondrosarcoma of bone. Am J Pathol 19:553–589

Michaels L (1987) Ear, nose and throat histopathology. Springer, Berlin, Heidelberg New York

18 Osteogenic Tumours

Osteoma

Osteoma is the commonest tumour to arise in the paranasal sinuses, and is said to occur in 1% of individuals (Mehta and Grewal 1963). It is seen more frequently in the frontal sinus, and less so in the ethmoids; it is rare in the sphenoid sinus and antrum.

There are two varieties: the ivory type showing predominantly lamellar bone with irregularly dispersed vascular spaces and some woven bone; and cancellous osteoma showing a predominantly lamellar structure. It has been stated that ivory or compact osteoma arises in bone preformed in membrane, and cancellous osteoma in bone derived from cartilage. However Atallah and Jay (1981), in their review of 23 sinus osteomata removed surgically, showed that all were composed of cancellous bone surrounded by peripheral dense compact bone. The proportions of each varied considerably, some tumours being largely composed of compact and others of cancellous bone. These authors suggested that the histological pattern is a reflection of the rate of bone growth rather than origin from membrane or cartilage. The osteomata also showed an age discrepancy, occurring at an earlier age in patients from the Middle East than in the European group, which showed maximum occurrence in the third and fourth decades.

The majority of sinus osteomata are asymptomatic and are found during routine sinus and skull radiography (Fig. 18.1). In those patients who do develop symptoms the commonest is facial pain, occurring with or without associated sinus infection. Large osteomata in the frontal or ethmoid sinuses frequently cause proptosis either by direct invasion of the orbit or by the formation of a mucocoele from a blocked fronto-nasal duct (Fig. 18.2). Occasionally upward extension into the anterior cranial fossa (Fig. 18.3) may lead to erosion of the dura and a fistula into the cranial cavity or brain, with an ensuing CSF leak, meningitis, aerocoele or brain abscess. This is especially likely to occur after surgery of the larger osteomata.

Multiple osteomata in the paranasal sinuses are found in Gardner's syndrome (Gardner and Plenk 1952) (Fig. 18.4). They are associated with benign tumours and cysts of the skin such as dermoids, and multiple intestinal polyps in the colon and rectum

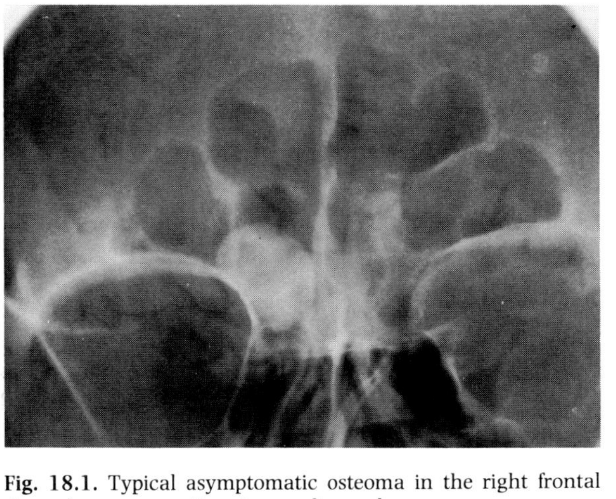

Fig. 18.1. Typical asymptomatic osteoma in the right frontal sinus shown on routine sinus radiography.

which show a marked tendency to malignant degeneration. One family reported by Gardner showed a history of nine deaths from large bowel carcinoma in five generations.

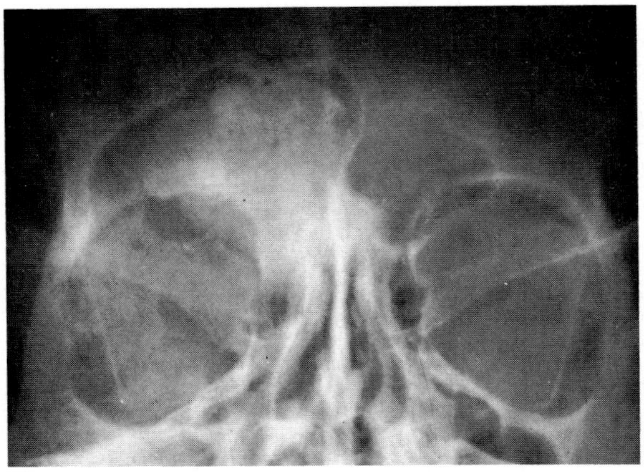

Fig. 18.2. Secondary mucocoele formation in the frontal sinus due to a large osteoma.

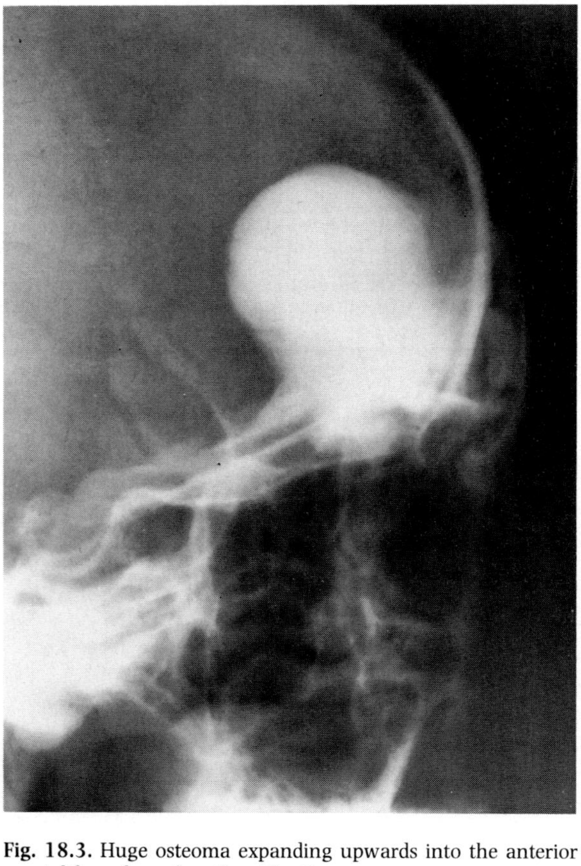

Fig. 18.3. Huge osteoma expanding upwards into the anterior cranial fossa from the ethmoid cells.

Radiologically osteoma is the easiest tumour to demonstrate in the sinuses. It is satisfactorily shown by plain radiography (Figs. 18.1, 18.4) and rarely are any more sophisticated techniques required. However, osteomata of the more cancellous variety may show relatively low density and their full extent is then better shown by CT.

Benign Osteoblastoma

This condition was first given the title osteogenic fibroma by Lichtenstein (1951) and described as giant osteoid osteoma by Dahlin and Johnson (1954). The more appropriate name of benign osteoblastoma was used by Jaffe (1956) and Lichtenstein (1956). The lesion is very rare in the sinuses, but involvement of the maxilla has been reported by Borello and Sedano (1967) and Kent et al. (1969); and Fu and Perzin (1974) described one tumour in the ethmoid region.

Microscopically the lesion consists of numerous trabeculae of osteoid or immature bone surrounded by osteoblasts, the intervening spaces being occupied by a highly vascular stroma containing multinucleate giant cells. A single example of this condition has been investigated in a 9-year-old female. The tumour was massive but well circumscribed (Fig. 18.5) and showed homogeneous density on CT that was marginally above soft tissue attenuation but consistent with the pattern of histology described for osteoblastoma.

Osteosarcoma

Although it is the commonest primary tumour of the bony skeleton, osteosarcoma is much less common in the nose and sinuses. Garrington et al. (1967) estimated that approximately 6.5% of all osteosarcomata arise in the jaws, the maxilla and sinuses being less frequently involved than the mandible. The disease may follow irradiation of the same bone and, as in the long bones, osteosarcoma may be a complication of Paget's disease (Fig. 18.6). The tumour affects an older age group than skeletal osteosarcoma, which predominates mainly in the second decade of life. In the jaw the tumour has a wider age distribution, with ages ranging from 15 to 50 (Garrington et al. 1967) and 30 to 69 (Windle-Taylor 1977).

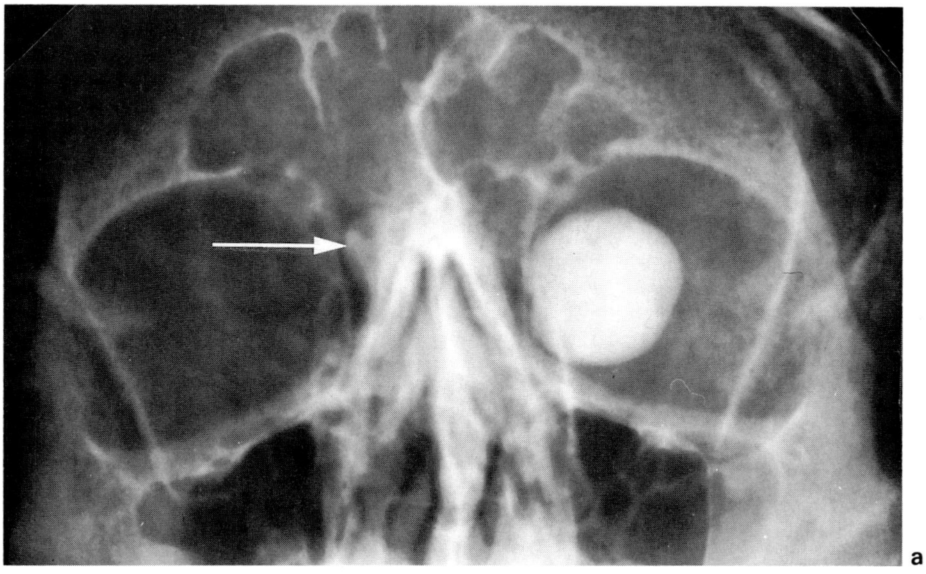

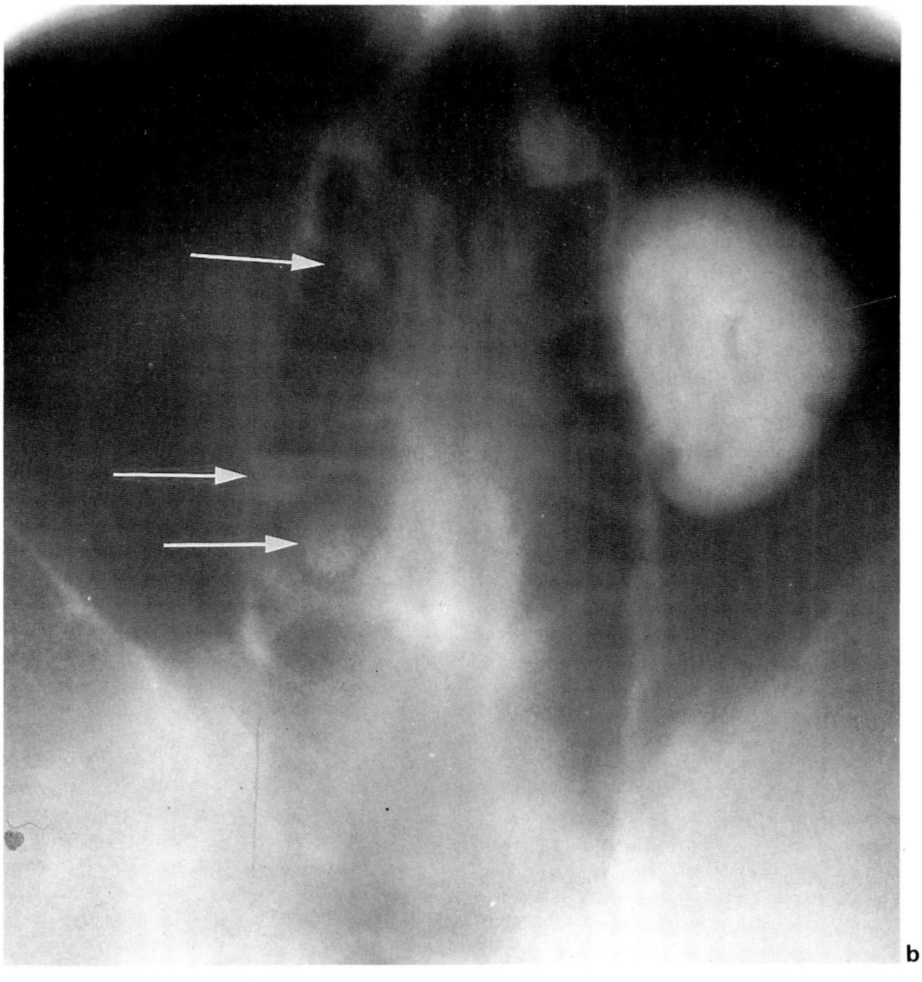

Fig. 18.4a,b. Gardner's syndrome. Plain radiograph (**a**) and axial hypocycloidal tomogram (**b**) showing one large ethmoid osteoma invading the orbit and several smaller osteomata in the ethmoid labyrinth (*arrows*). The patient had had a colectomy for large bowel polyps.

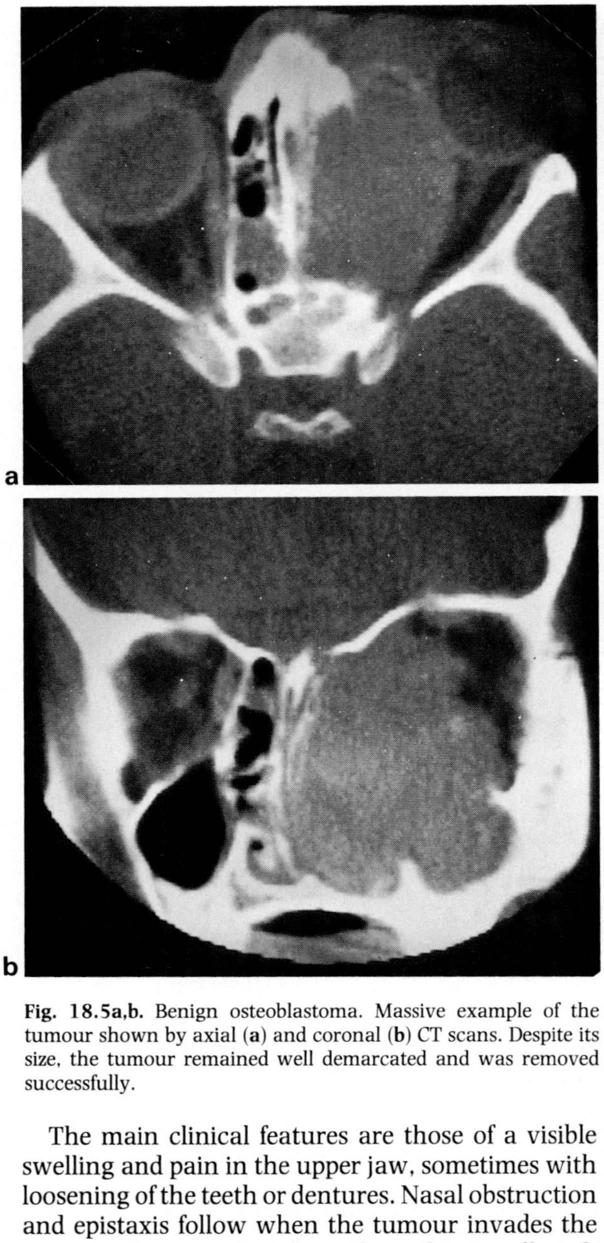

Fig. 18.5a,b. Benign osteoblastoma. Massive example of the tumour shown by axial (**a**) and coronal (**b**) CT scans. Despite its size, the tumour remained well demarcated and was removed successfully.

The main clinical features are those of a visible swelling and pain in the upper jaw, sometimes with loosening of the teeth or dentures. Nasal obstruction and epistaxis follow when the tumour invades the nose, and proptosis when the orbit is affected. Macroscopically the tumour presents the features of an irregularly calcified destructive tumour, which histologically is composed of spindle cell sarcoma with osteoid and immature neoplastic bone formation.

Radiology and Imaging

Seven patients with this condition have been investigated: five by CT, two by magnetic resonance and two by conventional techniques only. The majority of tumours (five patients) took origin in the maxilla, but one involved the sphenoid and another was located in the frontal sinus. Three patients were male and four female. The average age was 38 years and the age range 11–66 years.

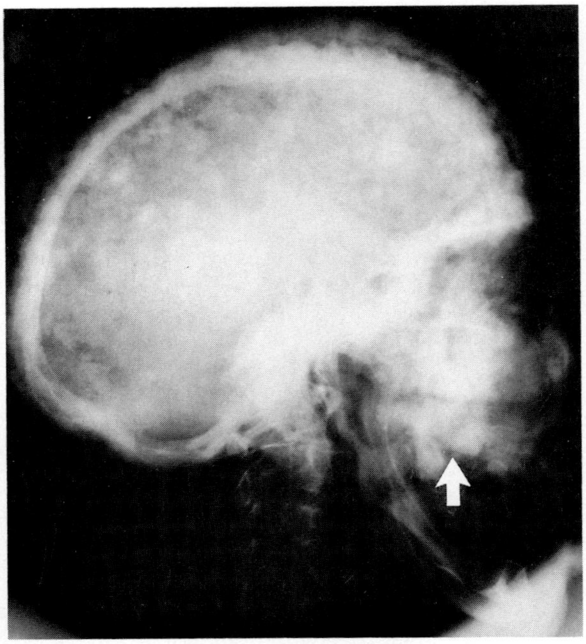

Fig. 18.6. Paget's disease of the skull, with osteosarcoma of the maxilla (*arrow*).

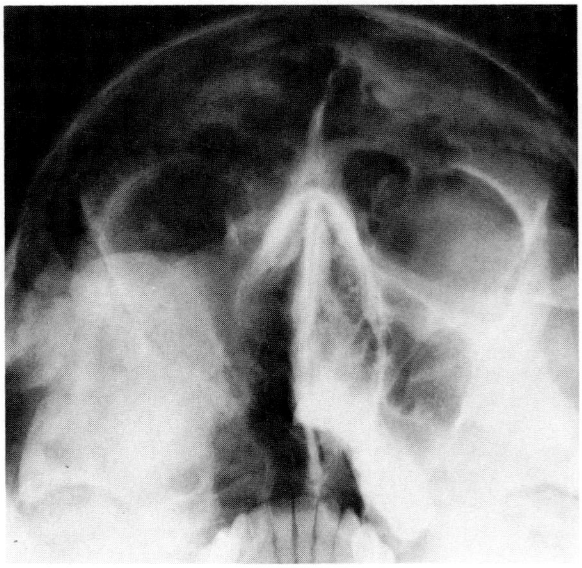

Fig. 18.7. Massive new bone formation in a recurrent osteosarcoma of the maxilla and zygoma.

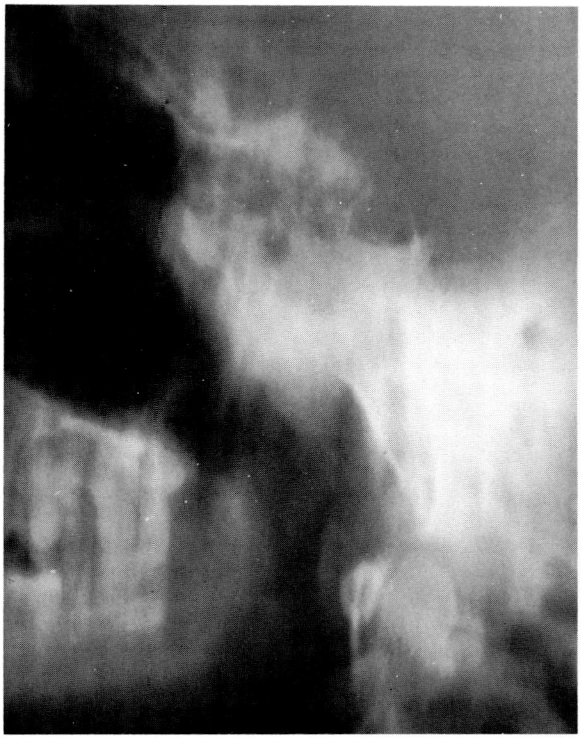

Fig. 18.8. Osteosarcoma of the sphenoids. Lateral tomogram showing new bone formation around the sphenoid sinus.

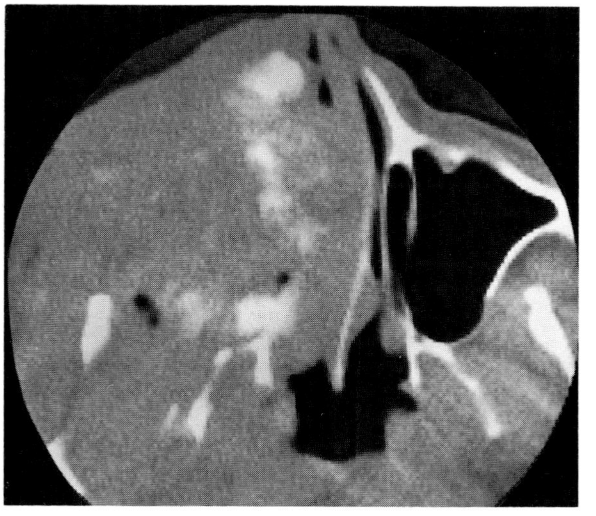

Fig. 18.9. Osteosarcoma of the maxilla. Axial CT scan showing massive tumour with bone destruction and areas of ill-defined calcification or new bone formation.

Basically these tumours present an osteolytic process as in other malignant sinus tumours, but they are characterised by new bone formation (Fig. 18.7), which was present in all but one of the patients examined. This is frequently ill defined and may be diffuse or nodular and of variable density. It gives the appearances both on plain radiography and on CT of being freshly formed (Figs. 18.8, 18.9). These areas may coexist with relatively well-defined nodular calcification which is radiologically indistinguishable from chondrosarcoma (Fig. 18.10).

It might be anticipated that in the presence of osteoid and new bone formation osteosarcoma would give low signal on magnetic resonance. This

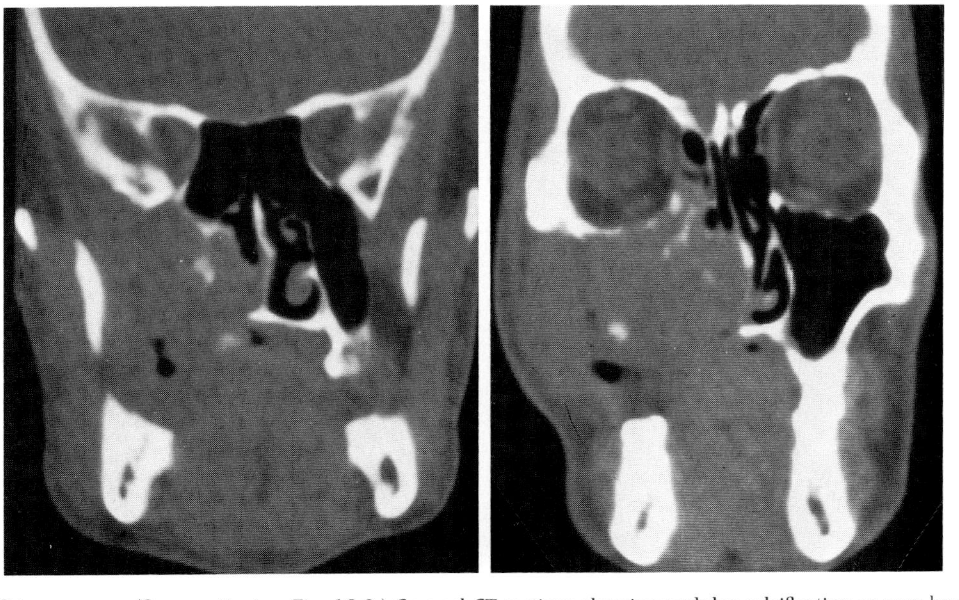

Fig. 18.10. Osteosarcoma. (Same patient as Fig. 18.9.) Coronal CT sections showing nodular calcification or new bone formation indistinguishable from a chondrosarcoma.

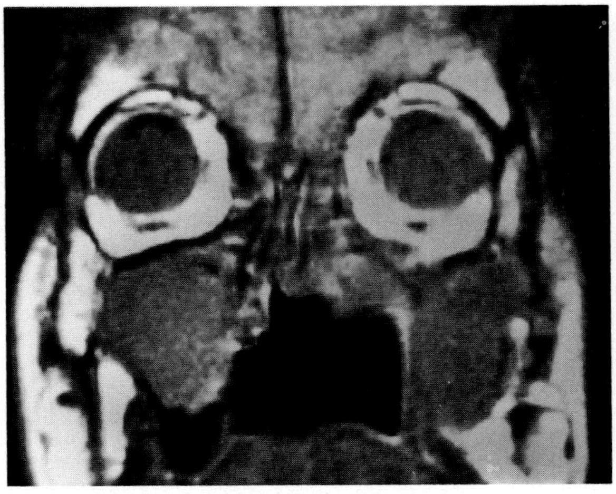

Fig. 18.11. Coronal magnetic resonance section of a recurrent osteosarcoma in the paranasal sinuses. The T_1-weighted scan shows no discrimination between tumour in the left antrum and inflammatory mucosa in the right.

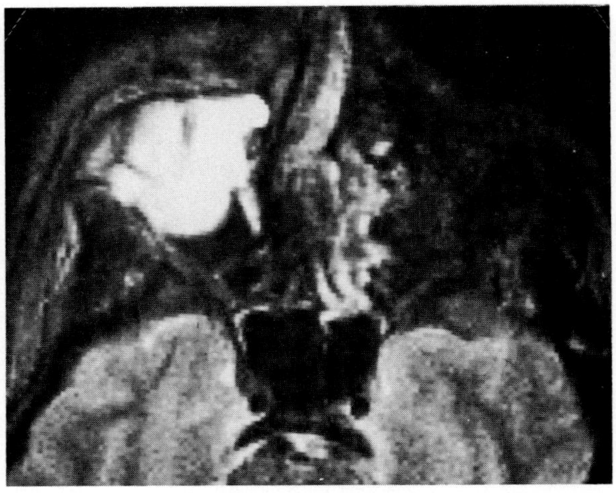

Fig. 18.12. Same patient as Fig. 18.11. Axial magnetic resonance section. On T_2-weighted spin echo sequences using a long time to echo and a long repetition time there is strong signal from inflamed mucosa in the right antrum and a very low-intensity signal from the left antrum containing recurrent osteosarcoma.

was so in one patient but the other showed moderately high signal on T_2-weighted spin echo sequences with areas of signal void corresponding to the nodular calcification shown on CT. One patient was scanned for a recurrent tumour. By using spin echo sequences with a long time to echo and a long repetition time it was possible to discriminate with total accuracy between tumour and inflammatory changes in the sinuses (Figs. 18.11, 18.12).

Ewing's Sarcoma

Ewing's sarcoma is an uncommon round cell tumour of bone which was first identified as a separate entity by Ewing in 1921. It is a locally aggressive neoplasm occurring predominantly in male patients in the first three decades of life and usually affecting the long bones. In the head and neck the tumour is rare. It may occur in the mandible and

a 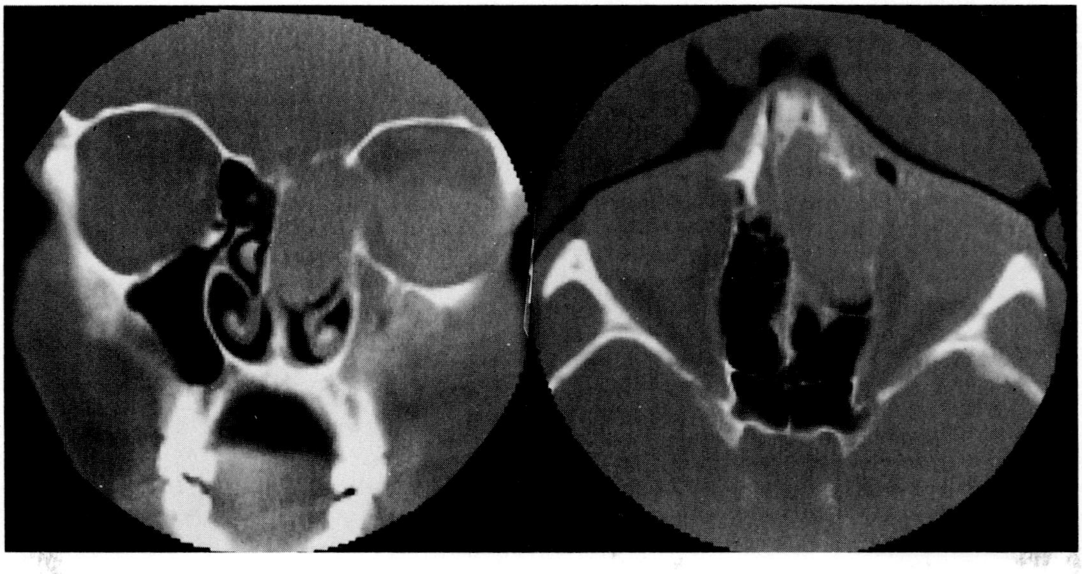 b

Fig. 18.13a,b. Coronal (a) and axial (b) CT sections showing Ewing's sarcoma in the anterior ethmoid cells invading the orbit and anterior cranial fossa.

sometimes in the maxilla, involving the maxillary antrum.

Brownson and Cook (1969) reported the first case to appear in an otolaryngological journal and found 10 cases in the general literature. Four additional patients with Ewing's sarcoma in the maxilla were reported by Ferlito (1978). This author described the tumour as a solitary lesion occurring between the ages of 5 and 20 years which could be identified by the presence of abundant glycogen in the cytoplasm of the neoplastic cells. All authors stressed the non-specific nature of the radiological changes, emphasising the lack of the "onion peel" character peculiar to Ewing's tumour in the long bones.

A more recent publication is that of Howard and Lund (1985), who described a case of primary Ewing's sarcoma arising from the ethmoid cells and extending via the cribriform plate to the anterior cranial fossa. The patient was a 14-year-old male who presented with recurrent swelling over the bridge of the nose associated with a firm mass at the medial canthus and swelling of the middle turbinate. On CT the anterior ethmoid cells and right upper nasal cavity were occupied by a well-defined soft tissue mass, which was beginning to expand into the orbit and anterior cranial fossa (Fig. 18.13). After surgical removal the tumour was shown to have the typical histological features of Ewing's sarcoma.

References

Atallah N, Jay MM (1981) Osteomas of the paranasal sinuses. J Laryngol Otol 95:291–309

Borello ED, Sedano HO (1967) Giant osteoid osteoma of the maxilla. Oral Surg Oral Med Oral Pathol 23:563–566

Brownson RJ, Cook RP (1969) Ewing's sarcoma of the maxilla. Ann Otol Rhinol Laryngol 78:1299–1304

Dahlin DC, Johnson EW (1954) Giant osteoid osteoma. J Bone Jt Surg [Am] 36:559–572

Ewing J (1921) Diffuse endothelioma of bone. Proc NY Pathol Soc 21:17–24

Ferlito A (1978) Primary Ewing's sarcoma of the maxilla. J Laryngol Otol 92:1007–1024

Fu Y, Perzin KH (1974) Non-epithelial tumours of the nasal cavity, paranasal sinuses and naso-pharynx. Cancer 33: 1289–1305

Gardner RJ, Plenk HP (1952) Hereditary pattern for multiple osteomas in a family group. Am J Genet 4:31–36

Garrington GE, Scofield HH, Comyn J, Hooker SP (1967) Osteosarcoma of the jaw. Analysis of 56 cases. Cancer 20:377–391

Howard D, Lund VJ (1985) Primary Ewing's sarcoma of the ethmoid bone. J Laryngol Otol 99:1019–1023

Jaffe HL (1956) Benign osteoblastoma. Bull Hosp Jt Dis 17:141–151

Kent JL, Castro HF, Girotti WR (1969) Benign osteoblastoma of the maxilla. Oral Surg Oral Med Oral Pathol 27:209–219

Lichtenstein L (1951) Classification of primary tumours of bone. Cancer 4:335–341

Lichtenstein L (1956) Benign osteoblastoma. Cancer 9:1044–1052

Mehta BS, Grewal GS (1963) Osteomas of the paranasal sinuses along with a case report of an orbito-ethmoidal osteoma. J Laryngol Otol 77:601–610

Windle-Taylor PC (1977) Osteosarcoma of the upper jaw. J Maxillofac Surg 5:62–68

Subject Index